Machine Learning in Sensors and Imaging

Machine Learning in Sensors and Imaging

Editor

Hyoungsik Nam

MDPI • Basel • Beijing • Wuhan • Barcelona • Belgrade • Manchester • Tokyo • Cluj • Tianjin

Editor
Hyoungsik Nam
Kyung Hee University
Korea

Editorial Office
MDPI
St. Alban-Anlage 66
4052 Basel, Switzerland

This is a reprint of articles from the Special Issue published online in the open access journal *Sensors* (ISSN 1424-8220) (available at: https://www.mdpi.com/journal/sensors/special_issues/ML-SI).

For citation purposes, cite each article independently as indicated on the article page online and as indicated below:

LastName, A.A.; LastName, B.B.; LastName, C.C. Article Title. *Journal Name* **Year**, *Volume Number*, Page Range.

ISBN 978-3-0365-3753-5 (Hbk)
ISBN 978-3-0365-3754-2 (PDF)

Contents

About the Editor

Hyoungsik Nam (Professor) received his B.S., M.S. and Ph.D. degrees in EECS from Korea Advance Institute of Science and Technology (KAIST), Daejon Korea, in 1996, 1998, and 2004. He joined Samsung Electronics as a senior engineer in 2005, where he had worked on Active-Matrix Liquid-Crystal Displays. He is currently an associate professor in the department of Information Display at Kyung Hee University, Seoul Korea. His current research interests are low power technologies, integrated circuits, signal/user interfaces for flat panel displays, and machine learning applications.

Preface to "Machine Learning in Sensors and Imaging"

With the recent spread of smartphones, smartwatches, and smartbands, people are carrying various sensors without realizing: a camera to capture an image, an infrared sensor to measure approach or heart rate, a microphone to sense a voice signal, an optical sensor or ultrasound sensor to sense a fingerprint, an accelerometer and a gravity sensor to measure movement or location, you name it. In addition to this, many CCTVs in our daily life and thermal sensors, which were widely distributed due to COVID-19, cannot be left out. A lot of data are created through this abundance of sensors, and various studies are being conducted to efficiently utilize such big data.

At the same time, machine learning greatly contributes to processing/analyzing this overflowing data and creating new applications. Machine learning has made remarkable progress in the name of deep learning thanks to the development of hardware and the accumulation of data from the late 2000s to the present. This Special Issue, Machine Learning in Sensors and Imaging (https://www.mdpi.com/journal/sensors/special_issues/ML-SI), of *Sensors* contains a variety of studies that apply machine learning to efficiently utilize the data obtained through sensors in various fields. We started accepting papers in August 2020 and collected a total of 15 research results by January 2022, and covers various fields that utilize the data obtained from sensors and machine learning technologies, including human activity recognition, fuzzy classification, failure detection, sensor-less estimation, automatic camera calibration, telescope control, object detection, wildfire assessment, shelf auditing, forest monitoring, road management, denoising, and touchscreen.

I was honored to participate in this Special Issue as a guest editor, and I look forward to contributing to the revitalization of machine learning research related to sensors, and opening up a new future.

Hyoungsik Nam
Editor

sensors

MDPI

Article

Computer Vision-Based Path Planning for Robot Arms in Three-Dimensional Workspaces Using Q-Learning and Neural Networks

Ali Abdi [1,2], Mohammad Hassan Ranjbar [2] and Ju Hong Park [1,*]

[1] Department of Convergence IT Engineering, Pohang University of Science and Technology (POSTECH), Pohang 37673, Korea; abdiali@postech.ac.kr

[2] School of Mechanical Engineering, College of Engineering, University of Tehran, Tehran 11155-4563, Iran; mhranjbar@ut.ac.ir

* Correspondence: juhpark@postech.ac.kr; Tel.: +82-54-279-8875

Abstract: Computer vision-based path planning can play a crucial role in numerous technologically driven smart applications. Although various path planning methods have been proposed, limitations, such as unreliable three-dimensional (3D) localization of objects in a workspace, time-consuming computational processes, and limited two-dimensional workspaces, remain. Studies to address these problems have achieved some success, but many of these problems persist. Therefore, in this study, which is an extension of our previous paper, a novel path planning approach that combined computer vision, Q-learning, and neural networks was developed to overcome these limitations. The proposed computer vision-neural network algorithm was fed by two images from two views to obtain accurate spatial coordinates of objects in real time. Next, Q-learning was used to determine a sequence of simple actions: up, down, left, right, backward, and forward, from the start point to the target point in a 3D workspace. Finally, a trained neural network was used to determine a sequence of joint angles according to the identified actions. Simulation and experimental test results revealed that the proposed combination of 3D object detection, an agent-environment interaction in the Q-learning phase, and simple joint angle computation by trained neural networks considerably alleviated the limitations of previous studies.

Keywords: path planning; Q-learning; neural network; YOLO algorithm; computer vision; robot arm; target reaching; obstacle avoidance

Citation: Abdi, A.; Ranjbar, M.H.; Park, J.H. Computer Vision-Based Path Planning for Robot Arms in Three-Dimensional Workspaces Using Q-Learning and Neural Networks. *Sensors* **2022**, *22*, 1697. https://doi.org/10.3390/s22051697

Academic Editor: Hyoungsik Nam

Received: 6 January 2022
Accepted: 20 February 2022
Published: 22 February 2022

1. Introduction

Intelligent robot arms can play a crucial role in automation. The extensive use of industrial [1], surgical [2], and home robots [3] are examples of applications in which robot arms have become indispensable. Many robot arms are synchronized to accomplish a task by using a program or are remotely controlled by human operators. Intelligent robot arms attached with numerous sensors and cameras have attracted considerable research attention [4]. These robots have powerful onboard processors, high memory capacity, and artificial intelligence (AI)-based algorithms. These features enable such robots to replicate human capabilities. Intelligent robot arms gather information regarding their environment to make decisions in real time.

Intelligent robot arms are also increasingly being used in numerous applications. For instance, in smart cities [5], these kinds of robots are used to scan buildings and generate automated three-dimensional (3D) reconstruction. In computer vision and computer graphics, 3D reconstruction is used to detail the shape and appearance of physical objects, define 3D profiles, and determine the 3D coordinates of any point on the profile. Furthermore, 3D reconstruction has applications in many fields, such as medicine, free-viewpoint video reconstruction, robotic mapping, city planning, gaming, virtual environments, virtual tourism, landslide inventory mapping, robot navigation, archaeology, augmented reality, reverse engineering, motion capture, gesture recognition, and hand tracking [6].

In addition to 3D reconstruction, intelligent robot arms can play an essential role in other applications. For example, in smart factories [7], intelligent robot arms can be used in production, manufacturing, assembly, and packing in various industries. Moreover, in smart hospitals, robot-assisted surgery allows doctors to perform many types of complex procedures with higher precision, flexibility, and control than is possible with conventional techniques. In smart homes, robot arms can assist people with disabilities, elderly individuals, and parents rearing children to considerably improve their quality of life. The accuracy, speed, and efficiency of robot arms enable them to perform daily work effortlessly.

Although intelligent robot arms can be used in many industries, developing robotic arm applications remains highly challenging. One crucial and difficult problem is to ensure that robot arm path planning is accurate, safe, and efficient [8]. Path planning refers to bringing the end-effector to the target without hitting obstacles. It depends on various algorithms that dictate the motion of the robot arm, and determine how a robot arm should approach, process, and orient itself for optimal productivity and collision avoidance. Numerous computer vision-based path planning approaches were reviewed. Given that the end-effector of robot arms is critical for achieving their goal, bringing the end-effector to the desired location with obstacle avoidance is a key challenge for these robots.

This study aimed to overcome certain limitations of current computer vision-based path planning of robot arms by using AI techniques to develop an intelligent robot arm with high performance, safety, and speed.

To this end, although we tried to address those limitations in our previous paper [9], some of them remain. This article that is, in fact, an extension of our previous study, tries to address the limitations of prior work. In the previous study, we developed a novel hybrid path planning method using Q-learning and neural networks. An action finding (active approach) and angle finding (passive approach) were the two components of the hybrid path planning system. The Q-learning algorithm was used in the active phase to determine a series of simple movements, such as going up, down, left, and right to reach a target cell in a two-dimensional (2D) grid workspace. In the passive phase, a neural network is trained to determine the joint angles of the robot arm with respect to the observed actions. According to our findings, this hybrid technique considerably improved the speed and reduced the complexity of the operation of the system. However, this study had the following limitations.

1. Although many applications require 3D movement, the scope of work was limited to the 2D workspace.
2. Finding a start, an obstacle, and a target point through their colors may negatively affect image processing accuracy depending on the ambient light. Obstacle detection based on shapes would be preferable.
3. Because the KNN algorithm was used in this study, a start, an obstacle, and a target cell were required to have distinct colors.
4. Only one obstacle could be located in the workspace unless distinct colors were used for each obstacle.

Therefore, in this study, we extended the scope of our previously proposed method from a 2D space to a 3D space and incorporated real-time object detection and localization for real-world applications.

In this article, the proposed path planning method comprises three phases: (1) detecting the spatial coordinates of a start, target, and obstacle object, (2) finding the optimal path from a start to a target object while avoiding an obstacle object, and (3) calculating the corresponding angles for the six joints of a robot arm. In the first stage, the spatial coordinates of a start, target, and obstacle object were recognized in a 3D workspace using a combined object detection technique and a neural network with two images captured using two cameras from different views. This stage required a training process based on the created dataset of the start, target, and obstacle shapes and their morphologies. In the next step, the Q-learning algorithm is used to determine the optimal actions in a gridded 3D workspace so that a robot arm could begin traveling from the start cell to the target cell without collision with an obstacle. In a Q-learning algorithm, states are represented by cells in a 3D workspace, and forward, backward, right, left, down, and up are defined as actions of the robot arms. Finally, in the next step, a trained neural network is used to calculate the angles for the six joints of a robot arm based on actions to place the end-effector in the required location in 3D space. The use of a neural network for this stage greatly reduces the calculation time and computing cost.

This approach exhibited high speed, low computational cost, and automated path generation for various situations. A real-time object detection algorithm is obtained by combining neural network and object detection. Furthermore, finding optimal actions by Q-learning algorithm and calculating robot arm joint angles with trained neural networks enabled us to create a precise, efficient, and fast method to address computer vision-based path planning issues for real-life applications. The remainder of this paper is organized as follows. In Section 2, the related works are presented. In Section 3, our novel computer vision-based path planning technique is proposed. In Section 4, the experimental results are presented and analyzed. In Section 5, the discussion is presented. Finally, in Section 6, the conclusion is presented.

2. Related Works

Prior to the development of a new computer vision-based path planning approach, numerous existing approaches were reviewed. Methods, such as probabilistic road map (PRM), artificial potential field (APF), rapidly exploring random tree (RRT), and reinforcement learning (RL)-based approaches, have been proposed. The most important of which are as follows.

Ka et al. presented a vision-based assistive robot arm assistance algorithm for a JACO robot in which a low-cost 3D depth-sensing camera and an improved inverse kinematic algorithm were used to enable semiautonomous or autonomous JACO operation [10]. Rai et al. proposed an autonomous robotic framework for academic, vocational, and training purposes. They used two webcams that provided the top and side views to consider the objects of various heights for positioning a robotic gripper at the center of the target [11]. Hsu et al. proposed a control design and implementation of an intelligent vehicle combined with a robotic manipulator and computer vision [12]. Chen et al. demonstrated the potential of combining augmented reality-based brain–computer interface and computer vision to control robotic arms. They employed hue, saturation, and lightness space (HSV) to provide the object's position and color in the 2D workspace. The objects they utilized were simply 2D disks with placements in a conventional gridding pattern [13]. Whang et al. used a popular object detection model, the faster R-CNN model, to detect nails and screws in construction waste recycling robots. Their result shows that the model's mean average precision (AP) for nails and screws was 0.891. This precision was well in their application [14]. Tebbe et al. demonstrated an innovative table tennis robot system with high precision vision detection and rapid robot response. They used a multi-camera calibration approach and iterative triangulation to reconstruct the 3D ball position with a 2.0 mm precision. They used classic image processing techniques and integrated color and background thresholding to detect the flying ball with higher velocities in real-time [15]. Sadhu et al. proposed an improvised FA that involved the Q-learning framework within itself for robot arms path planning. In this proposed Q-learning induced FA (QFA), the optimal parameter values for each firefly of a population were learned by the Q-learning strategy during the learning phase and applied thereafter during execution [16]. Wen et al. presented a new obstacle avoidance algorithm based on deep deterministic policy gradient (DDPG). Specifically, they proposed to use DDPG to plan the trajectory of a robot arm to realize obstacle avoidance [17]. Zhang et al. proposed a path planning method based on Q-learning for robot arm due to its simple and well-developed theory [18]. Huadong et al. analyzed the characteristics of obstacle avoidance path planning to improve the efficiency and accuracy of obstacle avoidance path planning [19]. Das et al. proposed a novel method of energy-efficient path planning of an industrial robot arm in a workspace with multiple obstacles using differential evolution (DE) algorithm [20]. Raheem et al. analyzed the shortest path and trajectory planning of a two-link robot arm with 2-DOF in the 2-D static known environment [21]. Chang et al. presented an automatic path planning of a six-axis robot for intelligent manufacturing based on network remote controlling and simulation [22]. Sugiura et al. determined an optimal path by computing the gradient of an equation using the APF method [23]. However, the algorithm may encounter local minima, rather than the absolute minimum, and therefore the shortest path may not be identified, as reported by Martínez et al. [24]. Kavraki et al. used Dijkstra's algorithm to calculate the shortest paths between nodes on a graph in the PRM method [25]. The path created by the sampling-based approach may not be optimal because the resultant path mainly depends on sample procedures, as reported by Hsu et al. [26]. Liu et al. reported that RRT can avoid precise environmental modeling and reduce calculations [27]. However, the convergence speed of motion planning is sluggish because of the random sampling and global uniform sampling technique of rapidly expanding random trees in redundant spaces. Therefore, producing the optimal path in a short period of time becomes challenging, as reported by Karaman et al. [28]. Prianto et al. used the soft actor-critic (SAC) deep-learning-based method for path planning. Because of the use of the entropy term in the goal function, the SAC exhibits high exploration capabilities for path planning [29]. Panov et al. investigated novel grid path planning with deep RL outcomes. Furthermore, they demonstrated the robust learning ability of a neural Q-learning agent on tiny maps and achieved promising results on new maps [30]. Low et al. suggested improved Q-learning and demonstrated its efficacy through experimental investigations [31]. Yu et al. proposed a neural-network-based path planning model for mobile robots based on hierarchical RL and compared this model to other algorithms. The results revealed the smoothness of the planned path and usable generalization in various scenarios by using an obstacle avoidance method based on non-uniform rational B-splines (NURBS) for robot arms [32]. It was also found that the NURBS method was highly effective for avoiding collisions [33].

These methods have distinct advantages and disadvantages. Certain methods have irregular paths, require preplanning, determine a non-optimal path, obtain cubic graphs, and are slow. Other methods have drawbacks, such as high complexity or are limited to certain conditions. Because robot arm performance cannot be generalized, some robot arm setups can become highly complex. The development of a novel path planning method that identifies and solves these limitations is essential for improving robot arm performance and production. The proposed approach meets this requirement with some advantages that will be presented in the next sections.

3. Methods

The proposed procedure can be summarized as follows: (1) detecting the coordinates of the bounding box of a start, target, and an obstacle by using an object detection algorithm. Then, converting them to their spatial coordinates using a neural network, (2) determining optimal actions using Q-learning, and (3) calculating the rotation angles of the joints of a six-degree-of-freedom robot arm using a neural network. This section describes these in detail.

3.1. Object Detection and Spatial Coordinates (Combined YOLO-Neural Networks 1)

In the first stage, the coordinates of the bounding boxes of a start, target, and obstacle are obtained in the 3D space. These coordinates can be automatically determined for real-time applications. In the proposed approach, an object detection algorithm and a neural network were combined. In an object detection algorithm, both object recognition and object localization in an image can be achieved. These algorithms can distinguish between a start, an obstacle, and a target object, and extract their bounding box coordinates. Because we required 3D coordinates, we extracted XYZ coordinates using two cameras with two planes, as displayed in Figure 1.

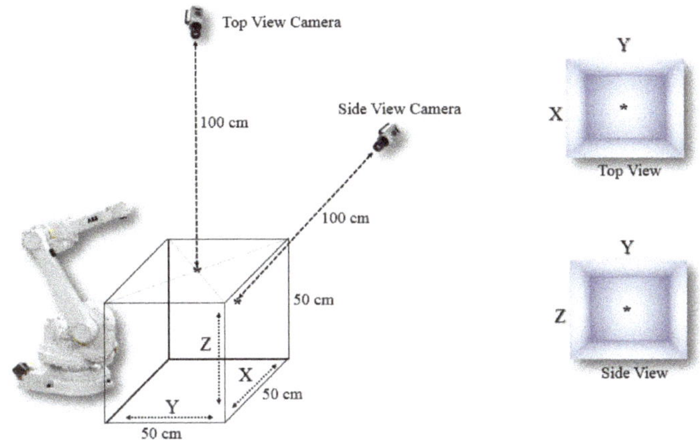

Figure 1. Workspace size and locations of cameras.

Many object detection algorithms, including SqueezeDet, MobileNet, R-CNN, fast R-CNN, mask R-CNN, single-shot detector, and you only look once (YOLO), have been proposed [34–38]. An algorithm can be evaluated from several perspectives, such as speed of detection and accuracy of identification. Speed is particularly important because targets may constantly change, so target detection should be performed in real time, allowing new paths to be computed. In [39], the accuracy and speed of various object detection algorithms were compared. The results revealed that the YOLO algorithm is accurate and operates at a high speed. Therefore, the YOLO algorithm was chosen.

In YOLO, object detection is regarded as a regression problem involving spatially separated bounding boxes and associated class probabilities. Full images are evaluated once, and a single neural network predicts bounding boxes and class probabilities based on those images alone. The use of a single network to monitor the detection process allows the network to be tuned to obtain the best results. We mainly used one variant, tiny-YOLOv4, and extracted the bounding boxes of a start, an obstacle, and the target to find the approximate coordinates of the bounding boxes of all objects with perspective correction (Figure 2). The cells within the bounding box of two perpendicular planes were classified as objects; thus, the cells of objects in the 3D space were extracted.

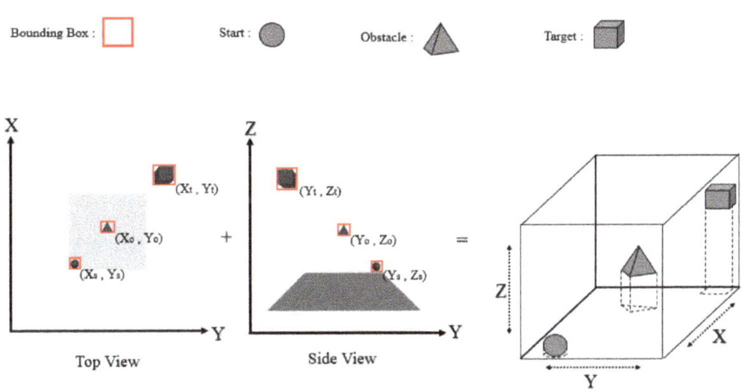

Figure 2. YOLO object detection in two views.

We used three objects to represent a start, an obstacle, and a target in this study. A start, an obstacle, and a target object were represented by a sphere, a pyramid, and a cube, respectively, as displayed in Figure 2. A dataset was created based on these three classes, and photographs were labeled using the YOLO standard labeling method (where each photograph is labeled with LabelImg, which is a free, open-source tool for graphically labeling images). Table 1 presents the details of the dataset.

Table 1. Details of the dataset.

Shape	Number of Data	Class
Pyramid	200	0
Sphere	200	1
Cube	200	2

In this stage, the spatial coordinates of each detected object are determined. To this end, we used a trained neural network to determine the X-, Y-, and Z-coordinates of these three classes. The inputs of the neural network were the class of objects, the center, and the sizes of the bounding boxes obtained by the YOLO algorithm.

A six-layer neural network topology with four hidden layers containing 16, 32, 64, and 16 neurons was used. The outputs were the spatial coordinates of the corresponding plates. Figure 3 depicts the neural network architecture and the mean absolute error.

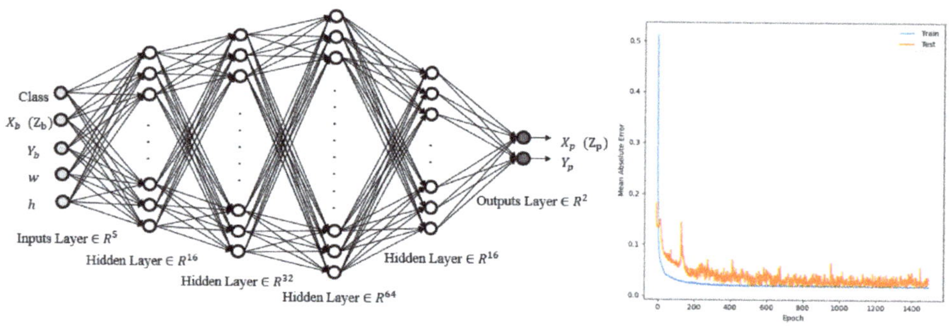

Figure 3. Structure of the first neural network and its mean absolute error convergence.

It is worth mentioning that we cannot use a simple mathematical problem to transform two 2D projected coordinates into a 3D coordinate. That is because many parameters influence the final accuracy of coordinates in real-world applications. Some of these factors are the focal length and distortion impact of the lens, camera sensor size, perspective effect, and positioning items behind another. When the focal length effect is paired with the perspective effect and the object is near the borders of an image, finding coordinates with high precision is a difficult operation using traditional mathematical approaches. Rather than a complex procedure, we proposed a combined YOLO-neural networks method to identify the coordinates of objects in pixel-based space with two cameras (YOLO) and convert the pixel-based space to 3D XYZ real-world coordinates (neural network). Using this method, we were able to extract 3D coordinates with acceptable precision in various positions of the objects. We could also reduce the detection time to roughly 0.04 s, allowing us to employ this approach in real-time applications.

We transformed the dimensions of the 3D workspace from 50 $\times$ 50 $\times$ 50 cm to 8 $\times$ 8 $\times$ 8 discrete cells, as displayed in Figure 4. This conversion simplified and accelerated the overall calculation process. According to the grid workspace and the bounding boxes of the start, obstacle, and target, the corresponding start, obstacle, and target cells can be identified. These labeled cells are used in the Q-learning algorithm, as explained in the following section.

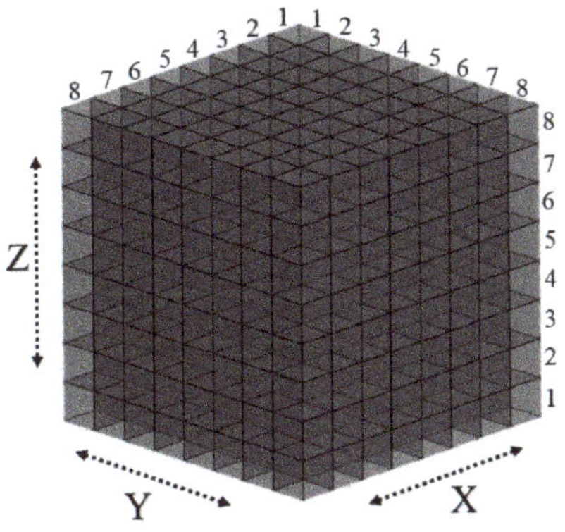

Figure 4. Gridded 3D workspace.

3.2. Action Finding (Q-Learning)

An action-finding process provides immediate action and state in the Q-learning technique of the cells after determining a start, an obstacle, and a target. In machine learning, Q-learning is an off-policy technique that focuses on how intelligent agents should function in a particular environment to optimize cumulative reward. An RL agent engages with its environment in discrete time steps. At each point t, the agent receives the current state $s_{\{t\}}$ and the reward $r_{\{t\}}$. As soon as action $a_{\{t\}}$ has been selected, it is conveyed to the environment. With each iteration, the environment changes to a new state ($s_{\{t+1\}}$), and the reward ($r_{\{t+1\}}$) associated with the change is computed (for each iteration). In RL, the projected total reward for the RL agent is maximized. This research utilized the Q-learning technique to identify a straightforward sequence of forward, backward, up, down, left, and right actions to maximize accumulated rewards. The method is inspired by the "windy grid world" problem. We divided the 3D workspace into several cells based on a designated resolution. As detailed in Sutton and Barto's original "windy grid world" problem [40], the main purpose was to travel from one cell to its adjacent cells until the target cell was reached. The difference between this technique and the RL-based methods mentioned in the background section is that in this method, a direct action is determined to move from a cell (state) to its neighboring cells instead of finding a sequence of joint angles for the same movement. This assumption simplifies and speeds up the action-finding process. Using this approach, state space and action space can be expressed as follows:

$$S = \{cell_1,\ cell_2,\ cell_3,\ \ldots,\ cell_{n \times m \times l}\}. \tag{1}$$

The overall width, length, and height are represented by n, m, and l, respectively. The higher the required resolution is, the more cells will be needed. The total number of members in the state space is $n \times m \times l$.

$$A = \{\ Up,\ Down,\ Left,\ Right,\ Forward,\ Backward\} \tag{2}$$

In this study, the action space has six members, namely forward, backward, up, down, left, and right, as shown by Equation (2). The state space and the action space are small, and an algorithmic model can be easily developed for the interaction of a robot arm within this reduced environment. This technique can drastically reduce the execution time of an algorithm and simultaneously decrease its complexity.

Before commencing the Q-learning process to discover the optimum path, the locations of the start, target, and obstacle were identified. These three positions were randomly arranged in each test in the workspace. In this technique, the 3D workspace was a portion of the total workspace available in front of the robot arm.

We selected forward, backward, up, down, left, and right as the total action space. Although the number of actions can be increased, the computing cost increases. Additional improvements in terms of resolution, defined as the magnitude of the end-effector motion in each step, may exponentially increase the search space, which increases the computational burden. However, reducing the actions may lead to path discovery failure. Hence, there is a trade-off between speed and efficiency on the one hand and the probability of success on the other. The time required to run the Q-learning code is directly proportional to the number of cells. Therefore, establishing an appropriate resolution and speed for a specific application considering the dimensions of the robot arm and objects is critical. We used a 3D grid workspace of $8 \times 8 \times 8$ cells (states) for this study, considering the sizes of the three objects and the speed of path planning. Therefore, this design implied a total of 512 cells (states) in which the end-effector could be placed. The start, target, and obstacle points were located within these cells. The Q-learning method was then used to determine the optimum actions from the start to the target cell under various reward and penalty scenarios.

Seijen revealed that in a "windy grid world" problem, Q-learning outperformed other algorithms, such as SARSA, to obtain the highest accumulated reward [41]. Therefore, we used Q-learning as the RL algorithm in the action-finding phase in this study. In this algorithm, if the robot arm successfully reaches the target cell, it earns a reward of 50 points. By contrast, if it reaches an obstacle cell, the agent receives a penalty of -100 points. Subsequently, all further acts result in a penalty of -1 point. The goal of the agent is simply to maximize its score. In the Q-learning algorithm, first, the best sequence of actions from the start cell to the target cell with avoidance of the obstacle cell is performed. A 3D grid workspace was set up as follows in this study:

- In an $8 \times 8 \times 8$ 3D grid workspace, six possible actions, namely forward, backward, up, down, left, and right, were considered.
- An agent starts from a randomly located start state and receives a reward of 50 points for reaching a randomly located target state.
- An agent receives a penalty of -100 points for reaching a randomly located obstacle cell.
- All other actions cause a penalty of -1 point.

This action-finding section created a 3D matrix of the robot arm's most rewarding actions in each cell. These best actions can be followed from a start cell to a target cell to create the best actions or optimal path sequence. In the next section, we describe the conversion of these actions to angles that rotate the joints of a robot arm.

3.3. 6-DOF Angle Finding (Neural Networks 2)

In the final stage, we need to transform the sequence of actions into a sequence of joint angles. One way is the use of inverse kinematics (IK). The idea behind IK is to calculate the joint angles of the robot arm for a given position of the end-effector. The joint configurations needed for each intermediate time step along the trajectory are calculated. However, using IK in each step makes the method extremely slow. Another way is the use of a trained neural network. The neural network does not need to be trained every time; instead, it is trained only one time, and this trained neural network is used every time during path planning.

In this study, an angle-finding process is performed by using a trained neural network to obtain the joint angles of each specified action in a particular cell. We trained a neural network for angle finding using RoboDK software because RoboDK can provide a 3D model of our robot arm. Because the joint angles required to move the end-effector of a robot arm in each direction are dependent on the present state of the end-effector, the current cell must be considered by the neural network at each stage of the process. Therefore, the neural network must be fed with the indices of the row, column, and height of the cell as well as the action that has been selected as inputs. The outputs of the neural network should be the joint angles that supply the location of the next cell corresponding to the action. Notably, because the input must be in numerical form, actions such as up, down, left, right, forward, and backward are labeled numerically as 1, 2, 3, 4, 5, and 6, respectively.

Furthermore, the number of outputs is proportional to the degrees of freedom (DOF) of the robot arm. The number of outputs in the robot arm in this research, with six joints, was therefore six.

We used a five-layer neural network topology with three hidden layers containing 4, 10, and 6 neurons. The number of hidden layers and neurons was determined empirically through trial and error during the training processes. Four inputs, namely the row, column, and height indices of a current cell, as well as an action to be performed, were considered. The robot arm used in this study, a model IRB 1600 with a 1.45 m arm length, had six DOF. The rectified linear unit (ReLU) is the activation function for input and output neurons because it is the most often implemented function and is not complex.

$$\sigma_{ReLU}(x) = \max\{0, x\} \tag{3}$$

First, we collected datasets to train a neural network according to the designed structure by collecting precise data with an end-effector placed in various grid cells. We considered six neighboring cells of the central cell, where the end-effector was located, to be six data points. These six cells were those orthogonal to the central cell, and reachable by moving up, down, left, right, backward, or forward. We collected 3072 data points for neural network training. Of these data points, 90% were used in the learning process, and the rest were used in the testing and validation process. These data points were sufficient to train a neural network properly with an 8 × 8 × 8 grid workspace. Because the research used 512 cells (8 × 8 × 8 grid), 3072 moves were possible, and each operation in a cell could be performed in six directions. Thus, we used the entire set of information. Because the dataset was sufficiently large to include all cell movements, the dataset was considered adequate. Six possible outcomes were possible when the robot was in cells on the edge of the workspace. The robot ignored actions to move outside the workspace.

Initially, we trained a neural network with weights with random values. The output value was then calculated for each training sample. The weights were then updated using a backpropagation method and a gradient descent process. This procedure was repeated until the weights reached their optimal levels and the error ranges were within the permitted limits. Figure 5 depicts the neural network architecture and the mean absolute error.

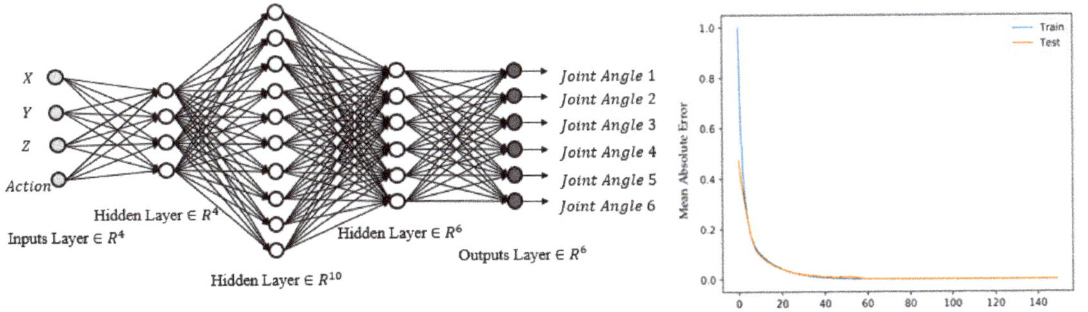

Figure 5. Structure of the second neural network and its mean absolute error convergence.

A multilayer perceptron with a ReLU activation function first proposed by Paul Werbos [42] was used in this study. In this approach, a neural network is trained only once and then used for determining angles. A lengthy training period is thus reduced to a single step before path planning. The trained neural network required only a few seconds for use and could be used repeatedly.

The proposed computer vision-based path planning method can be summarized as follows:

1. Capturing a snapshot of the 3D workspace with the two cameras;
2. Detecting a start, target, and obstacle cell using the YOLO object-detection algorithm

3. Obtaining the spatial coordinates of three objects using the first neural networks;
4. Using the Q-learning method to determine an optimal route from a starting point to a target point while avoiding obstacle collision.
5. Finding the joint angles of the discovered actions using the trained neural network;
6. Implementation in the actual or simulated world of the acquired joint angle sequence on the robot arm.

As displayed in Figure 6, we used a simple and fast technique in each phase of our process to provide an efficient path planning process. The use of low-quality images for the first stage could considerably speed up the picture analysis process because a start, an obstacle, and a target object shape could be identified from low-resolution photographs by using the algorithm. First, an object-detection method was used for object localization because it is one of the most accurate and rapid algorithms. Next, a trained neural network was used to obtain the spatial coordinates of these objects. We used the Q-learning algorithm to determine basic actions and determine the route with the highest reward. Next, the rapid conversion action of an experienced neural network was used for determining joint angles. In the next section, we discuss our testing of the effectiveness of the proposed method through simulation.

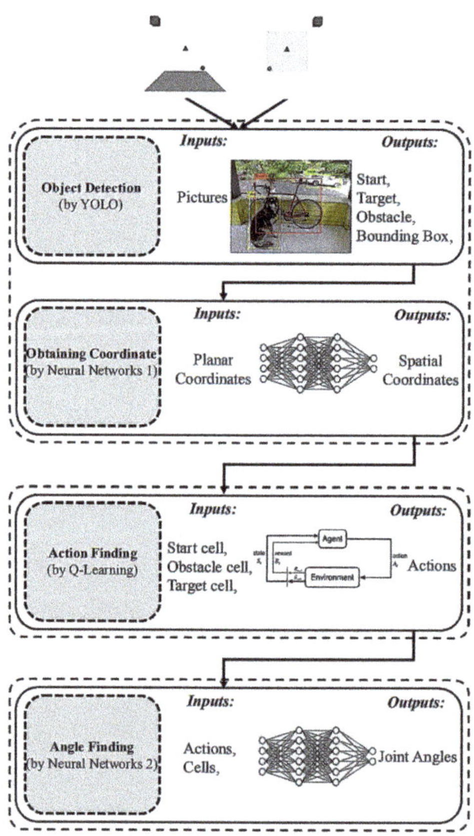

Figure 6. Method overview: 1. capturing pictures; 2. object detection by YOLO; 3. obtaining spatial coordinate by the first neural network; 4. action-finding by Q-learning; 5. angle-finding by the second neural network.

3.4. Simulation

To ensure that any proposed method is safe and provides desired results, simulation is essential before experimental implementation. First, we developed a 3D workspace by randomly locating a sphere, pyramid, and cube as a start, obstacle, and target object by using VPython, a library that allows users to create objects, such as spheres, cones, and other forms in 3D space. Two pictures displaying two views, namely the top and side views, were obtained. These two pictures are the inputs of the object-detection module. This module was configured to detect three classes of objects, namely a sphere, a pyramid, and a cube, in addition to finding the center of the bounding boxes as well as their dimensions.

For the next step, note that the outputs of the object detection (YOLO) algorithm were the inputs of the first neural network. Therefore, it has five inputs, namely (1) the class of an object, (2) the X- and (3) the Y- (or the Y- and the Z)-coordinates of the center of the bounding box, (4) the width, and (5) the length of the bounding box. It has two outputs, the X- and the Y- (or the Y and the Z) coordinates of an object in a spatial coordinate system. The coordinates of the center of the bounding box differ from the coordinates of the object in a spatial coordinate system because of the use of perspective pictures.

These spatial coordinates are then converted to a start, obstacle, and target cell. These cells are the inputs of the Q-learning algorithm, which finds the optimal path from the start cell to the target cell. In this simulation, we placed an obstacle object between the start and target objects to evaluate the performance of the Q-learning algorithm in terms of obstacle avoidance.

Next, the outputs of the Q-learning algorithm, including a list of actions and cell indexes, were used as inputs for the second neural network. Thus, four inputs, including the X-, Y-, Z-coordinates of the current cell and its corresponding actions, existed in this neural network. The outputs were six joint angles that indicated the next cell of the optimal path. Next, the end-effector of a robot arm in a RoboDK simulator started moving from a start cell and followed the generated optimal path according to the obtained sequence of joint angles. Figure 7 displays the results of the simulation.

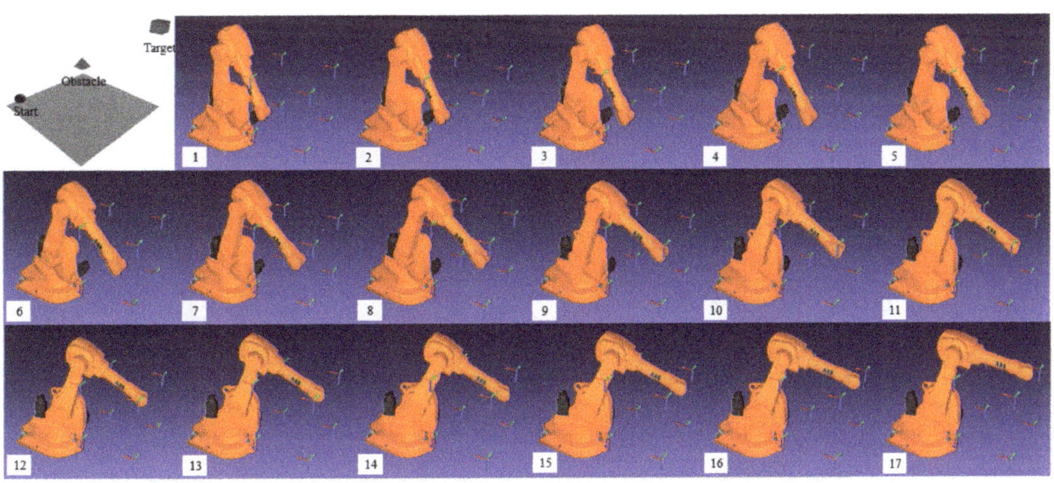

Figure 7. Simulation results in a RoboDK software: (**1–17**). The end-effector of a robot arm starts moving from a start point (Sphere) and avoid an obstacle point (Pyramid). Then reaches a target point (Cube).

4. Experiment Results

We tested the proposed method in a physical environment and verified its validity. The experimental setup (Figure 8) included a 6-DOF robot arm, two cameras, a computer, a 3D workspace, and three objects that represent a start, obstacle, and target point.

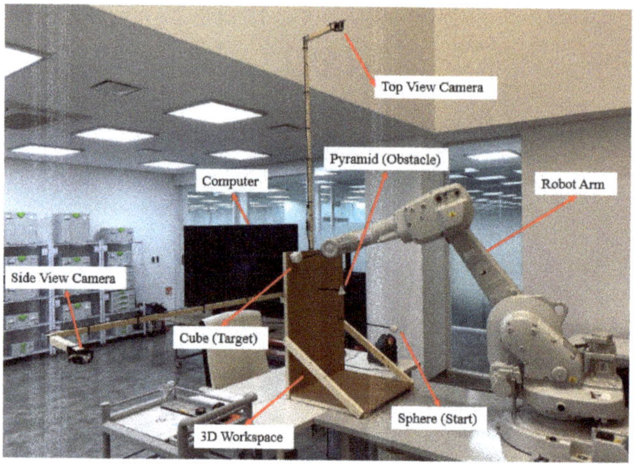

Figure 8. Experimental setup: cameras, robot arm, workspace, sphere (start), pyramid (obstacle), cube (target).

The test setup was similar to the simulated environment. The similarity was to enable comparison between the test results and simulation results. We installed two cameras and a robot arm, as was done in the simulation. Furthermore, a sphere (a start object), a pyramid (an obstacle object), and a cube (a target object) were placed in locations identical to those in the simulation. To ensure the physical test conditions were identical to those of the simulation, the specifications of the lens used in our cameras were identical to those of the virtual lens used in the simulation. Because of this similarity, the method presented in this article could be validated if physical robot test results were identical to the simulation result from the RoboDK software.

We placed a sphere (a start object) at the low-left corner, a pyramid (an obstacle object) in the middle, and a cube (a target object) at the upper-right corner of the 3D workspace. As displayed in Figure 9, the end-effector of the robot arm precisely followed the generated path, which was consistent with the results of the simulation test. Thus, the results of this physical test revealed that the proposed method exhibited satisfactory performance.

The results show that the robot follows the path well. This path was actually the best actions (optimal polices) of each cell (state) obtained by the Q-learning algorithm. When the Q-learning is running, it tries to find the best action that the agent can do in each state through interaction with the environment. Therefore, in each state, there is an action that is considered optimal action. In other words, the optimal policy is a 3D matrix ($8 \times 8 \times 8$) whose elements are a letter such as "U", "D", "L", "R", "B", "F" that represent Up, Down, Left, Right, Backward, Forward respectively. Figure 10, displays the optimal policy of our test. The letter "G" and "O" (is not found in the Figure 10) represent Goal (or Target) and Obstacle cell.

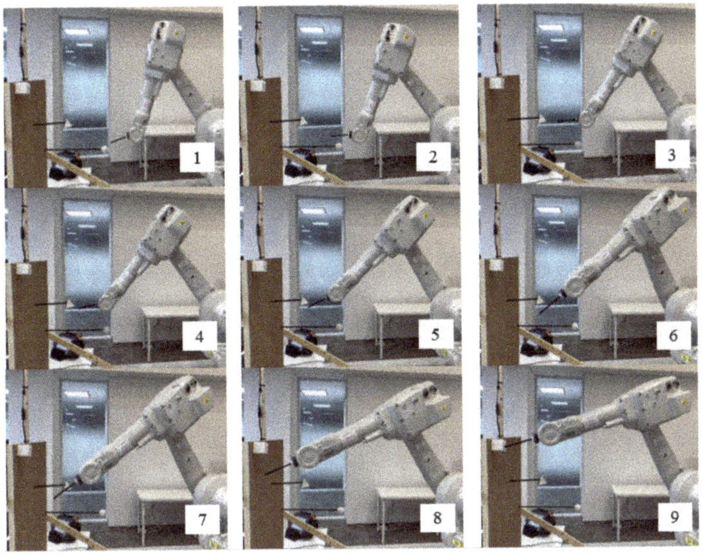

Figure 9. Real-world test: (**1**) The robot starts moving from the start point (Sphere). (**2–8**) The robot track the found path while avoiding the obstacle. (**9**) The robot reaches the target (Cube).

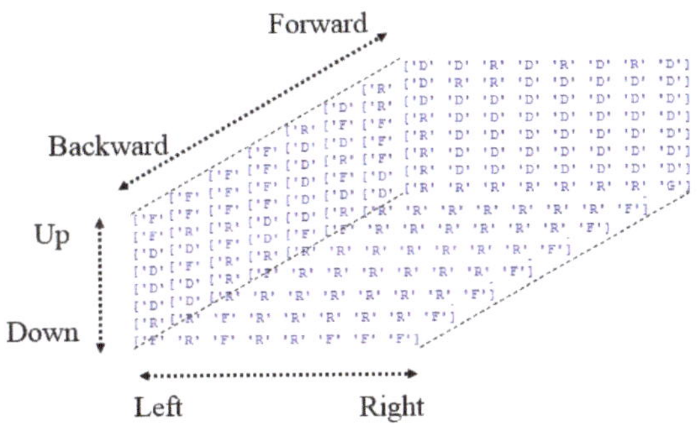

Figure 10. The optimal policy of each state (cell).

What is very important is that the same optimal policy matrix can be used for each start point if the location of the target and obstacle remains fixed. The extracted path from the Q-learning results for our test is {F,F,F,F,R,F,F,F,D,R,R,D,D,D,D,D,D,R,R,R,G}. This sequence of optimal actions begins from the start cell in which the sphere is placed. If we change the location of the sphere, we do not require to run Q-learning again; we can use the same optimal policy matrix to find a new sequence of actions from the new start cell. This make the path generation more straightforward and fast.

Another important point is that in the Q-learning algorithm, there is an optimal episode limit. When the episode limit is too high, it takes a long time for an agent to become an experienced one. However, reducing the episode limit may lead to gaining not enough experience by the agent, which means it cannot find the best actions for states. Hence, there is a trade-off between speed on the one hand and gaining enough experience on the other. In this test, the optimal episode limit is equal to 2000, which guarantees speed and enough experience. This number was achieved through trial and error. In the next section, we discuss our method in general and give some suggestions for its improvement.

5. Discussion

In this section, we discuss the proposed method's strengths and weaknesses. To this end, first, we compare our method with other methods in terms of speed which is a crucial feature in real-time path planning. To compare this method with other path planning methods, we refer to our previous study [9]. In that paper, we compared the speed of running time in our hybrid path planning method and other conventional path planning methods such as RL-based, APF, PRM, and RRT. In order to compare them, we ran a simple code of each method on the same computer to compute its running time. As we reported, our hybrid method required less running time to do the path planning process under the same conditions, which means it is faster than traditional methods due to using separate active and passive approaches. This comparison is discussed in detail in the previous work and can be referred to for more information.

In addition to strengths, this study has some limitations that should be taken into account in future studies. These limitations and our suggestions are presented in the following.

1. Although we tested the method using a physical robot arm, the physical setup was not exactly identical to the simulation because levitating multiple objects in the air and the middle of the workspace is difficult due to the use of fixtures and wood structures. Certain parts of a robot arm, such as the end-effector, may collide with fixtures. Therefore, in the future, the feasibility of using augmented reality (AR) for virtual test setup in physical robot tests should be considered. Furthermore, the efficacy of using a virtual sphere, pyramid, and cube instead of using physical objects that require fixtures should be evaluated. Such augmented tests can considerably reduce expenses and facilitate the rapid implementation of complex settings. Moreover, a more comprehensive analysis is possible than in an experimental test.
2. In future studies, trained deep Q-learning can also be incorporated into the proposed method to learn additional possible pathfinding to increase the speed of the action finding section. To use a deep Q-learning technique, a large dataset is required, which is currently not possible. This approach can drastically reduce the inference time, but concerns regarding high computing costs remain.
3. We used two cameras in two views which made our method less practical, especially for those applications that need a portable robot arm. One suggestion is that we use a single perspective picture taken by deep cameras then generate a path.
4. Our invention just takes shots at the beginning and a robot moves following a generated path. In order to be a realistic application, it should take many pictures (one per sec) and re-generate a path. Therefore, the system could be an intelligent real-time system.
5. We used a combined object detection-neural network method to calculate its spatial coordinates. It is suggested using 3D reconstruction methods to calculate objects' position.
6. To make the method more practical, further study may use actual objects such as a cup, pen, monitor, book, etc.

In the end, in order to make the proposed method more useful and practical, a list of useful applications is mentioned.

- It could be used for a harvesting robot to collect fruits from trees avoid obstacles.
- It could be used for a recycling robot to pick up bottles, cans, batteries, or other particular objects in a recycling factory in real-time.
- It could be used for warehouse robots to pick selected items from shelves or totes and place them into shipping containers to fulfill orders.
- It could be used for assembly line robots to pick components and place them at an appropriate location.

6. Conclusions

A novel computer vision approach was proposed for effective path planning by combining Q-learning and neural networks for robot arms. In the proposed approach, computer vision and neural networks were combined to obtain accurate spatial locations of a start, an obstacle, and a target object in real time. In the 3D workspace, a sphere, a pyramid, and a cube were used to represent the start, obstacle, and target points, respectively. Two images from two views were inputted into a trained YOLO algorithm to detect the aforementioned items and find bounding boxes for each detected object. A trained neural network converted the bounding boxes into spatial coordinates. Next, a Q-learning algorithm determined the optimal sequence of actions in a 3D workspace from a start cell to a target cell and simultaneously avoided obstacles. Next, a trained neural network converted the identified actions to the corresponding joint angles. Because the neural networks were trained before the path planning process, the method was fast. We tested this computer vision-based path planning algorithm through simulation and experimental methods. The results revealed that this research overcame the limitations of our previous research [9].

Future research directions are suggested as follows. First, the feasibility of using augmented reality (AR) for virtual test setup in physical robot tests could be considered. Second, the use of a trained deep Q-learning can also be considered to increase the speed of the action finding section. Third, the use of a single perspective picture taken by deep cameras could be considered to avoid using two cameras. Forth, instead of using pictures only one time, a live video can be used to make the method more real-time. Fifth, the use of 3D reconstruction methods to calculate objects' positions could be considered. Sixth, the use of actual objects such as a cup, pen, monitor, book, etc. could be considered to make the method more practical.

Author Contributions: A.A. raised the idea of 3D grid computer vision-based path planning, designed and trained the second neural network, performed the simulation on RoboDK software, surveyed a part of the background of this research, and wrote a part of the manuscript. M.H.R. designed and trained the object detection algorithm and first neural network, surveyed a part of the background of this research, and wrote a part of the manuscript. J.H.P. supervised and supported this study, prepared the experimental setup, and implemented the method on the real-world robot. All authors have read and agreed to the published version of the manuscript.

Funding: This research was supported by the MSIT (Ministry of Science and ICT), Korea, under the ICAN (ICT Challenge and Advanced Network of HRD) program (IITP-2021-2020-0-01822) supervised by the IITP (Institute of Information & Communications Technology Planning and Evaluation), and by the National Research Foundation of Korea (NRF) grant funded by the Korea government (MSIP; Ministry of Science, ICT and Future Planning) (No. 2019R1G1A1010859).

Institutional Review Board Statement: Not applicable.

Informed Consent Statement: Not applicable.

Data Availability Statement: Not applicable.

Acknowledgments: The authors appreciate the help of members of the Design Intelligence Lab.

Conflicts of Interest: The authors declare no conflict of interest.

Abbreviation

The following abbreviations are used in this manuscript:

APF	artificial potential field
PRM	probabilistic road maps
RRT	rapidly-exploring random tree
RL	reinforcement learning
DOF	degrees of freedom

References

1. Bugday, M.; Karali, M. Design optimization of industrial robot arm to minimize redundant weight. *Eng. Sci. Technol. Int. J.* **2019**, *22*, 346–352. [CrossRef]
2. Korayem, M.; Madihi, M.; Vahidifar, V. Controlling surgical robot arm using leap motion controller with Kalman filter. *Measurement* **2021**, *178*, 109372. [CrossRef]
3. Park, H.C.; Ahn, K.H.; Min, J.K.; Song, J.-B. 5 DOF Home Robot Arm based on Counterbalance Mechanism. *J. Korea Robot. Soc.* **2020**, *15*, 48–54. [CrossRef]
4. Chen, L.; Yang, H.; Liu, P. Intelligent robot arm: Vision-based dynamic measurement system for industrial applications. In *International Conference on Intelligent Robotics and Applications*; Springer: Berlin/Heidelberg, Germany, 2019.
5. Sharma, A.; Singh, P.K.; Hong, W.-C.; Dhiman, G.; Slowik, A. Introduction to the Special Issue on Artificial Intelligence for Smart Cities and Industries. *Scalable Comput. Pract. Exp.* **2021**, *22*, 89–91. [CrossRef]
6. Jezia, Y.; Samir, L.; Abdelmajid, B.A. Image-based 3D reconstruction precision using a camera mounted on a robot arm. *Int. J. Nonlinear Sci. Numer. Simul.* **2021**, *22*, 1–11. [CrossRef]
7. Sawangsri, W.; Suppasasawat, P.; Thamphanchark, V.; Pandey, S. Novel approach of an intelligent and flexible manufacturing system: A contribution to the concept and development of smart factory. In Proceedings of the 2018 International Conference on System Science and Engineering (ICSSE), New Taipei City, Taiwan, 28–30 June 2018.
8. Cheng, C.; Lv, X.; Zhang, J. Robot Arm Path Planning Based on Improved RRT Algorithm. In Proceedings of the 2021 3rd International Symposium on Robotics & Intelligent Manufacturing Technology (ISRIMT), Changzhou, China, 24–26 September 2021.
9. Abdi, A.; Adhikari, D.; Park, J.H. A novel hybrid path planning method based on q-learning and neural network for robot arm. *Appl. Sci.* **2021**, *11*, 6770. [CrossRef]
10. Ka, H.; Ding, D.; Cooper, R.A. Three dimentional computer vision-based alternative control method for assistive robotic manipulator. *Symbiosis* **2016**, *1*, 1–6.
11. Rai, N.; Rai, B.; Rai, P. Computer vision approach for controlling educational robotic arm based on object properties. In Proceedings of the 2014 2nd International Conference on Emerging Technology Trends in Electronics, Communication and Networking, Surat, India, 26–27 December 2014.
12. Hsu, Y.H.; Hsu, H.-Y.; Lin, J.-S. Control design and implementation of intelligent vehicle with robot arm and computer vision. In Proceedings of the 2015 International Conference on Advanced Robotics and Intelligent Systems (ARIS), Taipei, Taiwan, 29–31 May 2015.
13. Chen, X.; Huang, X.; Wang, Y.; Gao, X. Combination of augmented reality based brain-computer interface and computer vision for high-level control of a robotic arm. *IEEE Trans. Neural Syst. Rehabil. Eng.* **2020**, *28*, 3140–3147. [CrossRef] [PubMed]
14. Wang, Z.; Li, H.; Zhang, X. Construction waste recycling robot for nails and screws: Computer vision technology and neural network approach. *Autom. Constr.* **2019**, *97*, 220–228. [CrossRef]
15. Tebbe, J.; Gao, Y.; Sastre-Rienietz, M.; Zell, A. A table tennis robot system using an industrial kuka robot arm. In *German Conference on Pattern Recognition*; Springer: Berlin/Heidelberg, Germany, 2018.
16. Sadhu, A.K.; Konar, A.; Bhattacharjee, T.; Das, S. Synergism of firefly algorithm and Q-learning for robot arm path planning. *Swarm Evol. Comput.* **2018**, *43*, 50–68. [CrossRef]
17. Wen, S.; Chen, J.; Wang, S.; Zhang, H.; Hu, X. Path planning of humanoid arm based on deep deterministic policy gradient. In Proceedings of the IEEE International Conference on Robotics and Biomimetics (ROBIO), Kuala Lumpur, Malaysia, 12–15 December 2018; pp. 1755–1760.
18. Ji, M.; Zhang, L.; Wang, S. A path planning approach based on Q-learning for robot arm. In Proceedings of the 2019 3rd International Conference on Robotics and Automation Sciences (ICRAS), Wuhan, China, 1–3 June 2019.
19. Huadong, Z.; Chaofan, L.; Nan, J. A path planning method of robot arm obstacle avoidance based on dynamic recursive ant colony algorithm. In Proceedings of the 2019 IEEE International Conference on Power, Intelligent Computing and Systems (ICPICS), Shenyang, China, 12–14 July 2019.
20. Das, S.D.; Bain, V.; Rakshit, P. Energy optimized robot arm path planning using differential evolution in dynamic environment. In Proceedings of the 2018 Second International Conference on Intelligent Computing and Control Systems (ICICCS), Madurai, India, 14–15 June 2018.
21. Raheem, F.A.; Sadiq, A.T.; Abbas, N.A.F. Robot arm free Cartesian space analysis for heuristic path planning enhancement. *Int. J. Mech. Mechatron. Eng.* **2019**, *19*, 29–42.
22. Chang, W.-Y.; Lin, S.-Y.; Hsu, J.-W.; Hsu, B.-Y. Automatic path planning of robot for intelligent manufacturing based on network remoted controlling and simulation. In Proceedings of the 2019 4th Asia-Pacific Conference on Intelligent Robot Systems (ACIRS), Tianjin, China, 1–3 July 2019.
23. Sugiura, H.; Gienger, M.; Janssen, H.; Goerick, C. Real-time self collision avoidance for humanoids by means of nullspace criteria and task intervals. In Proceedings of the 2006 6th IEEE-RAS International Conference on Humanoid Robots, Genova, Italy, 4–6 December 2006.
24. Martínez, C.; Jiménez, F. Implementation of a Potential Field-Based Decision-Making Algorithm on Autonomous Vehicles for Driving in Complex Environments. *Sensors* **2019**, *19*, 3318. [CrossRef] [PubMed]
25. Kavraki, L.E.; Latombe, J.-C.; Motwani, R.; Raghavan, P. Randomized query processing in robot path planning. *J. Comput. Syst. Sci.* **1998**, *57*, 50–60. [CrossRef]

26. Hsu, D.; Latombe, J.-C.; Kurniawati, H. On the probabilistic foundations of probabilistic roadmap planning. *Int. J. Robot. Res.* **2006**, *25*, 627–643. [CrossRef]
27. Liu, Y.; Zuo, G. Improved RRT Path Planning Algorithm for Humanoid Robotic Arm. In Proceedings of the 2020 Chinese Control And Decision Conference (CCDC), Hefei, China, 22–24 August 2020.
28. Karaman, S.; Frazzoli, E. Sampling-based algorithms for optimal motion planning. *Int. J. Robot. Res.* **2011**, *30*, 846–894. [CrossRef]
29. Prianto, E.; Kim, M.; Park, J.-H.; Bae, J.-H.; Kim, J.-S. Path planning for multi-arm manipulators using deep reinforcement learning: Soft actor–critic with hindsight experience replay. *Sensors* **2020**, *20*, 5911. [CrossRef] [PubMed]
30. Panov, A.I.; Yakovlev, K.S.; Suvorov, R. Grid path planning with deep reinforcement learning: Preliminary results. *Procedia Comput. Sci.* **2018**, *123*, 347–353. [CrossRef]
31. Low, E.S.; Ong, P.; Cheah, K.C. Solving the optimal path planning of a mobile robot using improved Q-learning. *Robot. Auton. Syst.* **2019**, *115*, 143–161. [CrossRef]
32. Yu, J.; Su, Y.; Liao, Y. The path planning of mobile robot by neural networks and hierarchical reinforcement learning. *Front. Neurorobotics* **2020**, *14*, 63. [CrossRef] [PubMed]
33. Lai, T.-C.; Xiao, S.-R.; Aoyama, H.; Wong, C.-C. Path planning and obstacle avoidance approaches for robot arm. In Proceedings of the 2017 56th Annual Conference of the Society of Instrument and Control Engineers of Japan (SICE), Kanazawa, Japan, 19–22 September 2017.
34. Wu, B.; Wan, A.; Iandola, F.; Jin, P.H.; Keutzer, K. Squeezedet: Unified, small, low power fully convolutional neural networks for real-time object detection for autonomous driving. In Proceedings of the IEEE Conference on Computer Vision and Pattern Recognition Workshops, Honolulu, HI, USA, 21–26 July 2017.
35. Howard, A.G.; Zhu, M.; Chen, B.; Kalenichenko, D.; Wang, W.; Weyand, T.; Andreetto, M.; Adam, H. Mobilenets: Efficient convolutional neural networks for mobile vision applications. *arXiv* **2017**, arXiv:1704.04861.
36. Girshick, R.; Donahue, J.; Darrell, T.; Malik, J. Rich feature hierarchies for accurate object detection and semantic segmentation. In Proceedings of the IEEE Conference on Computer Vision and Pattern Recognition, Columbus, OH, USA, 23–28 June 2014.
37. He, K. Mask r-cnn. In Proceedings of the IEEE International Conference on Computer Vision, Venice, Italy, 22–29 October 2017.
38. Redmon, J.; Divvala, S.; Girshick, R.; Farhadi, A. You only look once: Unified, real-time object detection. In Proceedings of the IEEE Conference on Computer Vision and Pattern Recognition, Las Vegas, NV, USA, 27–30 June 2016.
39. CV-Tricks.com. Available online: https://cv-tricks.com/object-detection/faster-r-cnn-yolo-ssd/ (accessed on 12 December 2021).
40. Sutton, R.S.; Barto, A.G. *Reinforcement Learning: An Introduction*; MIT Press: Cambridge, MA, USA, 2018.
41. Van Seijen, H.; van Hasselt, H.; Whiteson, S.; Wiering, M. A theoretical and empirical analysis of Expected Sarsa. In Proceedings of the 2009 IEEE Symposium on Adaptive Dynamic Programming and Reinforcement Learning, Nashville, TN, USA, 31 March–1 April 2009.
42. Werbos, P. New Tools for Prediction and Analysis in the Behavioral Sciences. Ph.D. Thesis, Harvard University, Cambridge, MA, USA, 1974.

sensors

MDPI

Article

Comparing Sampling Strategies for Tackling Imbalanced Data in Human Activity Recognition

Fayez Alharbi [1,2,*], Lahcen Ouarbya [2] and Jamie A Ward [2]

[1] Computer Sciences and Information Technology College, Majmaah University, Al Majmaah 15341, Saudi Arabia
[2] Department of Computing, Goldsmiths, University of London, London SE14 6NW, UK; l.ouarbya@gold.ac.uk (L.O.); j.ward@gold.ac.uk (J.A.W.)
* Correspondence: fs.alharbi@mu.edu.sa

Abstract: Human activity recognition (HAR) using wearable sensors is an increasingly active research topic in machine learning, aided in part by the ready availability of detailed motion capture data from smartphones, fitness trackers, and smartwatches. The goal of HAR is to use such devices to assist users in their daily lives in application areas such as healthcare, physical therapy, and fitness. One of the main challenges for HAR, particularly when using supervised learning methods, is obtaining balanced data for algorithm optimisation and testing. As people perform some activities more than others (e.g., walk more than run), HAR datasets are typically imbalanced. The lack of dataset representation from minority classes hinders the ability of HAR classifiers to sufficiently capture new instances of those activities. We introduce three novel hybrid sampling strategies to generate more diverse synthetic samples to overcome the class imbalance problem. The first strategy, which we call the *distance-based method* (DBM), combines Synthetic Minority Oversampling Techniques (SMOTE) with Random_SMOTE, both of which are built around the k-nearest neighbors (KNN). The second technique, referred to as the noise detection-based method (NDBM), combines SMOTE Tomek links (SMOTE_Tomeklinks) and the modified synthetic minority oversampling technique (MSMOTE). The third approach, which we call the cluster-based method (CBM), combines Cluster-Based Synthetic Oversampling (CBSO) and Proximity Weighted Synthetic Oversampling Technique (ProWSyn). We compare the performance of the proposed hybrid methods to the individual constituent methods and baseline using accelerometer data from three commonly used benchmark datasets. We show that DBM, NDBM, and CBM reduce the impact of class imbalance and enhance $F1$ scores by a range of 9–20 percentage point compared to their constituent sampling methods. CBM performs significantly better than the others under a Friedman test, however, DBM has lower computational requirements.

Keywords: activity recognition; wearable sensors; imbalanced activities; sampling methods

Citation: Alharbi, F.; Ouarbya, L.; Ward, J.A. Comparing Sampling Strategies for Tackling Imbalanced Data in Human Activity Recognition. *Sensors* **2022**, *22*, 1373. https://doi.org/10.3390/s22041373

Academic Editor: Hyoungsik Nam

Received: 19 December 2021
Accepted: 27 January 2022
Published: 11 February 2022

1. Introduction

Human activity recognition (HAR) using body-worn or wearable sensors is an active research topic in mobile and ubiquitous computing [1]. Activity recognition is a useful tool because it provides information on an individual's behaviour that enables computing systems not only to monitor but also to analyse and assist with a range of day-to-day tasks [2,3].

Most HAR studies adopt a supervised learning approach [4]. Supervised learning typically requires immense amounts of labelled sensor data in order to train [2]. For such models to work well, the data are ideally recorded from a variety of real-word situations. Additionally, a diversity of sensor modalities and placements can help improve recognition performance [5,6].

Sensor data obtained from real-life settings is typically of poor quality (noisy) and frequently has missing data [7]. These issues arise due to factors such as bad or faulty placement of sensors, or sensor malfunctioning [8]. Similarly, sensor data may often be highly imbalanced due to significant individual variations, with limited labels for certain activities [9]. Further barriers to obtaining sufficient quantities of real-world data include the prohibitive cost of devices, issues related to privacy, or a desire to reduce battery consumption [10]. Sensor data from certain activities can be difficult to obtain because of the rare but critical nature of those activities, such as falls in the elderly [2] or heart failure [11].

For HAR to succeed as a viable technique, there is an urgent need for new approaches at making up for this shortfall in critical and underrepresented real-world data [2,12]. An important part of this is solving the class imbalance problem [13]. Imbalance can occur in both the between-class distribution and as within-class imbalance [14]. Between-class imbalance occurs, for example, when some activities are performed less often than others [15]. As a result, the sample sizes for these activities are smaller, so a supervised model might not have enough data to learn adequately. The related concept of within-class imbalance occurs when the same activity is performed in different ways by the same individual, yet there are insufficient examples from some of these for the model to generalise adequately [12].

There are usually two methods to solve class imbalance: data level (sampling) methods and algorithm level methods [16]. The data level approach involves changing a training set's class distribution by resampling. This might mean oversampling the minority classes, undersampling the majority classes, or a combination of both [17]. The algorithm level approach involves adjusting existing learning algorithms to focus more on the minority classes [16]. In this work we use the data level approach, which is less complex to configure and can be integrated with any learning algorithm.

The main contributions of the work are the following.

- We evaluate six sampling methods (SMOTE, Random_SMOTE, SMOTE_Tomeklinks, MSMOTE, CBSO, and ProWSyn) as solutions to the class imbalance problem across three commonly used datasets.
- We introduce three novel hybrid sampling approaches and show how these build on and improve upon their constituent methods. These are (1) DBM, a distance-based method that combines SMOTE and Random_SMOTE, (2) NDBM, a noise detection-based method that combines SMOTE_Tomeklinks and MSMOTE, and (3) CBM, a cluster-based method that combines CBSO and ProWSyn.
- We compare how useful the sampling methods are to improve the learning from imbalanced human activity data using both shallow and deep machine learning algorithms. Specifically, we test KNN, Logistic regression (LR), Random Forest (RF) and Support Vector Machine (SVM), and a Multilayer perceptron (MLP) [18,19]. We show that the sampling methods are only useful to improve the performance of the MLP compared to the other classifiers for imbalanced human activity data.

The remainder of the paper is organised as follows. Section 2, shows some of the existing work on class imbalance problem and techniques to deal with it in HAR. Section 3 provides background on the sampling methods used. Section 3.4 introduces the proposed method and Section 4 describes the datasets. Section 5 describes the data analysis and experimental setup. Section 6 introduces the experimental results, and Section 7 discusses the findings from these.

2. Related Work

Several authors have highlighted the importance of the class imbalance problem in HAR [2,12,20,21]. Ni et al. introduced a HAR system based on using the stacked denoising autoencoder (SDAE) to recognise static and dynamic ambulatory activities, such as standing and running, using accelerometers and gyroscopes [22]. The performance of their model dropped as the class (activity) distribution of samples became unbalanced. They used sampling techniques such as SMOTE and random undersampling to tackle the problem. Based on their experimental result, the sampling techniques were more successful than random undersampling at successfully treating imbalance and improving recognition performance. This is because the indiscriminate nature of the undersampling approach can lead to useful data being discarded.

Despite the promising results, Ni et al's. work is limited in that it oversamples the entire dataset. Implementing oversampling before splitting a dataset into different train and test partitions can result in information leakage from the original test data to the newly produced training data and this can then lead to overly optimistic classification performance [23]. In other words, the learning algorithm's performance might be less about its ability to generalise to the test data appropriately, than it is an indication of similar patterns in both train and test data due to information leakage. In the current work we avoid the information leakage problem by sampling exclusively on the training set.

Chen et al. [20] used data from accelerometers and gyroscopes to recognise activities such as walking, jogging, and jumping—again using an imbalanced dataset. They reported that the classifier always showed a good performance in recognising the majority class, whereas its performance was inadequate for the minority classes. Again, SMOTE was used to increase the count of underrepresented activities, leading to improved overall performance. One limitation of this work was that it only evaluated a single oversampling method. In contrast, our work evaluates a combination of different methods.

Inspired by the concept of data fusion, we introduce three hybrid sampling methods—DBM, NDBM, and CBM—which combine the outputs from different sampling methods. Fusion of diverse data sources and sensor modalities is a widely explored approach for improving recognition performance in HAR (e.g., [5,18,24,25]). Similarly, fusion of multiple, diverse, weak learners to produce a strong ensemble is a well-studied and effective approach in machine learning [26]. We hypothesise that by combining outputs from different sampling strategies we diversify the synthetic data and in turn improve the generalization ability of our learning models.

3. Sampling Methods

The underlying sampling methods used in this work can be categorised into three types: distance-based, noise detection-based, and cluster-based.

3.1. Distance-Based

SMOTE and Random_SMOTE both use distance-based algorithms to oversample the training data. SMOTE [27] takes an instance of the minority class x from the training set, and then computes its K nearest neighbours, identified as the shortest Euclidean distances between itself and other instances of the same class. To produce a synthetic sample, x_{new}, SMOTE randomly selects the K nearest neighbours from the minority class, e.g., x_k for the kth nearest neighbour, and computes the difference $x_k - x$. The new synthetic sample, x_{new}, is computed by multiplying this difference by a random number between 0 and 1 using Equation (1). The new synthetic instance x_{new} will lie along the line between x and x_k:

$$x_{new} = x + ||x - x_k|| \times rand\ (0,1) \tag{1}$$

Unfortunately, the linear design of SMOTE can lead to overfitting. Random_SMOTE [28] tackles this by opening up a much wider region for oversampling. For each minority instance, x, two minority samples x_1 and x_2 (with $x_{1,2} \neq x$) are randomly selected. A temporary synthetic sample x_{tmp} is then generated along the line between x_1 and x_2, as shown in Equation (2):

$$x_{tmp} = x_1 + ||x_2 - x_1|| \times rand\ (0,1) \tag{2}$$

The final synthetic sample x_{new} is then created along the line between x_{tmp} and the original sample x using Equation (3):

$$x_{new} = x + ||x_{tmp} - x|| \times rand\ (0,1) \tag{3}$$

3.2. Noise Detection-Based

Real world data contain noise from a variety of sources that can lead to poor recognition performance [29,30]. Frenay et al. [31] indicated that class noise (also known as label noise) is one of the most harmful noises in machine learning. This kind of noise can occur, for example, if a minority class sample is incorrectly labelled with a majority class label [16]. The SMOTE-Tomek Link [32] and modified synthetic minority oversampling technique (MSMOTE) algorithms are specifically designed to detect this kind of noise in order to minimize the risk of creating noisy synthetic samples [33].

SMOTE-Tomek Links oversamples using SMOTE on top of a Tomek link data cleaning step [32]. Tomek link works as follows: Consider two samples x_a and x_b belonging to different classes, where $d(x_a, x_b)$ is the Euclidean distance between x_a and x_b. A Tomek link is identified as an $(x_a x_b)$ pair if there is no sample z that meets the following conditions: $d(x_a, z) < d(x_a, x_b)$ or $d(x_b, z) < d(x_a, x_b)$. That is, x_a and x_b are each other's nearest neighbours [34]. Tomek links are therefore likely to be comprised of either boundary samples or noisy samples [30,35,36]. SMOTE-Tomek Links generates synthetic data in two steps [32]. First, the original minority training data are oversampled using SMOTE. Second, Tomek links are identified in the training data and removed to rebalance the data set.

MSMOTE is an improved version of SMOTE which first uses KNN to assign minority samples into three types: safe, border, and noise [33]. If a minority labelled sample is the same as the labels of its k near neighbours, then the sample is defined as 'safe'. If the labels are all different, then the sample is identified as 'noise'. Finally, if the sample is neither safe nor noise, it is classed as a 'border' sample. The second step of MSMOTE uses SMOTE to generate new samples. However, the random selection of neighbours is different depending on whether the sample is safe, border, or noise. For safe samples, MSMOTE will randomly choose the K nearest neighbours. For border samples, the algorithm only selects the nearest neighbour (i.e., $K = 1$). Noise samples are simply disregarded.

3.3. Cluster-Based

The cluster-based sampling methods include Cluster-Based Synthetic Oversampling (CBSO) and Proximity Weighted Synthetic Oversampling Technique (ProWSyn).

CBSO integrate clustering and SMOTE-it uses agglomerative clustering to first cluster minority samples with the aim of identifying those minority samples which are close to the majority samples border [37]. CBSO produces samples only in the neighbourhood of minority samples that are close to majority neighbours using SMOTE. For instance, in order to produce a new sample, CBSO will select a sample x from the minority class and randomly choose a minority sample x_k from x's cluster (using SMOTE Equation (1)), to produce a new sample.

ProWSyn is another cluster-based sampling method [38]. This algorithm computes the distance between minority class samples and majority class samples in order to assign greater weights to the minority samples. These weights are then used to assign greater significance to the minority samples during learning. ProWSyn operates in two steps: The first step splits the minority data into partitions (P) according to their distance from the class boundary. ProWSyn assigns a proximity level (L) to each partition. The level increases with distance from the boundary. When minority class samples are assigned to lower proximity levels, then they are considered more important for learning because they are close to the boundary. However, in cases where they are assigned higher proximity levels they are considered less important [38].

3.4. Proposed Hybrid Methods

All three proposed hybrid approaches, DBM, NDBM, and CBM, concatenate synthesized training data obtained from the constituent sampling methods. DBM combines SMOTE and Random_SMOTE. NDBM combines SMOTE_Tomeklinks and MSMOTE. Finally, CBM combines CBSO and ProWSyn. The three methods are evaluated as shown in Figure 1. Taking D as the original dataset, we first split D into D_{train} and D_{test}. D_{train} is then oversampled using the constituent methods. For more clarification, lets refer to the D_{train} as d and, for instance, if DBM is used to oversample d, it will be:

$$d_{DBM} = SMOTE(D_{train}, \alpha) + Random\_SMOTE(D_{train}, \alpha) \quad (4)$$

where α is the oversampling ratio. In case of using NDBM, it is denoted as:

$$d_{NDBM} = SMOTE\_TomekLinks(D_{train}, \alpha) + MSMOTE(D_{train}, \alpha) \quad (5)$$

For the CBM, it is referred to as:

$$d_{CBM} = ProWSyn(D_{train}, \alpha) + CBSO(D_{train}, \alpha) \quad (6)$$

We then concatenate the synthesized data to increase the size of D_{train}. The oversampled D_{train} is used to train a classifier, which is then evaluated on the left-aside D_{test}.

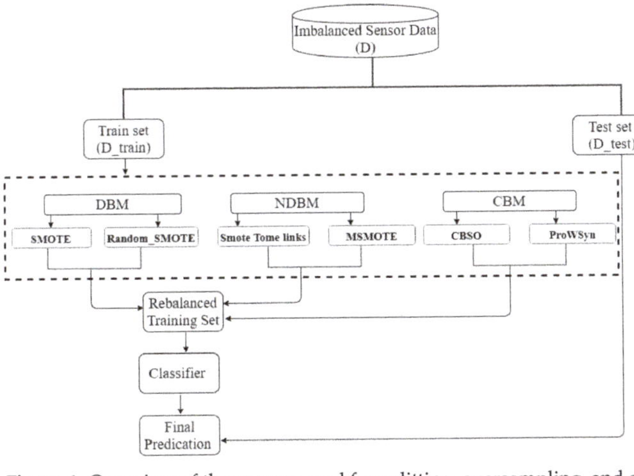

Figure 1. Overview of the process used for splitting, oversampling, and evaluating the data.

4. Datasets

We use three datasets that are widely used by HAR researchers: Opportunity [39], Physical Activity Monitoring (PAMAP2) [40], and Activities of Daily Living (ADL) [41]. Each of these comprise many individuals performing different types of human activity, including ambulation and daily living activities [42]. The ambulation activities are typically performed over a longer period of time, which comes in two difference forms: static (less repetitive) such as standing, or dynamic (more repetitive), for example, running. Shoaib et al. [43] describe these activities as 'simple' because they might be easily identified using a wrist-worn accelerometer placed at an individual wrist. Daily activities might consist of hand gestures such as waving hands or hand-to-mouth gestures (HMG), for example, eating or drinking [44]. Daily activities are not as repetitive as ambulatory dynamic activities, and these daily activities often are concurrent with each other due to their similar gestures such as eating, drinking, and brushing teeth [45]. Such activities are referred to as 'complex' because they are more challenging to identify using a single accelerometer compared to simple activities [43].

4.1. Opportunity

The Opportunity dataset was collected from 72 sensors, with different types of sensors integrated into the environment, objects and worn on participants' bodies [39]. Four participants performed daily living scenarios in a simulated kitchen environment. The dataset included around 6 h of recordings and was sampled at 30 Hz. The activities were annotated on two different levels: locomotion and gesture. For example, *cleaning up* and *open* door were labelled as gestures, with *sitting* and *lying* making up the locomotion subset. Here we focus solely on gesture activities. Figure 2 shows 17 activities categorised as gestures, including *Open Door1, Open Door2, Close Door1, Close Door2, Open Fridge, Close Fridge, Open Dishwasher, Close Dishwasher, Open Drawer1, Close Drawer1, Open Drawer2, Close Drawer2, Open Drawer3, Close Drawer3, Clean Table, Drink from Cup*, and *Toggle Switch*. The dataset contains several on-body and object sensors, but in this paper we use only the accelerometer in the lower right arm-worn inertial measurement unit (IMU).

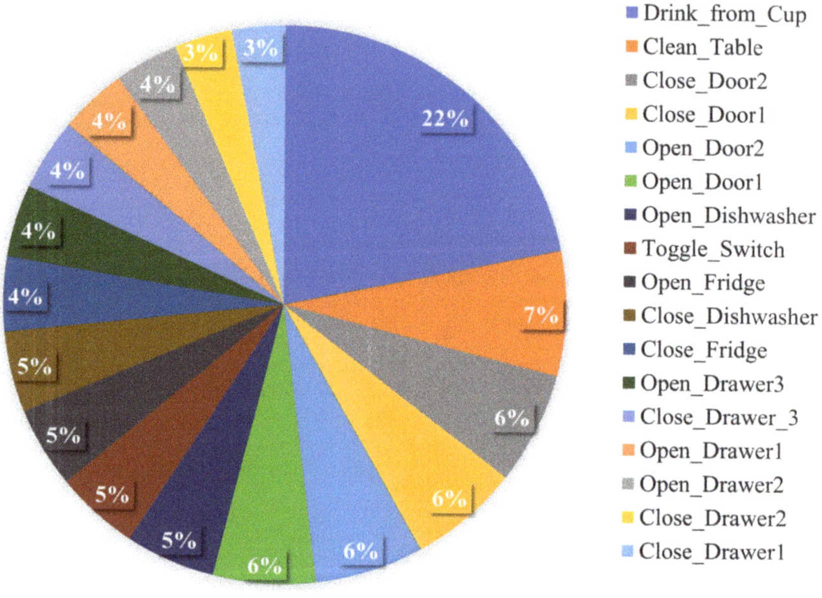

Figure 2. Activity distribution of the Opportunity dataset.

4.2. PAMAP2

The Physical Activity Monitoring dataset (PAMAP2) was collected from 9 participants who performed 12 activities for over 10 h and it was sampled at 100 Hz. Data were recorded by using IMUs placed on the hand, chest, and ankle [40]. Here we use only the accelerometer sensor of the hand-worn IMU. Figure 3 shows the activity distribution, and it can be seen that the dataset is imbalanced. It contains both simple and sporting activities such as walking, running, cycling, Nordic walking, and rope jumping. It also includes posture activities such as lying, sitting, and standing. Activities of daily living (ascending stairs, descending stairs), and households activities such as vacuum cleaning and ironing are also included.

4.3. ADL

The Activities of Daily Living (ADL) dataset is a public dataset collected using a single chest-worn wearable accelerometer on 15 participants [41]. The sampling rate of the accelerometer was 52 Hz. The participants performed seven daily living activities. The activities include *Working at Computer* (*WAC*), *Standing Up, Walking and Going Up/Downstairs* (*SWGUDS*), *Standing, Walking, Going Up/Downstairs* (*GUDS*), *Walking and Talking with Someone* (*WATWS*), and *Talking while Standing* (*TWS*). Figure 4 shows the activities distribution of the ADL dataset which indicates that the dataset is imbalanced.

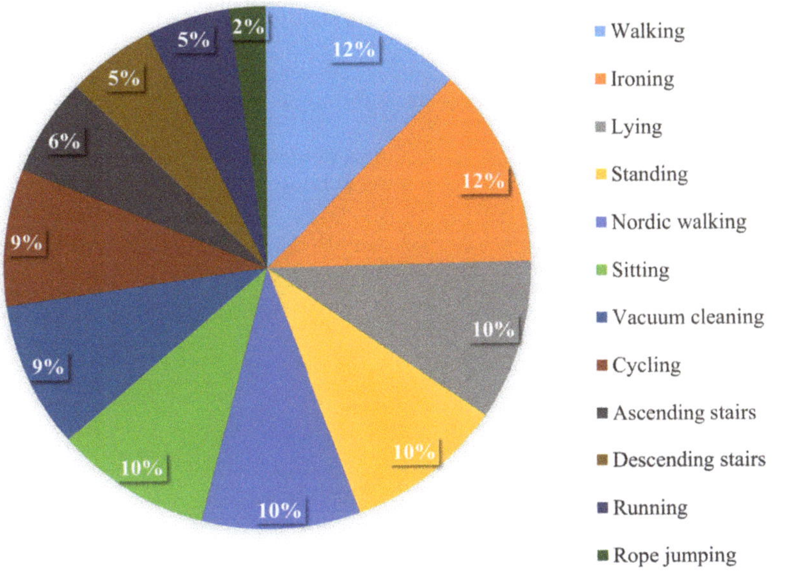

Figure 3. Activity distribution of the PAMAP2 dataset.

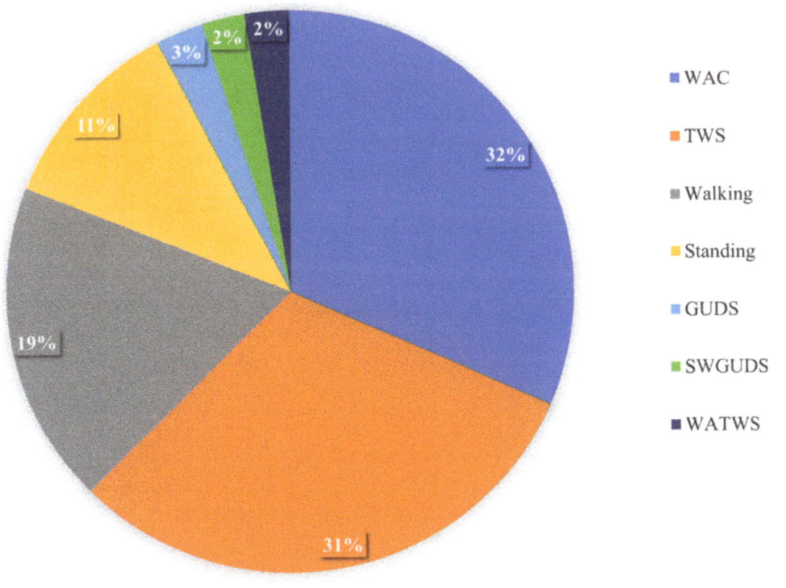

Figure 4. Activity distribution of the ADL dataset.

5. Data Analysis

5.1. Data Preprocessing

We explore how our proposed sampling methods might enhance a human activity model's performance in a real-life scenario where only a single 3-axis accelerometer is available. Although recognition performance is typically better when multiple sensors are used, in many scenarios access to multiple sensors is limited (e.g., in a single wrist-worn device) [10].

As a pre-processing step, we first calculate the Euclidean norm ($\sqrt{x^2 + y^2 + z^2}$) of each 3-axis sensor to ensure the data are invariant to shifting orientation of the sensors [46]. We then apply a non-overlapping sliding window to segment the data [24]. Table 1 provides more details such as the number of subjects, sampling rate, the window size, and sensor position we use.

We extract six time-domain features including mean, standard deviation, minimum, maximum, median, and range. The selected features are highlighted further in Table 2. These features are efficient as well as fast to compute [18].

Table 1. Datasets details.

Dataset	Number of Subjects	Sample Rate	Window Size (s)	Sensor Position	Number of Sensors Used
Opportunity	4	32	2	Right Arm	1 accelerometer
PAMAP2	8	100	3	Dominant Wrist	1 accelerometer
ADL	15	52	10	Chest	1 accelerometer

Table 2. Features description [18,43].

Feature	Description
Mean	It provides the average value of sensor data within a segment
Standard deviation	It describes how much sensor data are spread around the mean
Minimum	The minimum value of sensor data within a segment
Maximum	The maximum value of sensor data within a segment
Median	It finds the middle number of a sample within a segment
Range	The difference between the maximum and the minimum of sensor data within a segment

5.2. Parameters Setting

All of the evaluations in this work were carried out using a collection of shallow learning methods—specifically, SVM, LR, kNN, and RF—as well as a deep learning method based on MLP. The parameters for SVM, LR, and KNN were found using grid search (see Appendix C for details). For RF and MLP, we use the default settings provided by the Python implementation [47]. The MLP architecture that was used in the experiments is presented in Table 3.

Table 3. MLP architecture details.

Hidden Layers	Activation Function	Optimizer	Loss Function	Learning Rate	Regularization	Epochs
100	Relu	Adam	Cross-entropy	0.001	L2 penalty	200

Note that for brevity we include only the full results related to the overall best-performing classifier, MLP. The main findings using the remaining classifiers were broadly in agreement (as will be shown in Section 6.7). The full results for the remaining four classifiers are provided in Appendix B.

We also use the default settings on the Python implementation of our sampling algorithms—SMOTE, Random_SMOTE, SMOTE_TomekLinks, MSMOTE, CBSO, and ProWSyn. Number of neighbours and the number of samples to generate are common parameters among all sampling methods. In addition, other sampling methods use specific parameters, for example, ProWSyn utilizes number of levels. Ref. [37] provides more details about these parameters.

The percentage of samples to be created by a sampling method was set to 100%, which means that the number of minority samples in the training set will be equal to the number of majority samples in the training set after sampling.

5.3. Evaluation Method

Performance is measured using weighted F1 *score*, *recall*, and *precision* [12]. *Precision* records the proportion of class predictions that are correct, whereas *Recall* records the proportion of actual class samples that are correct [48]. The weighted F1 *score* used here weighs classes based on their sample proportion and is calculated as (e.g., [1]):

$$F1\,score = \sum_i 2 * w_i \frac{\times Precision_i \times Recall_i}{Precision_i + Recall_i} \tag{7}$$

Here, i corresponds to the class. $w_i = n_i/N$ corresponds to the proportion of class i and n_i is the number of samples of the class i. N is the total number of samples.

Pirttikangas et al. [49] suggested to combine all the data from different subject into one dataset. They augured this was because of the individual variation in body worn acceleration which is often dominates by strong commonalities among individuals in activity patterns.

Consequently, we followed Pirttikangas et al.'s [49] suggestion in this work and used 3-fold cross-validation to train the parameters in our analyses. We did not use more than 3 folds as most of the activities have a very low number of samples in some datasets. As an additional measure of reliability, our evaluations are all repeated 30 times using different random selections of data. The final results are presented as the mean (and standard deviation) of the F1 *score* over these repetitions. In the future we aim to use a leave-one-subject-out approach.

ANOVA and Friedman statistical tests were performed to discover whether there are significant differences in performance between the sampling methods across the five classifiers [50,51].

6. Results

Here we present the final results of each of the sampling methods and our hybrid methods vs the baseline (no resampling) case for each of the three datasets. We also present an evaluation of the power considerations for each method.

6.1. Distance-Based Method (DBM)

Table 4 shows the main results for the MLP classifier using DBM versus its constituent methods, SMOTE, and Random_SMOTE. The first thing to notice is a universal improvement when sampling is used compared to the baseline.

On the ADL dataset, the DBM F1 score is 92.59%, a 5.39 percentage point (pp) improvement over baseline, a 0.35% improvement over SMOTE, and a 1.52 pp improvement over Random_SMOTE. On Opportunity, DBM's F1 score is low (48.49%), however, this is a large 19.64 pp improvement on the baseline, and a 5 pp improvement over both constituent sampling methods. On PAMAP2, DBM F1 score is 80.15%, which is an 8.3 pp improvement on the baseline, and 5 pp on both constituent methods.

Table 4. Distance-based method results. Comparing the performance of MLP on DBM, SMOTE, and Random_SMOTE for multiple datasets. The reported mean of *F*1 score and (±standard deviation), recall, and precision are obtained from 30 repetitions. The *F*1 score, recall, and precision are in %. Highest scores are shown in bold.

Data	Method	*F*1 Score	Recall	Precision
ADL	Baseline	87.2 (±0.047)	87.03	89.02
	SMOTE	92.24 (±0.069)	91.44	94.21
	Random_SMOTE	91.07 (±0.086)	90.31	93.22
	DBM	**92.59** (±0.081)	**91.9**	**94.26**
Opportunity	Baseline	28.85 (±0.017)	34.1	29.57
	SMOTE	42.95 (±0.043)	42.45	45.73
	Random_SMOTE	42.74 (±0.04)	42.19	45.75
	DBM	**48.49** (±0.052)	**48.18**	**50.63**
PAMAP2	Baseline	71.85 (±0.081)	72.73	75.49
	SMOTE	74.73 (±0.055)	74.93	77.69
	Random_SMOTE	74.59 (±0.055)	74.64	77.83
	DBM	**80.15** (±0.046)	**80.23**	**81.93**

6.2. Noise Detection-Based Method (NDBM)

Table 5 demonstrates the MLP classifier performance of baseline, NDBM, SMOTE_TomekLinks, and MSMOTE across all datasets. Again, a large improvement is evident for all datasets when using sampling versus baseline.

On the ADL dataset, the NDBM *F*1 score is 93.7%, a 5.39 pp improvement over the baseline, and between 1–2 pp improvement over the constituent sampling methods. On the Opportunity dataset, NDBM performance is low (with *F*1 at 46.95%)—however, this is a dramatic 18.1 pp improvement over the baseline case. On the PAMAP2 dataset, the NDBM *F*1 score is 79.43%, a 7.58 pp improvement on the baseline, and 5 pp improvement over both constituent methods.

Table 5. Noise detection-based results. Comparing the performance of MLP for NDBM, MSMOTE, and SMOTE_TomekLinks on multiple datasets. The reported mean of *F*1 score and (±standard deviation), recall, and precision are obtained from 30 repetitions. The *F*1 score, recall, and precision are in %. Highest scores are shown in bold.

Data	Method	*F*1 Score	Recall	Precision
ADL	Baseline	87.2 (±0.047)	87.03	89.02
	SMOTE_TomekLinks	91.41 (±0.071)	90.52	93.56
	MSMOTE	90.7 (±0.067)	89.65	92.66
	NDBM	**92.7** (±0.065)	**91.69**	**94.77**
Opportunity	Baseline	28.85 (± 0.017)	34.1	29.57
	SMOTE_TomekLinks	42.89 (±0.039)	43.15	45.34
	MSMOTE	39.71 (±0.074)	39.58	42.07
	NDBM	**46.95** (±0.067)	**46.97**	**48.86**
PAMAP2	Baseline	71.85 (±0.081)	72.73	75.49
	SMOTE_TomekLinks	74.24 (±0.054)	74.51	77.13
	MSMOTE	73.73 (±0.059)	73.78	77.03
	NDBM	**79.43** (±0.054)	**79.46**	**81.35**

6.3. Cluster-Based Method (CBM)

Table 6 demonstrates the MLP classifier performance of baseline, CBM, CBSO, and ProWSyn across all datasets. Again, a clear improvement is evident for all datasets when using sampling versus baseline.

On the ADL dataset, the CBM *F*1 score is 92.96%, a 5.76 pp improvement over the baseline, and a 1.8–1.4 pp improvement over the constituent sampling methods. On the Opportunity dataset, CBM performance is low (with *F*1 at 48.87%)—however, this is a dramatic 20.02 pp improvement over the baseline case. On the PAMAP2 dataset, the CBM *F*1 score is 81.15%, a 9.13 pp improvement over the baseline, and a 5.29 pp and 6.56 pp improvement over CBSO and ProWSync, respectively.

Table 6. Cluster-based results. Comparing the performance of MLP using CBM, CBSO, and ProWsyn on multiple datasets. The reported mean of *F*1 scores and (±standard deviation), recall, and precision are obtained from 30 repetitions. The *F*1 score, recall, and precision are in %. Highest scores are shown in bold.

Data	Method	*F*1 Score	Recall	Precision
ADL	Baseline	87.2 (±0.047)	87.03	89.02
	CBSO	91.16 (±0.09)	90.22	93.66
	ProWSyn	91.56 (±0.091)	90.98	93.7
	CBM	**92.96** (0.087)	**91.93**	**95.29**
Opportunity	Baseline	28.85 (±0.017)	34.1	29.57
	CBSO	42.92 (±0.023)	42.96	45.12
	ProWSyn	42.78 (±0.055)	43.47	44.99
	CBM	48.87 (±0.045)	**48.82**	**50.67**
PAMAP2	Baseline	71.85 (±0.081)	72.73	75.49
	CBSO	75.69 (±0.042)	75.43	78.19
	ProWSyn	74.42 (±0.054)	74.4	77.5
	CBM	**80.98** (±0.051)	**80.9**	**82.54**

6.4. Comparing the Performance of the Proposed Sampling Approaches DBM, NDBM, and CBM

Table 7 combines the headline results from our proposed hybrid methods. On the ADL dataset, CBM outperforms the others, with an *F*1 score of 92.96%. On Opportunity, CBM (48.87% *F*1) narrowly outperforms DBM (48.49% *F*1) and NDBM (46.95% *F*1). Similarly, on PAMAP2, CBM (80.98% *F*1) outperforms DBM (80.15% *F*1) and NDBM (79.43% *F*1). The standard deviation across recognition scores for all evaluations is low throughout, with the maximum deviation being no more than 0.087.

Table 7. Comparing performance of DBM, NDBM, and CBM on multiple datasets. The reported mean of *F*1 scores and (±standard deviation), recall, and precision were obtained from 30 repetitions. The *F*1 score, recall, and precision are in %. Highest scores are shown in bold.

Data	Method	*F*1 Score	Recall	Precision
ADL	Baseline	87.2 (±0.047)	87.03	89.02
	DBM	92.59 (±0.081)	91.9	94.26
	NDBM	92.7 (±0.065)	91.69	94.77
	CBM	**92.96** (±0.087)	**91.93**	**95.29**
Opportunity	Baseline	28.85 (±0.017)	34.1	29.57
	DBM	48.49 (±0.052)	48.18	50.63
	NDBM	46.95 (±0.067)	46.97	48.86
	CBM	**48.87** (±0.045)	**48.82**	**50.67**
PAMAP2	Baseline	71.85 (±0.081)	72.73	75.49
	DBM	80.15 (±0.046)	80.23	81.93
	NDBM	79.43 (±0.054)	79.46	81.35
	CBM	**80.98** (±0.051)	**80.9**	**82.54**

6.5. Results for Minority Activities

For the Opportunity dataset, multiple activities were underrepresented, such as *Open_Fridge*, *Open_Drawer3*, and *Close_Drawer3*. Figure 5 indicates that the proposed DBM, NDBM, and CBM improve the *F*1 score of the MLP in recognising the underrepresented activities. Figure 5 also shows that without applying the sampling methods (baseline), the MLP classifier could not identify the *Open_Fridge* activity. By applying the proposed sampling methods, the MLP's ability to recognise underrepresented activities improved. For example, the *F*1 of the MLP's ability to classify the *Open_Fridge* activity improved by more than 10 pp using the DBM, NDBM, and CBM.

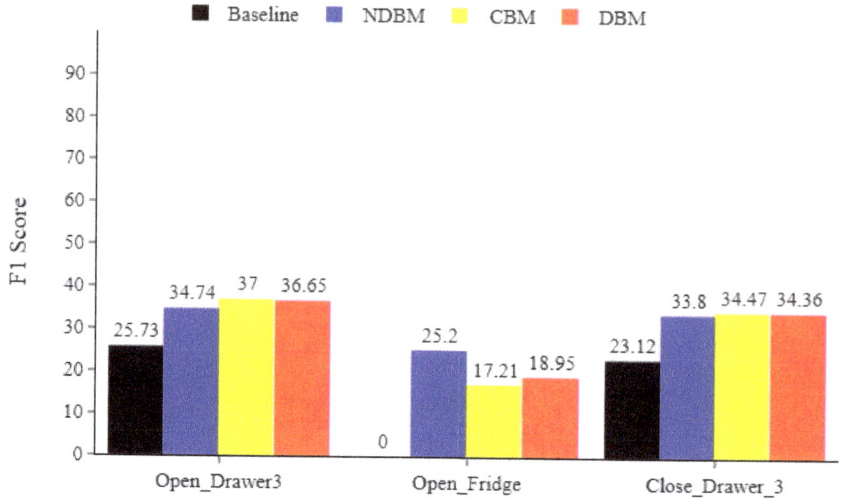

Figure 5. Opportunity minority classes. Comparing the impact of DBM, NDBM, and CBM on activity recognition performance, using MLP for the most underrepresented activities *Open_Fridge*, *Open_Drawer3*, and *Close_Drawer3*. The reported means of *F*1 scores are obtained from 30 repetitions. The *F*1 score is in %.

On the ADL dataset, Figure 6 also suggests that by applying the DBM, NDBM, and CBM MLP classifier, *F*1 score was improved by more than 10 pp and gained a significant advantage in identifying the underrepresented activities, including *Going Up/Downstairs* (*GUDS*), *Standing Up, Walking and Going Up/Downstairs* (*SWGUDS*), and *Walking and Talking with Someone* (*WATWS*).

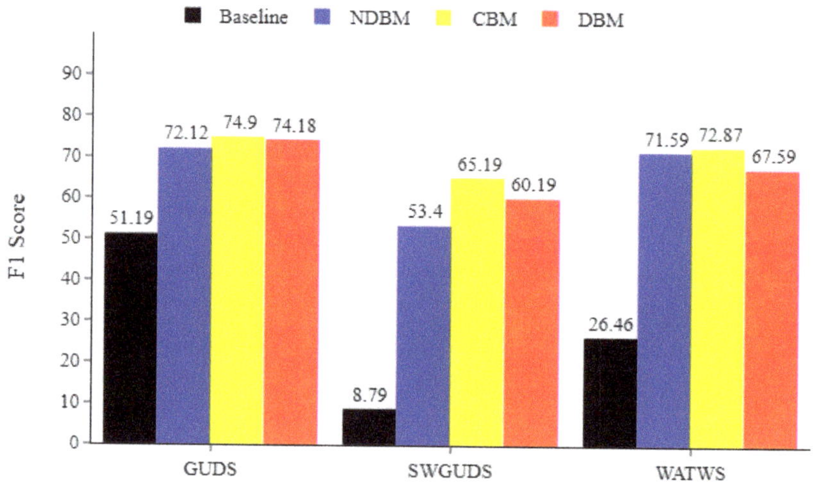

Figure 6. ADL minority classes. Comparing the impact of DBM, NDBM, and CBM on activity recognition performance, using MLP for the most underrepresented activities (*Going Up/Downstairs* (*GUDS*), *Standing Up, Walking and Going Up/Downstairs* (*SWGUDS*), and *Walking and Talking with Someone* (*WATWS*)). The reported means of *F*1 scores are obtained from 30 repetitions. The *F*1 score is in %.

Similarly, on the PAMAP2 dataset, Figure 7 implies that the MLP classifier was more capable of identifying the underrepresented activities, including rope jumping, running, descending stairs, and ascending stairs, when the proposed DBM, NDBM, and CBM were used. For example, the performance of the MLP improved on the *F*1 score by at least 4 pp when identifying the underrepresented rope jumping activity.

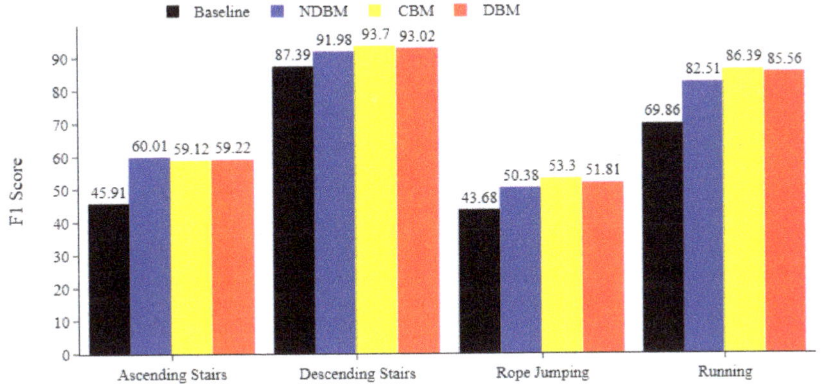

Figure 7. PAMAP2 minority classes. Comparing the impact of DBM, NDBM, and CBM on activity recognition performance, using MLP for the most underrepresented activities (ascending stairs, descending stairs, rope jumping, and running). The reported means of *F*1 scores are obtained from 30 repetitions. The *F*1 score is in %.

6.6. Run Times for DBM, NDBM, and CBM

Figure 8 offers a comparison for each proposed sampling method in terms of run times. The analysis was performed on a Fierce PC with 16 GB RAM, Intel Core i7-7700 processor with 3.60 GHz and using Ubuntu 16.04 LTS (64-bits). DBM demonstrated the best performance in terms of training time compared to NDBM and CBM.

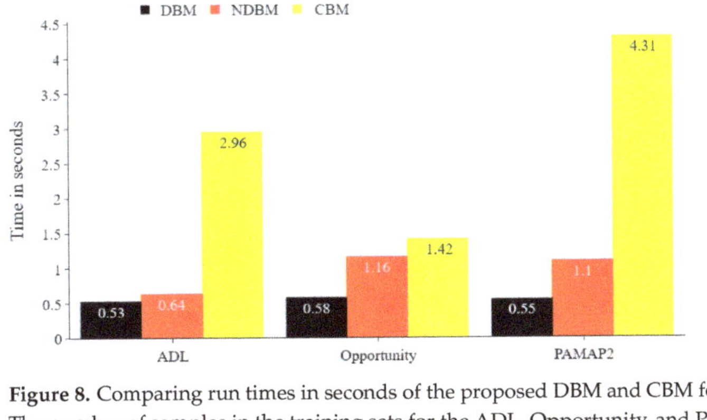

Figure 8. Comparing run times in seconds of the proposed DBM and CBM for all training datasets. The number of samples in the training sets for the ADL, Opportunity, and PAMAP2 datasets were 11,776, 1569, and 6450, respectively.

6.7. Statistical Analysis

A statistical analysis was performed to find out whether there are significant *F*1 performance differences between the nine sampling methods across five classifiers. The sampling methods analysed are SMOTE, Random_SMOTE, MSMOTE, SMOTE_TomeKLinks, CBSO, ProWSync, as well as the hybrid methods DBM, NDBM, and CBM. The classifiers are LR, RF, SVM, KNN, and MLP. The normality assumption is first estimated using the Anderson–Darling normality test on each sampling method and classifier combination [50,52]. This determines whether parametric statistical analysis, such as ANOVA, may be used in the case of normality, or a non-parametric method, such as the Friedman test, in the case of non-normality [53].

Table 8 shows the results of the Anderson–Darling normality test on sampling methods based on the five classifiers results for each dataset [52]. The mean $F1$ scores shown are obtained from 45 'samples', where one sample represents one sample method and classifier combination. On the PAMAP2 dataset, the Anderson–Darling p-value is more than 0.05 ($\alpha = 0.05$)—suggesting the null hypothesis of a normal distribution—and so an ANOVA could then be used. For the ADL- and Opportunity-based results, Table 8 shows a rejection of the null hypothesis ($p < 0.05$) [52]. This indicates that these dataset results are not normally distributed and, therefore, ANOVA cannot be applied [54]. In its place, we use a Friedman test [55].

Table 8. Anderson–Darling normality test on sampling methods based on the 5 classifiers results × 9 sampling methods (5 × 9 = 45 sample size) on each dataset. The p-value is less than 0.05 ($\alpha = 0.05$) for ADL and Opportunity which suggests that ADL and Opportunity are not normally distributed compared to PAMPA2.

Data	Mean	Standard Deviation	Sample Size	p-Value
ADL	0.8840	0.0399	45	0.0007
Opportunity	0.3773	0.0548	45	0.0000
PAMAP2	0.7272	0.0406	45	0.0680

6.7.1. ANOVA on PAMAP2

Table 9 reveals that the ANOVA test detected no statistical evidence to reject the null hypothesis ($p > 0.05$). In other words, when using PAMAP2, all sampling methods performed the same and none was found to perform significantly differently to the others.

6.7.2. Friedman Test on ADL and Opportunity

The Friedman test in Table 10 indicates that the p-values of the data are less than 0.05 ($\alpha = 0.05$) for the ADL and Opportunity datasets. Therefore, the null hypothesis is then violated. This means that there is a statistically significant difference across the sampling methods. In other words, one or more of the sampling methods can show different influences on these datasets.

Table 9. ANOVA for PAMAP2 dataset.

Data	Degrees of Freedom	Sum of Squares	Mean Square	F Value	p-Value
PAMAP2	8	0.0067	0.0008	0.4602	0.8757

Table 10. Friedman test results indicate that the p-value is less than 0.05 ($\alpha = 0.05$) for the ADL and Opportunity datasets. This means that one or more of the sampling methods is more effective than the others.

Data	Degrees of Freedom	Chi-Square	p-Value
ADL	8	21.8133	0.0053
Opportunity	8	24.2133	0.0021

Tables 11 and 12 display the ranks drawn from the Friedman test in the ADL and Opportunity datasets [55]. The test compares rankings across the five classifiers (rows) and nine sampling methods (columns). Ranking is conducted for each classifier row, with sampling methods ranked between 1 (lowest) and 9 (high). The tables then summarize the total ranks obtained for each column to obtain the overall ranking for each sampling method [55].

Both Tables 11 and 12 show that CBM has a consistently high ranking compared to the other sampling approaches across a range of classifiers. This supports our earlier finding that CBM is the highest performer.

Table 11. Friedman sum-of-ranks test on ADL-based results for all methods and classifiers. CBM is the overall highest ranking method.

Classifier	CBSO	NDBM	CBM	DBM	MSMOTE	Pro-WSyn	Random_SMOTE	SMOTE_TomekLinks	SMOTE
KNN	1	7	9	4	5	8	2	6	3
LR	1	8	3	9	2	5	6	7	4
MLP	3	8	9	7	1	5	2	4	6
RF	1	6	9	4	7	8	3	5	2
SVM	1	8	7	9	2	3	6	4	5
Sum of ranks	7	37	**37**	33	17	29	19	26	20

Table 12. Friedman sum-of-ranks test on Opportunity-based results for all methods and classifiers. CBM is the overall highest ranking method.

Classifier	CBSO	NDBM	CBM	DBM	MSMOTE	Pro-WSyn	Random_SMOTE	SMOTE_TomekLinks	SMOTE
KNN	5	6	9	7	1	4	8	3	2
LR	5	9	7	8	1	2	6	4	3
MLP	5	7	9	8	1	3	2	4	6
RF	4	5	8	3	1	9	7	6	2
SVM	2	7	8	9	1	4	3	5	6
Sum of ranks	21	34	**41**	35	5	22	26	22	19

7. Discussion and Future Work

Prior studies such as [2,4] have highlighted the lack of works that address and investigate the impact of the class imbalance problem in human activity recognition. Our present study fills this gap by proposing three approaches, DBM, NDBM, and CBM, to reduce the class imbalance and substantially improve human activity recognition (HAR) performance. We found that the proposed hybrid sampling methods worked better than applying any single sampling algorithm across three different HAR datasets. The benefit of the proposed approaches is that they generate more diverse samples, and thus improve the generalisability of the learning algorithm.

The cluster-based method (CBM) reveals consistently better performance than distance (DBM) or noise detection (NDBM)-based methods. A Friedman test additionally showed the statistical superiority of CBM over the other methods for two datasets, ADL and Opportunity, across five different classifiers.

Minority classes in particular benefit from using CBM, as shown in Figures 5–7. This indicates that CBM would be a good choice when working with imbalanced HAR data involving activities similar to those found in Opportunity and ADL.

DBM, which is a combination of distance-based SMOTE and Random_SMOTE, provides the next highest performing combination. The main advantage of DBM, however, is that it uses significantly less computational resources than CBM. We suggest this method in instances where the training data suffer from small sample size and there is limited computational power. The main issue with DBM compared to NDBM and CBM is more likely to introduce noisy samples. The DBM does not perform any data filtering such as clustering processes prior to oversampling the data. One way to improve the DBM is to use a technique to assess the similarities between the synthetic samples and the training data samples (i.e., the original data), then to use only the most similar synthetic samples to the original sample in order to oversample the training data. For example, one can use the SMOTE and Random_SMOTE approaches to generate synthetic samples from the original training samples and use an efficient similarity metric such as Euclidean distance to compare the generated synthetic data to the original training samples and then use only the most similar synthetic samples and disregard the least similar. Our reason for this is that it might ensure that high-quality synthetic samples are used to oversample the training data.

Moreover, the key problem with the NDBM is that is relies on sampling methods that eliminate some samples during the oversampling process. This can lead to losing some valuable information of the activities. Therefore, we suggest that CBM be used by researchers to ensure they can be able to oversample the imbalance human activity data without losing any information.

To minimise complexity, we opted to use the default settings for most of the classifiers and sampling algorithms. Despite this, we believe that the general findings of the work regarding the influence of sampling on HAR still hold, and we have left further optimisation for future work.

One limitation of our proposed solutions is the choice of features. We chose to adopt time-domain features because these are efficient and fast to compute. This work might be extended by expanding on the feature set to incorporate, say, frequency domain features [12]. Additionally, we only considered data from a single accelerometer and a single location per dataset. How our sampling strategies might perform using an expanded feature set and a larger number of sensors will be the focus of future work.

A further area of future work will be to investigate more complex ensembles of sampling methods, e.g., combing distance with cluster-based methods. One challenge here will be to reduce the computational complexity of the clustering method, while preserving its ability to produce diverse samples.

Author Contributions: F.A. conceived the idea and research methodology as well as developing the proposed approaches, performing the experiments, and writing. J.A.W. contributed to the research by supervising and directing the research and analysis, then reviewing and editing the paper. L.O. contributed by reviewing and providing feedback. All authors have read and agreed to the published version of the manuscript.

Funding: Jamie A. Ward is funded by a Leverhulme supported APEX award from The British Academy, Royal Academy of Engineering and Royal Society (R156910).

Acknowledgments: Fayez Alharbi was supported by Majmaah University through Saudi Arabia Cultural Bureau in United Kingdom.

Conflicts of Interest: The authors declare no conflict of interest.

Appendix A

Table A1. Parammeters settting of SVM, LR and KNN on multiple dataset. Ref. [47] provides description about the parameters that we used.

Algorithms	Parameters	ADL	Opportunity	PAMAP2
SVM	gamma	0.1	0.1	0.1
	C	20	20	20
	kernel	rbf	rbf	rbf
	max_iter	−1	−1	−1
	decision_function_shape	ovr	ovr	ovr
LR	multi_class	multinomial	multinomial	multinomial
	solver	newton-cg	sag	sag
	max_iter	250	250	250
	C	2	2	2
	penalty	L2	L2	L2
KNN	n_neighbors	3	5	3
	algorithm	auto	auto	auto

Appendix B

Table A2. Comparing the performance of the baseline classifiers on multiple datasets. The reported mean of *F*1 scores and (±standard deviation) were obtained from 30 repetitions. The *F*1 score is as %.

Data	Classifier	*F*1 Score
ADL	KNN	85.63 (±0.043)
	LR	84.51 (±0.026)
	MLP	87.2 (±0.047)
	RF	82.76 (±0.037)
	SVM	90.76 (±0.037)
Opportunity	KNN	31.36 (±0.052)
	LR	26.03 (±0.012)
	MLP	28.85 (±0.017)
	RF	33.15 (±0.032)
	SVM	34.04 (±0.012)
PAMAP2	KNN	69.44 (±0.033)
	LR	64.81 (±0.094)
	MLP	71.85 (±0.081)
	RF	71.72 (±0.057)
	SVM	75.18 (±0.06)

Appendix C

We exhibited the *F*1 score of the baseline classifiers including the SVM, RF, LR and KNN in order to compare the influence of the sampling methods in improving their *F*1 score. The sampling methods were the proposed DBM, NDBM and CBM as well as the existing methods including, SMOTE, Random_SMOTE, SMOTE_Tomeklinks, MSMOTE, CBSO, and ProWSyn.

The below figures compared the *F*1 scores of the SVM, RF, LR, and KNN on the Opportunity, PAMAP2, and ADL datasets. For more details about the dataset, see Section 4.

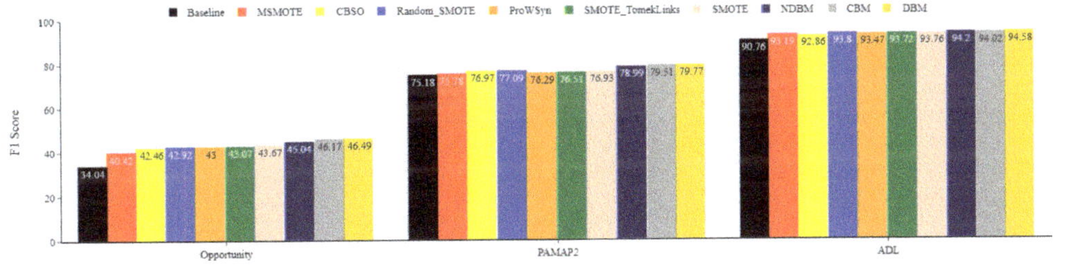

Figure A1. *F*1 score of baseline (SVM), the proposed method, and the sampling methods for the Opportunity, PAMAP2, and ADL datasets. The reported means of *F*1 scores were obtained from 30 repetitions.

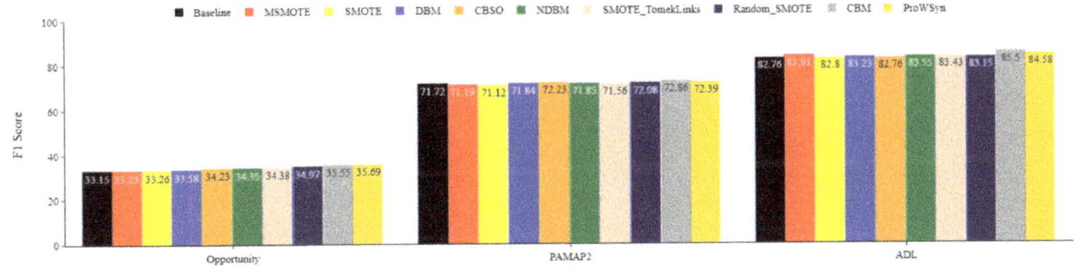

Figure A2. *F*1 score of baseline (RF), the proposed method, and the sampling methods for the Opportunity, PAMAP2, and ADL datasets. The reported means of *F*1 scores were obtained from 30 repetitions.

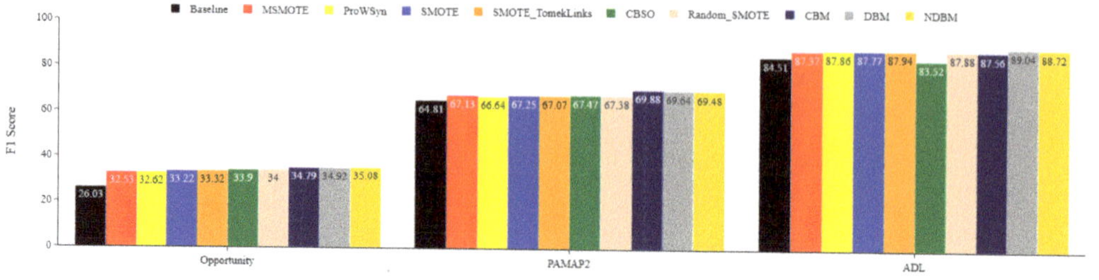

Figure A3. $F1$ score of baseline (LR), the proposed method, and the sampling methods for the Opportunity, PAMAP2 and ADL datasets. The reported means of $F1$ scores were obtained from 30 repetitions.

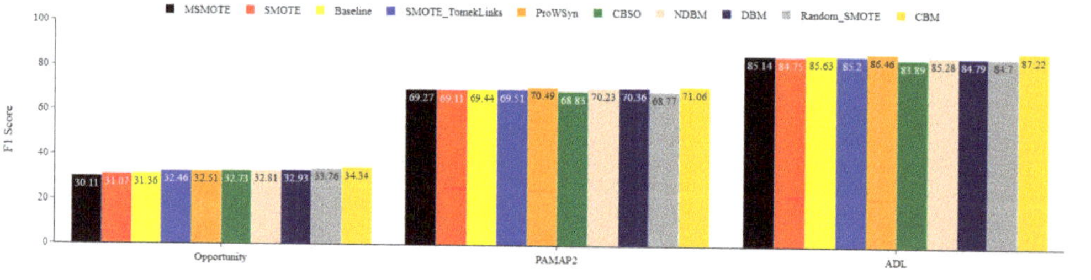

Figure A4. $F1$ score of baseline (KNN), the proposed method, and the sampling methods for the Opportunity, PAMAP2, and ADL datasets. The reported means of $F1$ scores were obtained from 30 repetitions.

References

1. Ordóñez, F.J.; Roggen, D. Deep Convolutional and LSTM Recurrent Neural Networks for Multimodal Wearable Activity Recognition. *Sensors* **2016**, *16*, 115. [CrossRef] [PubMed]
2. Chen, K.; Zhang, D.; Yao, L.; Guo, B.; Yu, Z.; Liu, Y. Deep Learning for Sensor-Based Human Activity Recognition: Overview, Challenges, and Opportunities. *ACM Comput. Surv.* **2021**, *54*, 1–40. [CrossRef]
3. Bao, L.; Intille, S.S. *Activity Recognition from User-Annotated Acceleration Data BT-Pervasive Computing*; Springer: Berlin/Heidelberg, Germany, 2004; pp. 1–17.
4. Wang, J.; Chen, Y.; Gu, Y.; Xiao, Y.; Pan, H. SensoryGANs: An Effective Generative Adversarial Framework for Sensor-based Human Activity Recognition. In Proceedings of the 2018 International Joint Conference on Neural Networks (IJCNN), Rio de Janeiro, Brazil, 8–13 July 2018; pp. 1–8.
5. Ward, J.A.; Lukowicz, P.; Troster, G.; Starner, T.E. Activity Recognition of Assembly Tasks Using Body-Worn Microphones and Accelerometers. *IEEE Trans. Pattern Anal.* **2006**, *28*, 1553–1567. [CrossRef]
6. Van Laerhoven, K.; Gellersen, H.W. Spine versus porcupine: A study in distributed wearable activity recognition. In Proceedings of the Eighth International Symposium on Wearable Computers, Arlington, VA, USA, 31 October–3 November 2004; Volume 1, pp. 142–149. [CrossRef]
7. Catellier, D.J.; Hannan, P.J.; Murray, D.M.; Addy, C.L.; Conway, T.L.; Yang, S.; Rice, J.C. Imputation of missing data when measuring physical activity by accelerometry. *Med. Sci. Sport. Exerc.* **2005**, *37*, S555–S562. [CrossRef] [PubMed]
8. Saeed, A.; Ozcelebi, T.; Lukkien, J. Synthesizing and Reconstructing Missing Sensory Modalities in Behavioral Context Recognition. *Sensors* **2018**, *18*, 2967. [CrossRef]
9. Ramamurthy, S.R.; Roy, N. Recent trends in machine learning for human activity recognition—A survey. In *Interdisciplinary Reviews: Data Mining and Knowledge Discovery*; Wiley Online Library: Hoboken, NJ, USA, 2018; Volume 8, pp. 1–11.
10. Lago, P.; Matsuki, M.; Inoue, S. Achieving Single-Sensor Complex Activity Recognition from Multi-Sensor Training Data. *arXiv* **2020**, arXiv:2002.11284.
11. Anliker, U.; Ward, J.; Lukowicz, P.; Troster, G.; Dolveck, F.; Baer, M.; Keita, F.; Schenker, E.; Catarsi, F.; Coluccini, L.; et al. AMON: A wearable multiparameter medical monitoring and alert system. *IEEE Trans. Inf. Technol. Biomed.* **2004**, *8*, 415–427. [CrossRef]
12. Bulling, A.; Blanke, U.; Schiele, B. A tutorial on human activity recognition using body-worn inertial sensors. *ACM Comput. Surv.* **2014**, *46*, 1–33. [CrossRef]
13. Kwon, H.; Tong, C.; Haresamudram, H.; Gao, Y.; Abowd, G.D.; Lane, N.D.; Plötz, T. IMUTube: Automatic Extraction of Virtual on-Body Accelerometry from Video for Human Activity Recognition. In *Proceedings of the ACM on Interactive, Mobile, Wearable and Ubiquitous Technologies*; ACM: New York, NY, USA, 2020; Volume 4. [CrossRef]

14. Kaur, H.; Pannu, H.S.; Malhi, A.K. A systematic review on imbalanced data challenges in machine learning: Applications and solutions. *ACM Comput. Surv.* **2019**, *52*, 4. [CrossRef]
15. Nweke, H.F.; Teh, Y.W.; Mujtaba, G.; Alo, U.R.; Al-garadi, M.A. Multi-sensor fusion based on multiple classifier systems for human activity identification. *Human-Centric Comput. Inf.* **2019**, *9*, 1. [CrossRef]
16. He, H.; Garcia, E.A. Learning from imbalanced data. *IEEE Trans. Knowl. Data Eng.* **2009**, *21*, 1263–1284.
17. Guo, H.; Li, Y.; Shang, J.; Gu, M.; Huang, Y.; Gong, B. Learning from class-imbalanced data: Review of methods and applications. *Expert Syst. Appl.* **2017**, *73*, 220–239.
18. Cook, D.J.; Krishnan, N.C. *Activity Learning: Discovering, Recognizing and Predicting Human Behavior from Sensor Data*; John Wiley & Sons: Hoboken, NJ, USA, 2015.
19. Yang, J.; Nguyen, M.N.; San, P.P.; Li, X.L.; Krishnaswamy, S. Deep Convolutional Neural Networks on Multichannel Time Series for Human Activity Recognition. In Proceedings of the Twenty-Fourth International Joint Conference on Artificial Intelligence, Buenos Aires, Argentina, 25–31 July 2015.
20. Chen, Y.; Shen, C. Performance Analysis of Smartphone-Sensor Behavior for Human Activity Recognition. *IEEE Access* **2017**, *5*, 3095–3110. [CrossRef]
21. Nweke, H.F.; Teh, Y.W.; Al-garadi, M.A.; Alo, U.R. Deep learning algorithms for human activity recognition using mobile and wearable sensor networks: State of the art and research challenges. *Expert Syst. Appl.* **2018**, *105*, 233–261. [CrossRef]
22. Ni, Q.; Fan, Z.; Zhang, L.; Nugent, C.D.; Clel, I.; Zhang, Y.; Zhou, N. Leveraging wearable sensors for human daily activity recognition with stacked denoising autoencoders. *Sensors* **2020**, *20*, 5114. [CrossRef]
23. Vandewiele, G.; Dehaene, I.; Kovács, G.; Sterckx, L.; Janssens, O.; Ongenae, F.; De Backere, F.; De Turck, F.; Roelens, K.; Decruyenaere, J.; et al. Overly optimistic prediction results on imbalanced data: A case study of flaws and benefits when applying over-sampling. *Artif. Intell. Med.* **2021**, *111*, 101987. [CrossRef]
24. Sousa Lima, W.; Souto, E.; El-Khatib, K.; Jalali, R.; Gama, J. Human Activity Recognition Using Inertial Sensors in a Smartphone: An Overview. *Sensors* **2019**, *19*, 3213. [CrossRef]
25. Akbari, A.; Jafari, R. Transferring activity recognition models for new wearable sensors with deep generative domain adaptation. In Proceedings of the IPSN 2019-18th International Conference on Information Processing in Sensor Networks, Montreal, QC, Canada, 16–18 April 2019; pp. 85–96.
26. Dietterich, T.G. Ensemble methods in machine learning. In Proceedings of the International Workshop on Multiple Classifier Systems, Cagliari, Italy, 21–23 June 2000; pp. 1–15.
27. Chawla, N.V.; Bowyer, K.W.; Hall, L.O.; Kegelmeyer, W.P. SMOTE: Synthetic minority over-sampling technique. *J. Artif. Intell. Res.* **2002**, *16*, 321–357. [CrossRef]
28. Dong, Y.; Wang, X. A new over-sampling approach: Random-SMOTE for learning from imbalanced data sets. *Lect. Notes Comput. Sci.* **2011**, *7091*, 343–352.
29. Zhu, X.; Wu, X. Class Noise vs. Attribute Noise: A Quantitative Study. *Artif. Intell. Rev.* **2004**, *22*, 177–210. [CrossRef]
30. He, H.; Ma, Y. *Imbalanced Learning: Foundations, Algorithms, and Applications*, 1st ed.; Wiley-IEEE Press: Hoboken, NJ, USA, 2013.
31. Frenay, B.; Verleysen, M. Classification in the Presence of Label Noise: A Survey. *IEEE Trans. Neural Netw. Learn. Syst.* **2014**, *25*, 845–869. [CrossRef] [PubMed]
32. Batista, G.E.A.P.A.; Prati, R.C.; Monard, M.C. A study of the behavior of several methods for balancing machine learning training data. *ACM SIGKDD Explor. Newsl.* **2004**, *6*, 20–29. [CrossRef]
33. Hu, S.; Liang, Y.; Ma, L.; He, Y. MSMOTE: Improving classification performance when training data is imbalanced. In Proceedings of the 2nd International Workshop on Computer Science and Engineering: WCSE 2009, Qingdao, China, 28–30 October 2009.
34. Kubat, M.; Matwin, S. Addressing the Curse of Imbalanced Training Sets: One-Sided Selection. In Proceedings of the Fourteenth International Conference on Machine Learning, Nashville, TN, USA, 8–12 July 1997; pp. 179–186.
35. Saez, J.A.; Luengo, J.; Stefanowski, J.; Herrera, F. SMOTE-IPF: Addressing the noisy and borderline examples problem in imbalanced classification by a re-sampling method with filtering. *Inf. Sci.* **2015**, *291*, 184–203. [CrossRef]
36. Hoens, T.R.; Chawla, N.V. Imbalanced datasets: From sampling to classifiers. In *Foundations, Algorithms, and Applications: In Imbalanced Learning*; Wiley: Hoboken, NJ, USA, 2013; pp. 43–59. [CrossRef]
37. Kovács, G. smote-variants: A Python Implementation of 85 Minority Oversampling Techniques. *Neurocomputing* **2019**, *366*, 352–354. [CrossRef]
38. Barua, S.; Islam, M.; Murase, K. ProWSyn: Proximity Weighted Synthetic Oversampling Technique for Imbalanced Data Set Learning. *Adv. Knowl. Discov. Data Min.* **2013**, *7819*, 317–328 .
39. Chavarriaga, R.; Sagha, H.; Calatroni, A.; Digumarti, S.T.; Tröster, G.; Millán, J.D.; Roggen, D. The Opportunity challenge: A benchmark database for on-body sensor-based activity recognition. *Pattern Recognit. Lett.* **2013**, *34*, 2033–2042. [CrossRef]
40. Reiss, A.; Stricker, D. Creating and benchmarking a new dataset for physical activity monitoring. In Proceedings of the 5th International Conference on PErvasive Technologies Related to Assistive Environments, Crete, Greece, 6–8 June 2012.
41. Casale, P.; Pujol, O.; Radeva, P. Personalization and user verification in wearable systems using biometric walking patterns. *Pers. Ubiquitous Comput.* **2012**, *16*, 563–580. [CrossRef]
42. Roggen, D.; Calatroni, A.; Rossi, M.; Holleczek, T.; Förster, K.; Tröster, G.; Lukowicz, P.; Bannach, D.; Pirkl, G.; Ferscha, A.; et al. Collecting complex activity datasets in highly rich networked sensor environments. In Proceedings of the 2010 Seventh International Conference on Networked Sensing Systems (INSS), Kassel, Germany, 15–18 June 2010; pp. 233–240.

43. Shoaib, M.; Bosch, S.; Incel, O.D.; Scholten, H.; Havinga, P.J.M. Complex human activity recognition using smartphone and wrist-worn motion sensors. *Sensors* **2016**, *16*, 426. [CrossRef]
44. Tang, Q.; Vidrine, D.J.; Crowder, E.; Intille, S.S. Automated detection of puffing and smoking with wrist accelerometers. In Proceedings of the PervasiveHealth 2014: 8th International Conference on Pervasive Computing Technologies for Healthcare, Oldenburg, Germany, 20–23 May 2014; pp. 80–87.
45. Lara, O.D.; Labrador, M.A. A survey on human activity recognition using wearable sensors. *IEEE Commun. Surv. Tutor.* **2013**, *15*, 1192–1209. [CrossRef]
46. Shoaib, M.; Bosch, S.; Scholten, H.; Havinga, P.J.M.; Incel, O.D. Towards detection of bad habits by fusing smartphone and smartwatch sensors. In Proceedings of the 2015 IEEE International Conference on Pervasive Computing and Communication Workshops, St. Louis, MO, USA, 23–27 March 2015; pp. 591–596.
47. Pedregosa, F.; Varoquaux, G. Scikit-learn: Machine learning in Python. *J. Mach. Learn. Res.* **2011**, *12*, 2825–2830.
48. Sokolova, M.; Lapalme, G. A systematic analysis of performance measures for classification tasks. *Inf. Process. Manag.* **2009**, *45*, 427–437. [CrossRef]
49. Pirttikangas, S.; Fujinami, K.; Nakajima, T. Feature Selection and Activity Recognition from Wearable Sensors. In Proceedings of the Third International Conference on Ubiquitous Computing Systems UCS'06, Seoul, Korea, 11–13 October 2006; Springer: Berlin/Heidelberg, Germany, 2006; pp. 516–527. [CrossRef]
50. Ghorbani, R.; Ghousi, R. Comparing Different Resampling Methods in Predicting Students' Performance Using Machine Learning Techniques. *IEEE Access* **2020**, *8*, 67899–67911. [CrossRef]
51. Shahi, A.; Deng, J.D.; Woodford, B.J. A streaming ensemble classifier with multi-class imbalance learning for activity recognition. In Proceedings of the 2017 International Joint Conference on Neural Networks (IJCNN), Anchorage, AK, USA, 14–19 May 2017; pp. 3983–3990.
52. Jantschi, L.; Bolboaca, S.D. Computation of Probability Associated with Anderson-Darling Statistic. *Mathematics* **2018**, *6*, 88. [CrossRef]
53. Demšar, J. Statistical Comparisons of Classifiers over Multiple Data Sets. *J. Mach. Learn. Res.* **2006**, *7*, 1–30.
54. Fisher, R.A. *Statistical Methods and Scientific Inference*; Hafner Publishing Co.: Oxford, UK, 1956; p. viii 175.
55. Kaur, A.; Kaur, I. An empirical evaluation of classification algorithms for fault prediction in open source projects. *J. King Saud-Univ.-Comput. Inf. Sci.* **2018**, *30*, 2–17. [CrossRef]

sensors

MDPI

Article

Fuzzy Overclustering: Semi-Supervised Classification of Fuzzy Labels with Overclustering and Inverse Cross-Entropy

Lars Schmarje [1,*], Johannes Brünger [1], Monty Santarossa [1], Simon-Martin Schröder [1], Rainer Kiko [2] and Reinhard Koch [1]

1 Multimedia Information Processing Group, Kiel University, 24118 Kiel, Germany; jbr@informatik.uni-kiel.de (J.B.); msa@informatik.uni-kiel.de (M.S.); sms@informatik.uni-kiel.de (S.-M.S.); rk@informatik.uni-kiel.de (R.K.)

2 Laboratoire d'Océanographie de Villefranche, Sorbonne Université, 06230 Villefranche-sur-Mer, France; rainer.kiko@obs-vlfr.fr

* Correspondence: las@informatik.uni-kiel.de

Abstract: Deep learning has been successfully applied to many classification problems including underwater challenges. However, a long-standing issue with deep learning is the need for large and consistently labeled datasets. Although current approaches in semi-supervised learning can decrease the required amount of annotated data by a factor of 10 or even more, this line of research still uses distinct classes. For underwater classification, and uncurated real-world datasets in general, clean class boundaries can often not be given due to a limited information content in the images and transitional stages of the depicted objects. This leads to different experts having different opinions and thus producing fuzzy labels which could also be considered ambiguous or divergent. We propose a novel framework for handling semi-supervised classifications of such fuzzy labels. It is based on the idea of overclustering to detect substructures in these fuzzy labels. We propose a novel loss to improve the overclustering capability of our framework and show the benefit of overclustering for fuzzy labels. We show that our framework is superior to previous state-of-the-art semi-supervised methods when applied to real-world plankton data with fuzzy labels. Moreover, we acquire 5 to 10% more consistent predictions of substructures.

Keywords: semi-supervised; fuzzy; deep learning; noisy; real-world; plankton; marine

Citation: Schmarje, L.; Brünger, J.; Santarossa, M.; Schröder, S.-M.; Kiko, R.; Koch, R. Fuzzy Overclustering: Semi-Supervised Classification of Fuzzy Labels with Overclustering and Inverse Cross-Entropy. *Sensors* **2021**, *21*, 6661. https://doi.org/10.3390/s21196661

Academic Editor: Hyoungsik Nam

Received: 24 August 2021
Accepted: 2 October 2021
Published: 7 October 2021

1. Introduction

Over the past years, we have seen the successful application of deep learning to many underwater computer vision problems [1–4]. Automatic analysis of underwater data allows us to monitor ecological changes by evaluating large amounts of for example plankton data [5,6]. While it is relatively easy to create a lot of underwater image data, its analysis is time-consuming and thus expensive because the annotation requires trained taxonomists. The possible reasons for this issue include the huge amounts of data, the high imbalance between classes and the variability of annotations [7].

In underwater classification, domain experts often differ in their annotations [7–9]. This issue arises due to the following reasons: Firstly, automatically captured underwater images often have a lower quality than images taken manually by humans. This difference in quality arises for example due to the underwater lighting conditions and no manual corrections to e.g., insufficient sharpness or not centering the target inside the focus. For example the analyis of benthic images can suffer from these issues [8,9]. Even in the best scenario, a single image generally does not contain most of the information needed for a clear identification (e.g., three-dimensional configuration, minute morphological details, fluorescence). Secondly, intermediate stages actually exist between classes [10]. For example, in Figure 1 we show two different physical appearances (puff & tuft) of trichodesmium, while the dataset also contains intermediate stages between these two classes.

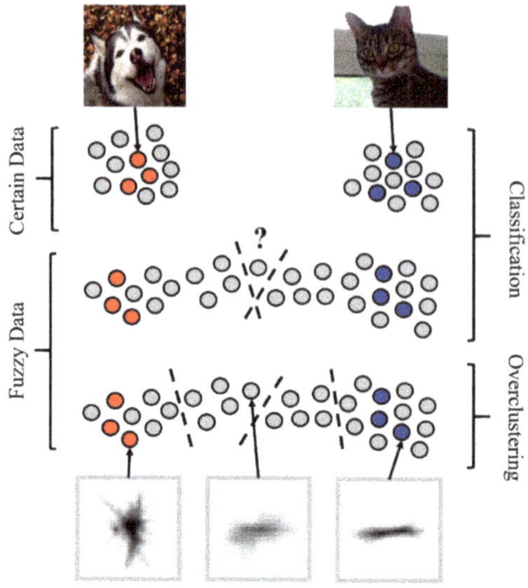

Figure 1. Illustration of `fuzzy` data and overclustering—The grey dots represent unlabeled data and the colored dots labeled data from different classes. The dashed lines represent decision boundaries. For `certain` data, a clear separation of the different classes with one decision boundary is possible and both classes contain the same amount of data (**top**). For `fuzzy` data determining a decision boundary is difficult because of intermediate datapoints between the classes (**middle**). These `fuzzy` datapoints can often not be easily sorted into one consistent class between annotators. If you overcluster the data, you get smaller but more consistent substructures in the `fuzzy` data (**bottom**). The images illustrate possible examples for `certain` data (cat & dog) and `fuzzy` plankton data (trichodesmium puff and tuft). The center plankton image was considered to be trichodesmium puff or tuft by around half of the annotators each. The left and right plankton image were consistently annotated.

This issue of different annotations is also known as *intra-* and *inter-observer* variability [11] and is common in many biological and medical application fields [8,9,12–17]. Even in a curated dataset [1], we quote Tarling et al. who state "there will very likely be inaccuracies, bias, and even inconsistencies in the labeling which will have affected the training capacity of the model and lead to discrepancies between predictions and ground truths" [18]. When aggregating multiple annotations per image, we call the resulting label `fuzzy` if we have different annotations between experts (non-zero variance), and `certain` if all annotations agree with each other. The mathematical formulation of a `fuzzy` label would be a unknown soft probability distribution l for k classes. The distribution $l \in (0,1)^k$ can only be approximated with a high cost e.g., by averaging over multiple annotations.

Semi- and Self-Supervised Learning are promising approaches to decrease the needed amount of annotated data by a factor of 10 or even more [19–21]. These approaches leverage unlabeled data in addition to the normal labeled data to improve the training. A common strategy is to define a pretext task like image rotation prediction [22] or mutual information maximization [23] for pretraining. A broad overview of current trends, ideas and methods in semi-, self- and unsupervised learning is available in [24]. However, this research mainly focuses on established curated classification datasets such as STL-10 [25]. In these datasets, a clear distinction between classes such as cats and dogs are given. The hard partitioning of intermediate morphologies is not appropriate and does not allow the identification of substructures. We show that state-of-the-art semi-supervised algorithms are not well suited to handle `fuzzy` labels. These algorithms expect only `certain` labels as shown in the upper part of Figure 1. If we apply previous semi-supervised algorithms to `fuzzy` data which include `fuzzy` images, these algorithms arbitrarily assign undecidable images to one class (middle part of Figure 1).

Noisy labels are a common data quality issue and are discussed in the literature [11,26,27]. The fuzziness of labels is known as a special case of label noise that exist "due to subjectiveness of the task for human experts or the lack of experience in annotator[s]" [26]. In contrast to us, most methods [28–30] and literature surveys [11,26,27] interpret `fuzzy` labels as corrupted labels. We argue that `fuzzy` labels are valid signals derived from ambiguous images and that it is important to discover the substructures for real-world data handling [12–17].

Geng proposed to learn the label distribution to handle `fuzzy` data [31] and the idea was extended to the application of real-world images [32]. However, these methods are not semi-supervised and therefore depend on large labeled datasets. A variety of methods was proposed to handle `fuzzy` data in a semi-supervised learning approach [33–35]. These methods use lower-dimensional features spaces in contrast to images as input. Liu et al. proposed to use independent predictions of multiple networks as pseudo-labels for the estimation of the label distribution for photo shot-type classification [36]. We argue that the true label distribution is difficult to approximate and thus difficult to evaluate. We do not learn the label distribution but use clustering to identify substructures.

We propose *Fuzzy Overclustering* (FOC) which separates the `fuzzy` data into a larger number of visual homogeneous clusters (lower part, Figure 1) which can then be annotated very efficiently [10]. We will show on a Plankton dataset that state-of-the-art semi-supervised algorithms perform worse on `fuzzy` data in comparison to our method FOC which explicitly considers `fuzzy` images. Moreover, we will show that this leads to 5 to 10% more self-consistent predictions of plankton data.

One main idea is to rephrase the handling of `fuzzy` labels as a semi-supervised learning problem by using a small set of `certain` images and a large number of `fuzzy` images that are treated as unlabeled data. This approach allows us to use the idea of overclustering from semi-supervised literature [23,37] and apply it to `fuzzy` data. The difference to previous work is that we use overclustering not only to improve classification accuracy on the labeled data but improve the clustering and therefore the identification of substructures of `fuzzy` data. We show that overclustering allows us to cluster the `fuzzy` images in a more meaningful way by finding substructures and therefore allowing experts to analyze `fuzzy` images more consistently in the future.

We show the benefits of our method mainly on a plankton dataset which highlights the benefit for underwater classification. However, the issue of `fuzzy` labels is neither limited to plankton data nor to underwater classification. On a synthetic dataset, we show a proof-of-concept for the generalizability of our model to other datasets.

Our key contributions are:

- We identify an issue of semi-supervised algorithms that they do not work well with `fuzzy` labels. However, such `fuzzy` labels occur regularly in underwater image classification e.g due to high natural variation of depicted objects which leads to a high inter- and intraobserver variability.
- We propose a novel framework for handling `fuzzy` labels with a semi-supervised approach. This framework uses overclustering to find substructures in `fuzzy` data and outperforms common state-of-the-art semi-supervised methods like FixMatch [38] on `fuzzy` plankton data.
- We propose a novel loss, *Inverse Cross-entropy* (CE^{-1}), which improves the overclustering quality in semi-supervised learning.
- We achieve 5 to 10% more self-consistent predictions on `fuzzy` plankton data.

2. Method

Our framework Fuzzy Overclustering (FOC) aims at creating an overclustering for `fuzzy` labels by using an auxiliary classification and not the other way round like previous literature [23,37]. In this section, we describe our framework in general and explain important parts in detail in the following subsections. We use the following notation for the given semi-supervised classification task. Our training data consists of the two subsets

X_l and X_u. X_l is a labeled image dataset with images $x \in X_l$ and corresponding labels y. X_u is an unlabeled image dataset, i.e., there is/exists no label for images $x \in X_u$.

We generate three inputs x_1, x_2, x_3 based on one image $x \in X_l \cup X_u$ depending on the availability of the corresponding label y. If y is not available, the images x_1 and x_2 are augmented views of x and x_3 is an augmented version of a random image $x' \in X_l \cup X_u$. If y is available, x_1 is an augmented view of x, x_2 is a supervised augmentation (see Section 2.3) and x_3 an inverse example. For the inverse example, we choose an image $x' \in X_l$ with a different label y' ($y! = y'$). We use an augmented version of this image as third input $x_3 = g_3(x')$ with augmentation g_3. We constraint the ratio from unlabeled to labeled data to a fixed ratio r to improve the run time of the model (see Section 2.4). The inputs are processed by a neural network Φ which is composed of a backbone like ResNet50 [39] and linear output prediction layers. Following [23], we call this linear predictors *heads* and use them either as normal or overclustering heads. As output we use the soft-max classifications of these normal and overclustering heads. If k_{GT} is the number of ground-truth classes a normal head outputs a probability for each of the k_{GT} classes. The overclustering head has k output nodes with $k > k_{GT}$ and give probabilities for more clusters than ground-truth classes (overclustering). Both type of heads are therefore fully connected layers with softmax activation but of different output size. We can average the training over multiple independent heads per type as shown in [23]. We use the notation Φ_{n_i} or Φ_{o_i} for the i-th normal or overclustering head respectively. An overview about the general pseudo code of FOC including the loss calculation is given in Algorithm 1.

For both heads the loss is different but can be written as the weighted sum of an unsupervised and a supervised loss as follows:

$$\mathcal{L} = \lambda_s \mathcal{L}_s + \lambda_u \mathcal{L}_u \tag{1}$$

$\mathcal{L}_s$ is cross-entropy ($\mathcal{L}_{CE}$) for the normal head and our novel CE^{-1} loss ($\mathcal{L}_{CE^{-1}}$) for the overclustering head (see Section 2.1). For both heads $\mathcal{L}_u$ is the mutual information loss $\mathcal{L}_{MI}$ (see Section 2.2). An illustration of the complete pipeline is given in Figure 2. We initialize our backbones with pretrained weights and can therefore directly use RGB images as input. For further implementation details see Section 3.2.

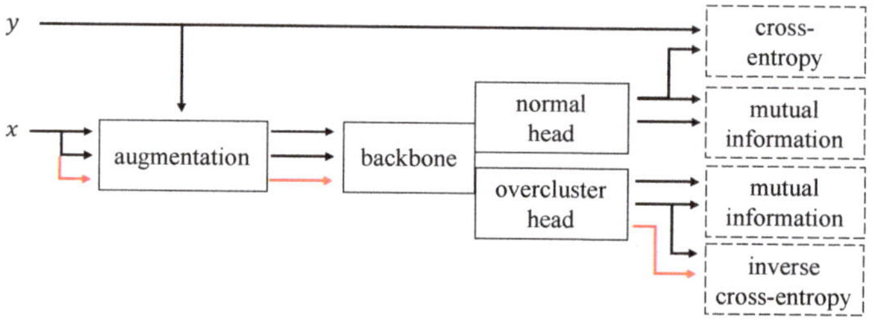

Figure 2. Overview of our framework FOC for semi-supervised classification—The input image is x and the corresponding label is y. The arrows indicate the usage of image or label information. Parallel arrows represent the independent copy of the information. The usage of the label for the augmentations is described in Section 2.3. The red arrow stands for an inverse example image x' with a different label than y. The output of the normal and the overclustering head have different dimensionalities. The normal head has as many outputs as ground-truth classes exist (k_{GT}) while the overclustering head has k outputs with $k > k_{GT}$. The dashed boxes on the right side show the used loss functions. More information about the losses inverse cross-entropy and mutual information can be found in Sections 2.1 and 2.2 respectively.

If we use FOC with $\lambda_s = 0$ and without supervised augmentations our model is comparable to the pretext task of Invariant Information Clustering (IIC) [23]. We can use this configuration as a warm-up to pretrain the weights. During the evaluation, we will refer to using the pretext task for IIC and the warm-up of FOC synonymously. Our framework FOC can also be used to perform standard unsupervised clustering. The details about unsupervised clustering and a comparison to previous literature is given in the supplementary.

Algorithm 1: Pseudocode for our method Fuzzy Overclustering

Data: Batch of images of size b from labeled image data X_l and unlabeled image data X_u
Result: calculate loss value for one given batch for a network Φ with n normal and overclustering heads
L: matrix of size $b \times 2n$;
`/* iterate over batch */`
for $i \leftarrow 0$ **to** b **do**
  $x \leftarrow i$-th image in batch;
  if *label y for image x_i available* **then**
    $x_1 \leftarrow g_1(x)$ with random augmentation g_1;
    `/* Supervised augmentation defined in Section 2.3 */`
    $x_2 \leftarrow g_2(x)$ with supervised augmentation g_2;
    `/* Inverse example defined in Section 2 */`
    $x_3 \leftarrow g_3(x')$ with random augmentation g_3 and inverse example x';
  else
    $x_1 \leftarrow g_1(x)$ with random augmentation g_1;
    $x_2 \leftarrow g_2(x)$ with random augmentation g_2;
    $x_3 \leftarrow g_3(x')$ with random augmentation g_3 and random image x';
  end
  `/* iterate over heads */`
  for $j \leftarrow 0$ **to** n **do**
    calculate forward pass for outputs Φ_{n_j} and Φ_{o_j};
    `/* CE loss for normal head */`
    L[i,j] $\leftarrow \mathcal{L}_{CE}(\Phi_{n_j}(x_i), l_i)$ with l_i;
    `/* CE`$^{-1}$` loss for overclustering head */`
    L[i,j+n] $\leftarrow \mathcal{L}_{CE^{-1}}(x_1, x_2, x_3)$ with Equation (2) ;
  end
end
`/* calculate loss */`
$\mathcal{L}_s \leftarrow$ average supervised loss across heads and batch from L;
$\mathcal{L}_u \leftarrow$ unsupervised MI loss across batch with Equations (3) and (4);
$\mathcal{L} \leftarrow \lambda_s \mathcal{L}_s + \lambda_u \mathcal{L}_u$;

2.1. *Inverse Cross-Entropy (CE^{-1})*

Inverse Cross-Entropy is a novel supervised loss for an overclustering head and one of the key contributions of this work. The loss is needed to use the label information for an overclustering head. For normal heads, we can use cross-entropy (CE) to penalize the divergence between our prediction and the label. We can not use CE directly for the overclustering heads since we have more clusters than labels and no predefined mapping between the two. However, we know that the inputs x_1/x_2 and x_3 should not belong to the same cluster. Therefore, our goal with CE^{-1} is to define a loss that pushes their output distributions (e.g., $\Phi(x_1)$ and $\Phi(x_3)$) apart from each other.

Let us assume we could define a distribution that $\Phi(x_3)$ should not be. In short, an inverse distribution $\Phi(x_3)^{-1}$. If we had such a distribution we could use CE to penalize the divergence for example between $\Phi(x_1)$ and $\Phi(x_3)^{-1}$.

One possible and easy solution for an inverse distribution is $\Phi(x_3)^{-1} = 1 - \Phi(x_3)$. For a binary classification problem, $\Phi(x_3)^{-1}$ can even be interpreted as a probability distribution again. This is not the case for a multi-class classification problem. We could use a function like softmax to cast $\Phi(x_3)^{-1}$ into a probability distribution but decided against it for three reasons. Firstly, we would penalize correct behavior. For example in a three class problem with $\Phi_1(x_1) = 0.5 = \Phi_2(x_1)$ and $\Phi_3(x_3) = 1$ we only get $CE(\Phi(x_1), \Phi(x_3)^{-1}) = 0$ if $\Phi(x_3)^{-1}$ is not a probability distribution. Otherwise either $\Phi_1(x_3)^{-1}$ or $\Phi_2(x_3)^{-1}$ have to be real smaller than 1. Secondly, we are still minimizing the entropy of $\Phi(x_1)$ which leads to more confident predictions in semi-supervised learning [19,20,40–43]. The proof is given in the supplementary. Thirdly, it is easier and in practice, it is not needed. For the input $i = (x_1, x_2, x_3)$, we define the cross-entropy inverse loss $\mathcal{L}_{CE^{-1}}$ as shown in Equation (2).

$$\begin{aligned} \mathcal{L}_{CE^{-1}}(i) &= 0.5 \cdot CE^{-1}(\Phi(x_1), \Phi(x_3)) \\ &\quad + 0.5 \cdot CE^{-1}(\Phi(x_2), \Phi(x_3)) \text{, with} \\ CE^{-1}(p, q) &= -\sum_{c=1}^{k} p(c) \cdot ln(1 - q(c)) \,. \end{aligned} \tag{2}$$

2.2. Mutual Information (MI)

For the unlabeled data, we use the loss proposed by Ji et al. because it is calculated directly on the output clusters [23]. Therefore similar images are pulled to the same clusters while CE^{-1} pushes different images apart. For this purpose, we want to maximize the mutual information between two output predictions $\Phi(x_1), \Phi(x_2)$ with x_1, x_2 images which should belong to the same cluster and $\Phi : X \to [0,1]^k$ a neural network with k output dimensions. We can interpret $\Phi(x)$ as the distribution of a discrete random variable z given by $P(z = c|x) = \Phi_c(x)$ for $c \in \{1, \ldots, k\}$ with $\Phi_c(x)$ the c-th output of the neural network. With z, z' such random variables we need the joint probability distribution for $P_{cc'} = P(z = c, z' = c')$ for the calculation of the mutual information $I(z, z')$. Ji et al. propose to approximate the matrix P with the entry $P_{cc'}$ at row c and column c' by averaging over the multiplied output distributions in a batch of size n [23]. Symmetry of P is enforced as shown in Equation (3).

$$P = \frac{Q + Q^T}{2} \text{ with } Q = \frac{1}{n}\sum_{i=1}^{n} \Phi(x_i) \cdot \Phi(x_i')^T \tag{3}$$

We can maximize our objective $I(z, z')$ with the marginals $P_c = P_{c'} = P(z = c)$ given as sums over the rows or columns as shown in Equation (4).

$$I(z, z') = \sum_{c=1}^{k} \sum_{c'=1}^{k} P_{cc'} \cdot ln \frac{P_{cc'}}{P_c \cdot P_{c'}} \tag{4}$$

2.3. Supervised Augmentations

In the unsupervised pretraining, we use the same image x to create the two inputs $x_1 = g_1(x)$ and $x_2 = g_2(x)$ based on the augmentations g_1 and g_2. Otherwise, without supervision, it is difficult to determine similar images. However, if we have the label y for x we can use a secondary image $x' \in X_l$ with the same label to mock an ideal image transformation to which the network should be invariant. In this case we can create $x_2 = g_2(x')$ based on the different image. We call this *supervised augmentation*.

2.4. Restricted Unsupervised Data

Unlabeled data has a small impact on the results but drastically increases the runtime in most cases. The increased runtime is caused by the facts that we often have much more unlabeled data than labeled data and that a neural network runtime is normally linear in the number of samples it needs to process. However, unlabeled data is essential for our proposed framework and we can not just leave it out. We propose to restrict the unlabeled data to a fixed upper-bound ratio r in every batch and therefore the unlabeled data per epoch. Detailed examples and experiments are given in the supplementary. It is important to notice that we restrict only the unlabeled data per batch/epoch. While for one epoch the network will not process all unlabeled data, over time all unlabeled data will be seen by the network. We argue that the impact on training time negatively outweighs the small benefit gained from all unlabeled data per epoch.

3. Experiments

We conducted our experiments mainly on a real-world plankton dataset. We used the common image classification dataset STL-10 as a comparison with only `certain` labels and a synthetic dataset for a proof-of-concept for the generalizability to other datasets. We compare ourselves to previous work and make several ablations. Additional results like unsupervised clustering, more detailed ablations and further details are given in the supplementary material.

3.1. Datasets

While the issue of `fuzzy` labels is present in multiple datasets [12–17], they are not well suited for evaluations. If we want to quantify the performance on `fuzzy` labels, we need a dataset with very good `fuzzy` ground-truth. This can only be achieved with a high cost e.g., by multiple annotations and thus is often not feasible. For all used datasets, we ensure that the labeled training data only consists of `certain` images and that the `fuzzy` images are used as unlabeled data. If we include `fuzzy` labels in the labeled data which is used as guidance during training, this will lead to worse performance as illustrated in the ablations (Table 3).

3.1.1. Plankton

The plankton dataset contains diverse grey-level images of marine planktonic organisms. The images were captured with an Underwater Vision Profiler 5 [44] and are hosted on EcoTaxa [45]. In the citizen science project PlanktonID (https://planktonid.geomar.de/en (accessed on 6 October 2021)), each sample was classified multiple times by citizen scientists. The data for the PlanktonID project is a subset of the data available on EcoTaxa [45]. It was presorted to contain a more balanced representation of the available classes. The dataset consists of 12,280 images in originally 26 classes. We merged minor and similar classes so that we ended up with 10 classes. The class no-fit represents a mixture of left-over classes. The merging was necessary because some classes had too few images for current state-of-the-art semi-supervised approaches. After this process, a class imbalance is still present with the smallest class containing about 4.16% and the largest class 30.37% of all samples. We use the mean over all annotations as the `fuzzy` label. The citizen scientists agree on most images completely. We call these images and their label `certain`. However, about 30% of the data has as least one disagreeing annotation. We call these images and their label `fuzzy` and use the most likely class as ground-truth if we need a hard label for evaluation. The `fuzzy` labeled images are used only as unlabeled data. More details about the mapping process, the number of used samples and graphical illustrations are given in the supplementary.

3.1.2. STL-10

STL-10 is a common semi-supervised image classification dataset [25] and a subset of ImageNet [46]. It consists of 5000 training samples and 8000 validation samples depicting everyday objects. Additionally, 100,000 unlabeled images are provided that may belong to the same or different classes than the training images. In contrast to the plankton and synthetic dataset, no labels are provided for the unlabeled data and no `fuzzy` datapoints exist. We use this dataset only to illustrate the difference in the performance of FOC to previous semi-supervised methods.

3.1.3. Synthetic Circles and Ellipses (SYN-CE)

This dataset is a mixture of circles and ellipses (bubbles) on a black background with different colors. The 6 ground-truth classes are blue, red and green circles or ellipses. An image is defined as `certain` if the hue of the color is 0 (red), 120 (green) or 240 (blue) and the main axis ratio of the bubble is 1 (circle) or 2 (ellipse). Every other datapoint is considered `fuzzy` and the ground-truth label l is calculated as the product of the interpolation of the color p_c and the geometry p_g distribution. More details are in the supplementary. The dataset consists of 1800 `certain` and 1000 `fuzzy` labeled images for train, validation and unlabeled data split. We look at three subsets: *Ideal*, *Real* and *Fuzzy*. The *Ideal* subset uses the maximal class of the `fuzzy` label l as a ground-truth class and represents the ideal case that we certainly know the most likely label to each image. For the *Real* subset, the ground-truth classes in randomly picked with the distribution of the `fuzzy` label l and represent the real or common case. For example due to only one annotation, the percentage that the label corresponds to the actual most likely class is linear to the `fuzzy` label. The *Fuzzy* subset only uses `certain` labeled images as training data and represent a cleaned training dataset. We will show that this handling of `fuzzy` labels leads to a higher classification performance in comparison to the *Real* dataset in Section 3.5.1. The *Ideal* and the *Real* subset can be evaluated on the unlabeled data of the *Fuzzy* subset with some overlap in the images.

3.2. Implementation Details

As a backbone for our framework, we used either a ResNet34 variant [23] or a standard ResNet50v2 [39]. The heads are single fully connected layers with a softmax activation function. Following [23], we use five randomly initialized copies for each type of head and repeat images per batch three times for more stable training. We alternated between training the different types of heads. The inputs are either sobel-filtered images or color images for pretrained networks. For the ResNet34 backbone, we use CIFAR20 (20 superclasses in CIFAR-100 [47]) weights and for the ResNet50v2 backbone ImageNet [46] weights. We use in general $\lambda_s = 1 = \lambda_u$ and an unlabeled data restriction of $r = 0.5$. We call our Framework FOC-Light if we use $\lambda_u = 0$ and no warm-up. This means we do not use the loss introduced by [23] and therefore also do not have to use their stabilization methods like repetitions. During the pretext task or warm-up and the main training, we train the framework with Adam and an initial learning rate of 1×10^{-4} for 500 epochs. When switching from the pretext task to fine-tuning, we train only the heads for 100 epochs with a learning rate of 1×10^{-3} before switching to the lower learning rate of 1×10^{-4}. The number of outputs for the overclustering head should be about 5 to 10 times the number of classes. The exact number is not crucial because it is only an upper bound for the framework. We use 70 for STL-10 and 60 for the plankton dataset. We selected all hyperparameters heuristically based on the STL-10 dataset and did not change them for the plankton dataset. We used the recommended hyperparameters by the original authors for the previous methods. We compared with the following methods Semantic Clustering by Adopting Nearest neighbors (SCAN) [48], Information Invariant Clustering (IIC) [23], Mean-Teacher [49], Pi(-Model) [29], Pseudo-label [50] and FixMatch [38]. More detailed descriptions are given in the supplementary.

3.3. Metrics

The evaluation protocols vary slightly depending on the used output and dataset. The used data splits training, validation and unlabeled are defined above in Section 3.1.

On STL-10, we calculate accuracy of the validation data. Accuracy is the portion of true positive and true negatives from the complete dataset.

$$\text{Accuracy} = \frac{\text{TP} + \text{TN}}{\text{TP} + \text{TN} + \text{FP} + \text{FN}} \tag{5}$$

TP, TN, FP and FN are the true positive, true negative, false positive and false negative respectively. We calculate these values per class and then sum the up before calculating the accuracy (micro averaging). For the overclustering head, we need to find a mapping between the output clusters and the given classes. We calculate this mapping based on the majority class in each cluster on the training data as in [23].

On the `fuzzy` plankton and synthetic datasets, we evaluate the macro-f1 score on the unlabeled data. We calculate the macro F1-Score i.e., the average of the F1-scores per class due to the skewed class distribution.

$$\text{F1-Score} = \frac{2\text{TP}}{2\text{TP} + \text{FP} + \text{FN}} \tag{6}$$

Mind that a micro averaged F1-Score would be in our case the same as the above defined accuracy. We use the unlabeled data as evaluation dataset because the `fuzzy` images, in which we are interested, are only included in the unlabeled data split by definition. The mapping for the overclustering head is calculated based on the unlabeled data split because we expect human experts to be involved in this process for the identification of substructures. The best unlabeled results of the `fuzzy` Plankton and Synthetic dataset are reported based on the validation metrics.

If not stated otherwise, we report the maximum score for the overclustering and the normal head and the average and standard deviation over 3 independent repetitions.

3.4. Results

3.4.1. State-of-the-Art Comparison

We compare the state-of-the-art methods on `certain` and `fuzzy` data in Table 1.

Table 1. Comparison of state-of-the-art on certain and fuzzy data—We use STL-10 as a `certain` dataset and the Plankton data as a `fuzzy` dataset. We report the Accuracy for STL-10 and the F1-Score for the Plankton data due to class imbalance. It is important to notice that STL-10 is a curated dataset while the Plankton dataset still contains the `fuzzy` images. For more details about the metrics see Section 3.3. The results of previous methods are reported in the original paper or the original authors code was used to replicate the results. The best results are marked bold. Legend: † A MLP used for fine-tuning. ‡ Used only 1000 labels instead of 5000. * Unsupervised method.

		Type of Data	
Method	**Network**	**Certain**	**Fuzzy**
SCAN * [48]	ResNet18	76.80 ± 1.10	37.64 ± 3.56
IIC [23]	ResNet34	85.76 ± 1.36	65.47 ± 1.86
IIC † [23]	ResNet34	88.8	66.81 ± 1.85
Mean-Teacher [49]	Wide ResNet28	78.577 ± 2.39 ‡	72.85 ± 0.46
Pi [29]	Wide ResNet28	73.77 ± 0.82 ‡	74.34 ± 0.58
Pseudo-label [50]	Wide ResNet28	72.01 ± 0.83 ‡	75.04 ± 0.52
FixMatch [38]	Wide ResNet28	**94.83 ± 0.63** ‡	76.28 ± 0.27
FOC-Light (Ours)	ResNet50	–	72.79 ± 2.99
FOC (Ours)	ResNet50	86.12 ± 1.22	**76.79 ± 1.18**

We see that FOC reaches a performance of about 86% on `certain` data but is not able to reach the performance of FixMatch. FixMatch outperforms FOC by a clear margin of nearly 8% while using a fifth of the labels. This performance is expected as FOC does not focus like the others on classifying certain but `fuzzy` data. If we look at the less curated `fuzzy` Plankton dataset, we see that FOC outperforms all all methods by a small margin. All previous methods focus on certain and curated data and we see this leads to a huge performance degeneration if they are applied to `fuzzy` data. FixMatch reaches in both datasets the best performance except for our method FOC. We conclude that the overclustering from FOC is the key for handling `fuzzy` data because it allows more flexibility during training. Previous semi-supervised methods did not consider the issue of inter- and intraobserver variability and thus are worse than FOC in classifying `fuzzy` data.

If we use FOC-Light without the loss and stabilization of [23] the F1-Score drops slightly to 75% but the used GPU hours can be decreased from 58 to 4 h. We conclude that the overclustering head is more suitable for handling fuzzy real-world data as we assumed at the beginning. Moreover, we see that the combination of cross-entropy and our novel loss CE^{-1} can also successfully train an overclustering head.

3.4.2. Consistency

Up to this point, we analyzed classification metrics based on the 10 ground-truth classes but the quality of substructures was not evaluated. We can judge the consistency of each image within its cluster with the help of experts as a quality measure. An image is consistent if an expert views it as visually similar to the majority of the cluster. The consistency is calculated by dividing the number of consistent images by all images. The consistency over all classes or per class for FOC and FixMatch is given in Table 2 and raw numbers are provided in the supplementary. We provide a comparison based on all data and without the no-fit class because this class contains a mixture of different plankton entities. Visual similarity is therefore difficult to judge because it can only be defined by not being similar the other nine classes. Based on the F1-Score, FixMatch and FOC perform similarly but if we look at the consistency we see that FOC is more than 5% more consistent than FixMatch. If we exclude the class no-fit from the analysis, FOC reaches a consistency of around 86% in comparison to 77% from FixMatch. For both sets, our method FOC reaches a higher average consistency per cluster and lower standard deviation. This means the clusters produced by FOC are more relevant in practice because there are fewer low-quality clusters which can not be used. Overall, this higher consistency can lead to faster and more reliable annotations.

Table 2. Consistency comparison on plankton dataset—The consistency is rated by experts over the complete data and a subset without the class no-fit. The score is given overall as as average per cluster with standard deviation and is described in Section 3.4.2. The best results are marked bold.

Method	All Data		Ignore Class No-Fit	
	Overall	Per Cluster	Overall	Per Cluster
FixMatch [38]	82.56	78.78 ± 28.22	77.11	69.61 ± 29.41
FOC (Ours)	**87.80**	**79.66 ± 18.88**	**86.31**	**86.41 ± 13.68**

3.4.3. Qualitative Results

We illustrate some qualitative results of FOC in Figure 3. All images in a cluster are visually similar, even the probably wrongly assigned images (red box). For the images in the first row, the annotators are certain that the images belong to the same class. In the second row, annotators show a high uncertainty of assignment between the two variants of the same biological object. This illustrates the benefit of overclustering since visual similar items are in the same cluster even for uncertain annotations. In a consensus process for the second row, experts could decide if the cluster should be the puff, tuft or a new borderline class. Moreover, this clustering could be beneficial for monitoring the current imaging process. We provide more randomly selected results in the supplementary.

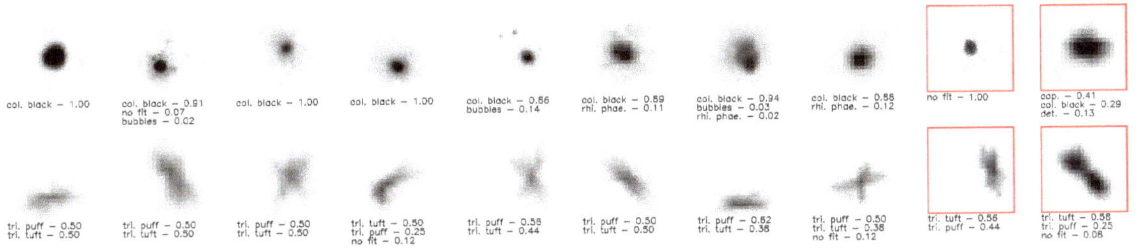

Figure 3. Qualitative results for unlabeled data—The results in each row are from the same predicted cluster. The three most important `fuzzy` labels based on the citizen scientists' annotations are given below the image. The last two items with the red box in each row show examples not matching the majority of the cluster.

3.5. Ablation Studies

3.5.1. SYN-CE

We compare our framework with some previous methods on the three subsets of SYN-CE in Table 3. All semi-supervised method reach almost a F1-Score of 100% on the unlabeled `fuzzy` data for the subset *Ideal*. In real-world data, it is unlikely that we have the real `fuzzy` ground-truth labels. It is more likely that we have uncertain/wrong labels for training and validation or no labels at all for `fuzzy` data like in the subsets *Real* or *Fuzzy*. In both cases, we see that our method reaches a superior performance with up to 10% higher F1-Score. While FOC-Light is only slightly better in comparison to the other semi-supervised methods on the *Real* subset it is comparable to the complete framework on the *Fuzzy* dataset. This is one indication that CE^{-1} is one of the key components for successfully training the overclustering heads. We see the F1-Score on the *Fuzzy* subset is around 10% higher than on the *Real* subset. We conclude that FOC can also generalize to other datasets. We conclude that these results supports our idea of separating `certain` and `fuzzy` data during training because we do not need to potentially falsely approximate the real `fuzzy` ground-truth label like in the *Real* subset.

Table 3. Comparison to state-of-the-art on SYN-CE datasets—Each column represent a subset of the dataset SYN-CE. The results are F1-Scores which were calculated on the unlabeled data which include the fuzzy labels. All results within a one percent margin of the best result are marked bold.

Method	Ideal	Real	Fuzzy
Mean-Teacher [49]	97.11 ± 0.78	73.23 ± 2.49	66.57 ± 16.27
Pi [29]	$\mathbf{98.44 \pm 0.28}$	72.74 ± 2.43	77.69 ± 5.02
Pseudo-label [50]	$\mathbf{98.17 \pm 0.30}$	75.70 ± 1.98	89.48 ± 1.94
FixMatch [38]	$\mathbf{98.32 \pm 0.01}$	71.81 ± 1.06	$\mathbf{93.82 \pm 1.83}$
FOC-Light (Ours)	$\mathbf{97.46 \pm 4.39}$	78.77 ± 7.83	$\mathbf{94.29 \pm 0.87}$
FOC (Ours)	$\mathbf{97.72 \pm 4.52}$	$\mathbf{83.86 \pm 4.21}$	$\mathbf{94.15 \pm 0.29}$

3.5.2. Loss & Network

In Table 4 multiple ablations for STL-10 and the plankton dataset are given. The scores are averaged across the different output heads of our framework. Based on these tables, we illustrate the impact of the warm-up, the initialization and the usage of the MI and CE^{-1} loss for our framework. The normal accuracy can be improved by about 10% when using the unsupervised warm-up on the STL-10 dataset. On the plankton dataset, the impact is less but tends to give better results of some percent. Warm-up in combination with the MI loss leads to a performance which is not more than 10% worse than the full setup for all ablations except for one. For this exception, CE^{-1} is needed to stabilize the overclustering performance due to the poor initialization with CIFAR-20 weights. We attribute this worse performance to the initialization and not the different backbone because on STL-10 the CIFAR-20 initializations of the ResNet34 backbone outperform the ImageNet weights of the ResNet50v2 backbone. We believe the positive effects of ImageNet weights for its subset STL-10 and the better network are negated by the different loss.

IIC is similar to FOC with warm-up and no additional losses but we train also train an overclustering head for handling `fuzzy` data. Taking this into consideration, we achieve an 8 to 11% better F1-Score than IIC. A special case is FOC-light which does only use the CE^{-1} loss and therefore no stabilization method proposed in [23]. This decreases gpu memory usage and runtime and results in a total decrease of the GPU hours from 58 to 4 h. Overall, our novel loss CE^{-1} improves the overclustering performance regardless of the dataset and the weight initialization by 10% on STL-10 and up to 7% on the plankton dataset. We see that CE^{-1} is a key component for training an overclustering head and can even be trained without the stabilization of the warm-up and the MI loss.

Table 4. Ablation study—The second to fourth column indicates if a warm-up, the MI loss or our CE^{-1}loss were used respectively. The fifth column indicates if CIFAR-20 (C), ImageNet (I) or no (–) weights were used. Sobel filtered images are used as input for no weights. The Top1 and Top3 results are marked bold respectively. * Original authors code. † A MLP used for fine-tuning.

Method	Warm	MI	CE^{-1}	Weight	Accuracy	
					Overcluster	Normal
FOC		X		–	70.92 ± 2.42	76.39 ± 0.05
IIC * [23]	X			–		85.76
FOC	X	X		–	73.88 ± 0.21	82.01 ± 5.31
FOC	X	X	X	–	82.59 ± 0.06	86.49 ± 0.01
FOC	X	X	X	C	84.36 ± 0.64	78.59 ± 7.40
FOC	X	X	X	I	83.57 ± 0.10	85.21 ± 0.03

(a) STL-10

Method	Warm	MI	CE^{-1}	Weight	F1-Score	
					Overcluster	Normal
IIC [23]	X			–	–	66.63
IIC † [23]	X			–	–	69.92
FOC				C	31.45 ± 6.02	39.35 ± 1.30
FOC		X		C	29.82 ± 2.98	60.65 ± 0.02
FOC		X	X	C	70.11 ± 1.99	64.10 ± 0.13
FOC	X			C	23.95 ± 2.63	58.71 ± 2.07
FOC	X	X		C	69.36 ± 0.05	56.59 ± 0.04
FOC	X	X	X	C	70.68 ± 0.10	58.09 ± 0.03
FOC				I	29.88 ± 2.75	54.92 ± 0.03
FOC-Light			X	I	74.93 ± 0.22	73.64 ± 0.06
FOC		X		I	72.70 ± 0.36	64.78 ± 0.04
FOC		X	X	I	73.93 ± 0.29	64.84 ± 0.03
FOC	X			I	73.93 ± 0.29	64.84 ± 0.03
FOC	X	X		I	69.64 ± 1.04	66.56 ± 0.08
FOC	X	X	X	I	74.01 ± 3.17	65.17 ± 0.18

(b) plankton dataset

4. Conclusions

In this paper, we take the first steps to address real-world underwater issues with semi-supervised learning. Our presented novel framework FOC can handle `fuzzy` labels via overclustering. We showed that overclustering can achieve better results than previous state-of-the-art semi-supervised methods on `fuzzy` plankton data. The additional overclustering output is a key difference to previous work to achieve this superior performance.

While on `certain` data FOC is not state-of-the-art by a clear margin of over 10%, it slighlty outperforms all other methods on the `fuzzy` plankton data. These beneficial effects have to be verified on other `fuzzy` datasets and with more semi-supervised algorithms in the future. Due to better performance of FOC on `fuzzy` data, we expect a similar outcome. We illustrated the visual similarity on qualitative results from these predictions and results in 5 to 10% more consistent predictions. We showed that CE^{-1} is the key component for training the overclustering head.

Supplementary Materials: The following are available at https://www.mdpi.com/article/10.3390/s21196661/s1, The details about unsupervised clustering and a comparison to previous literature.

Author Contributions: Conceptualization, L.S., J.B., M.S., S.-M.S., R.K. (Rainer Kiko) and R.K. (Reinhard Koch); methodology, L.S, J.B., M.S. and S.-M.S; software, L.S.; validation, L.S.; formal analysis, L.S.; investigation, L.S., J.B., M.S., S.-M.S. and R.K. (Rainer Kiko); resources, L.S. and R.K. (Rainer Kiko); data curation, L.S., S.-M.S. and R.K. (Rainer Kiko); writing—original draft preparation, L.S., J.B., M.S., S.-M.S., R.K. (Rainer Kiko) and R.K. (Reinhard Koch); writing—review and editing, L.S., J.B., M.S., S.-M.S., R.K. (Rainer Kiko) and R.K. (Reinhard Koch); visualization, L.S., J.B., M.S., S.-M.S., R.K. (Rainer Kiko) and R.K. (Reinhard Koch); supervision, R.K. (Reinhard Koch); project administration, Not applicable; funding acquisition, Not applicable. All authors have read and agreed to the published version of the manuscript.

Funding: We acknowledge funding of L. Schmarje by the ARTEMIS project (Grant number 01EC1908E) funded by the Federal Ministry of Education and Research (BMBF, Germany). We acknowledge funding of M. Santarossa by the KI-SIGS project (Grant number FKZ 01MK20012E) funded by the Federal Ministry for Economic Affairs and Energy (BMWi, Germany). S-M Schöder was supported by the "CUSCO—Coastal Upwelling System in a Changing Ocean" project (Grant number 03F0813) funded by the Federal Ministry of Education and Research (Germany). R Kiko also acknowledges support via a "Make Our Planet Great Again" grant of the French National Research Agency within the "Programme d'Investissements d'Avenir"; reference "ANR-19-MPGA-0012". Funds to conduct the PlanktonID project were granted to R Kiko and R Koch (CP1733) by the Cluster of Excellence 80 "Future Ocean" within the framework of the Excellence Initiative by the Deutsche Forschungsgemeinschaft (DFG) on behalf of the German federal and state governments. This work was supported by Land Schleswig-Holstein through the Open Access Publikationsfonds Funding Program.

Institutional Review Board Statement: Not applicable.

Informed Consent Statement: Not applicable.

Data Availability Statement: The used STL-10 dataset is introduced in [25]. The raw image plankton data is hosted on EcoTaxa [45] and the annotations were created in the project PlanktonID https://planktonid.geomar.de/de (accessed on 6 October 2021). The annotations can be requested from the original data owners. The source code is available at https://github.com/Emprime/FuzzyOverclustering (accessed on 6 October 2021). The used data is available at https://doi.org/10.5281/zenodo.5550918 (accessed on 6 October 2021).

Acknowledgments: We thank our colleagues, especially Claudius Zelenka, for their helpful feedback and recommendations on improving the paper. Moreover, we are grateful for all citizen scientist which participated in PlanktonID and the team of PlanktonID for providing us with their data. We thank Xu Ji, Ting Chen, Kihyuk Sohn and Wouter Van Gansbeke for answering our questions regarding their respective work.

Conflicts of Interest: The authors declare no conflict of interest.

References

1. Saleh, A.; Laradji, I.H.; Konovalov, D.A.; Bradley, M.; Vazquez, D.; Sheaves, M. A realistic fish-habitat dataset to evaluate algorithms for underwater visual analysis. *Sci. Rep.* **2020**, *10*, 14671. [CrossRef] [PubMed]
2. Gómez-Ríos, A.; Tabik, S.; Luengo, J.; Shihavuddin, A.S.M.; Krawczyk, B.; Herrera, F. Towards highly accurate coral texture images classification using deep convolutional neural networks and data augmentation. *Expert Syst. Appl.* **2019**, *118*, 315–328. [CrossRef]
3. Thum, G.W.; Tang, S.H.; Ahmad, S.A.; Alrifaey, M. Toward a highly accurate classification of underwater cable images via deep convolutional neural network. *J. Mar. Sci. Eng.* **2020**, *8*, 924. [CrossRef]

4. Knausgård, K.M.; Wiklund, A.; Sørdalen, T.K.; Halvorsen, K.T.; Kleiven, A.R.; Jiao, L.; Goodwin, M. Temperate fish detection and classification: A deep learning based approach. *Appl. Intell.* **2021**. [CrossRef]
5. Lombard, F.; Boss, E.; Waite, A.M.; Uitz, J.; Stemmann, L.; Sosik, H.M.; Schulz, J.; Romagnan, J.B.; Picheral, M.; Pearlman, J.; et al. Globally consistent quantitative observations of planktonic ecosystems. *Front. Mar. Sci.* **2019**, *6*, 196. [CrossRef]
6. Giering, S.L.C.; Cavan, E.L.; Basedow, S.L.; Briggs, N.; Burd, A.B.; Darroch, L.J.; Guidi, L.; Irisson, J.O.; Iversen, M.H.; Kiko, R.; et al. Sinking Organic Particles in the Ocean—Flux Estimates From in situ Optical Devices. *Front. Mar. Sci.* **2020**, *6*, 834. [CrossRef]
7. Addison, P.F.E.; Collins, D.J.; Trebilco, R.; Howe, S.; Bax, N.; Hedge, P.; Jones, G.; Miloslavich, P.; Roelfsema, C.; Sams, M.; et al. A new wave of marine evidence-based management: Emerging challenges and solutions to transform monitoring, evaluating, and reporting. *ICES J. Mar. Sci.* **2018**, *75*, 941–952. [CrossRef]
8. Durden, J.M.; Bett, B.J.; Schoening, T.; Morris, K.J.; Nattkemper, T.W.; Ruhl, H.A. Comparison of image annotation data generated by multiple investigators for benthic ecology. *Mar. Ecol. Prog. Ser.* **2016**, *552*, 61–70. [CrossRef]
9. Schoening, T.; Bergmann, M.; Ontrup, J.; Taylor, J.; Dannheim, J.; Gutt, J.; Purser, A.; Nattkemper, T.W. Semi-automated image analysis for the assessment of megafaunal densities at the Artic deep-sea observatory HAUSGARTEN. *PLoS ONE* **2012**, *7*, e38179. [CrossRef]
10. Schröder, S.M.; Kiko, R.; Koch, R. MorphoCluster: Efficient Annotation of Plankton images by Clustering. *Sensors* **2020**, *20*, 3060. [CrossRef]
11. Karimi, D.; Dou, H.; Warfield, S.K.; Gholipour, A. Deep learning with noisy labels: Exploring techniques and remedies in medical image analysis. *Med. Image Anal.* **2020**, *65*, 101759. [CrossRef] [PubMed]
12. Brünger, J.; Dippel, S.; Koch, R.; Veit, C. 'Tailception': Using neural networks for assessing tail lesions on pictures of pig carcasses. *Animal* **2019**, *13*, 1030–1036. [CrossRef] [PubMed]
13. Schmarje, L.; Zelenka, C.; Geisen, U.; Glüer, C.C.; Koch, R. 2D and 3D Segmentation of Uncertain Local Collagen Fiber Orientations in SHG Microscopy. In *DAGM German Conference of Pattern Regocnition*; Springer: Berlin/Heidelberg, Germany, 2019; Volume 11824 LNCS, pp. 374–386. [CrossRef]
14. De Fauw, J.; Ledsam, J.R.; Romera-Paredes, B.; Nikolov, S.; Tomasev, N.; Blackwell, S.; Askham, H.; Glorot, X.; O'Donoghue, B.; Visentin, D.; et al. Clinically applicable deep learning for diagnosis and referral in retinal disease. *Nat. Med.* **2018**, *24*, 1342–1350. [CrossRef] [PubMed]
15. Karimi, D.; Nir, G.; Fazli, L.; Black, P.C.; Goldenberg, L.; Salcudean, S.E. Deep Learning-Based Gleason Grading of Prostate Cancer From Histopathology Images—Role of Multiscale Decision Aggregation and Data Augmentation. *IEEE J. Biomed. Health Inform.* **2020**, *24*, 1413–1426. [CrossRef]
16. dos Reis, F.J.C.; Lynn, S.; Ali, H.R.; Eccles, D.; Hanby, A.; Provenzano, E.; Caldas, C.; Howat, W.J.; McDuffus, L.A.; Liu, B.; et al. Crowdsourcing the general public for large scale molecular pathology studies in cancer. *EBioMedicine* **2015**, *2*, 681–689. [CrossRef]
17. Culverhouse, P.; Williams, R.; Reguera, B.; Herry, V.; González-Gil, S. Do experts make mistakes? A comparison of human and machine identification of dinoflagellates. *Mar. Ecol. Prog. Ser.* **2003**, *247*, 17–25. [CrossRef]
18. Tarling, P.; Cantor, M.; Clapés, A.; Escalera, S. Deep learning with self-supervision and uncertainty regularization to count fish in underwater images. *arXiv* **2021**, arXiv:2104.14964.
19. Berthelot, D.; Carlini, N.; Cubuk, E.D.; Kurakin, A.; Sohn, K.; Zhang, H.; Raffel, C. ReMixMatch: Semi-Supervised Learning with Distribution Alignment and Augmentation Anchoring. *arXiv* **2019**, arXiv:1911.09785.
20. Zhai, X.; Oliver, A.; Kolesnikov, A.; Beyer, L. S4L: Self-Supervised Semi-Supervised Learning. In Proceedings of the IEEE International Conference on Computer Vision, Seoul, Korea, 27 October–2 November 2019; pp. 1476–1485.
21. Chen, T.; Kornblith, S.; Norouzi, M.; Hinton, G. A Simple Framework for Contrastive Learning of Visual Representations. *arXiv* **2020**, arXiv:2002.05709.
22. Gidaris, S.; Singh, P.; Komodakis, N. Unsupervised Representation Learning by Predicting Image Rotations. *arXiv* **2018**, arXiv:1803.07728.
23. Ji, X.; Henriques, J.F.; Vedaldi, A.; Ji, X.; Henriques, J.F.; Vedaldi, A. Invariant information clustering for unsupervised image classification and segmentation. In Proceedings of the IEEE International Conference on Computer Vision, Seoul, Korea, 27 October–2 November 2019; pp. 9865–9874.
24. Schmarje, L.; Santarossa, M.; Schroder, S.M.; Koch, R. A Survey on Semi-, Self-and Unsupervised Learning for Image Classification. *IEEE Access* **2021**, *9*, 82146–82168. [CrossRef]
25. Coates, A.; Ng, A.; Lee, H. An analysis of single-layer networks in unsupervised feature learning. In Proceedings of the Fourteenth International Conference on Artificial Intelligence and Statistics, Fort Lauderdale, FL, USA 11–13 April 2011; pp. 215–223.
26. Algan, G.; Ulusoy, I. Image Classification with Deep Learning in the Presence of Noisy Labels: A Survey. *Knowl.-Based Syst.* **2021**, *215*, 106771. [CrossRef]
27. Song, H.; Kim, M.; Park, D.; Lee, J. Learning from Noisy Labels with Deep Neural Networks: A Survey. *arXiv* **2020**, arXiv:1406.2080.
28. Nguyen, D.T.; Mummadi, C.K.; Ngo, T.P.N.; Nguyen, T.H.P.; Beggel, L.; Brox, T. SELF: Learning to Filter Noisy Labels with Self-Ensembling. *arXiv* **2019**, arXiv:1910.01842.
29. Laine, S.; Aila, T. Temporal ensembling for semi-supervised learning. *arXiv* **2016**, arXiv:1610.02242.
30. Li, J.; Socher, R.; Hoi, S.C.H. DivideMix: Learning with Noisy Labels as Semi-supervised Learning. *arXiv* **2020**, arXiv:2002.07394.
31. Geng, X. Label distribution learning. *IEEE Trans. Knowl. Data Eng.* **2016**, *28*, 1734–1748. [CrossRef]

32. Gao, B.B.; Xing, C.; Xie, C.W.; Wu, J.; Geng, X. Deep Label Distribution Learning With Label Ambiguity. *IEEE Trans. Image Process.* **2017**, *26*, 2825–2838. [CrossRef] [PubMed]
33. Liu, J.; Ma, Y.; Qu, F.; Zang, D. Semi-supervised Fuzzy Min–Max Neural Network for Data Classification. *Neural Process. Lett.* **2020**, *51*, 1445–1464. [CrossRef]
34. Kowsari, K.; Bari, N.; Vichr, R.; Goodarzi, F.A. FSL-BM: Fuzzy Supervised Learning with Binary Meta-Feature for Classification. In *Future of Information and Communication Conference*; Springer: Cham, Switzerland, 2018; pp. 655–670.
35. El-Zahhar, M.M.; El-Gayar, N.F. A semi-supervised learning approach for soft labeled data. In Proceedings of the 2010 10th International Conference on Intelligent Systems Design and Applications, Cairo, Egypt, 29 November–1 December 2010; pp. 1136–1141.
36. Liu, Y.; Liang, X.; Tong, S.; Kumada, T. Photo Shot-Type Disambiguation by Multi-Classifier Semi-Supervised Learning. In Proceedings of the 2019 IEEE International Conference on Image Processing (ICIP), Taipei, Taiwan, 22–25 September 2019; pp. 2466–2470.
37. Caron, M.; Bojanowski, P.; Joulin, A.; Douze, M. Deep clustering for unsupervised learning of visual features. In Proceedings of the European Conference on Computer Vision (ECCV), Munich, Germany, 8–14 September 2018; pp. 132–149.
38. Sohn, K.; Berthelot, D.; Li, C.L.; Zhang, Z.; Carlini, N.; Cubuk, E.D.; Kurakin, A.; Zhang, H.; Raffel, C. FixMatch: Simplifying Semi-Supervised Learning with Consistency and Confidence. *arXiv* **2020**, arXiv:2001.07685.
39. He, K.; Zhang, X.; Ren, S.; Sun, J. Identity Mappings in Deep Residual Networks. In Proceedings of the European Conference on Computer Vision (ECCV), Amsterdam, the Netherlands, 8–16 October 2016; pp. 630–645.
40. Grandvalet, Y.; Bengio, Y. Semi-supervised learning by entropy minimization. *Adv. Neural Inf. Process. Syst.* **2005**, 367, 529–536.
41. Xie, Q.; Dai, Z.; Hovy, E.; Luong, M.T.; Le, Q.V. Unsupervised Data Augmentation for Consistency Training. *arXiv* **2019**, arXiv:1904.12848.
42. Miyato, T.; Maeda, S.I.; Koyama, M.; Ishii, S. Virtual adversarial training: A regularization method for supervised and semi-supervised learning. *IEEE Trans. Pattern Anal. Mach. Intell.* **2018**, *41*, 1979–1993.
43. Berthelot, D.; Carlini, N.; Goodfellow, I.; Papernot, N.; Oliver, A.; Raffel, C.A. Mixmatch: A holistic approach to semi-supervised learning. *arXiv* **2019**, arXiv:1905.02249.
44. Picheral, M.; Guidi, L.; Stemmann, L.; Karl, D.M.; Iddaoud, G.; Gorsky, G. The Underwater Vision Profiler 5: An advanced instrument for high spatial resolution studies of particle size spectra and zooplankton. *Limnol. Oceanogr. Methods* **2010**, *8*, 462–473. [CrossRef]
45. Picheral, M.; Colin, S.; Irisson, J.O. EcoTaxa, a Tool for the Taxonomic Classification of Images. 2017. Available online: https://ecotaxa.obs-vlfr.fr/ (accessed on 6 October 2021).
46. Krizhevsky, A.; Sutskever, I.; Hinton, G.E. Imagenet classification with deep convolutional neural networks. *Adv. Neural Inf. Process. Syst.* **2012**, *60*, 1097–1105. [CrossRef]
47. Krizhevsky, A.; Hinton, G. Learning Multiple Layers of Features from Tiny Images. Technical Report. 2009. Available online: https://www.cs.toronto.edu/~kriz/learning-features-2009-TR.pdf (accessed on 6 October 2021).
48. Van Gansbeke, W.; Vandenhende, S.; Georgoulis, S.; Proesmans, M.; Van Gool, L. Scan: Learning to classify images without labels. In *European Conference on Computer Vision*; Springer: Cham, Switzerland, 2020; pp. 268–285.
49. Tarvainen, A.; Valpola, H. Mean teachers are better role models: Weight-averaged consistency targets improve semi-supervised deep learning results. *arXiv* **2017**, arXiv:1703.01780.
50. Lee, D.H. Pseudo-label: The simple and efficient semi-supervised learning method for deep neural networks. In *Workshop on Challenges in Representation Learning*; ICML: Atlanta, GA, USA, 2013; Volume 3, p. 2.

sensors

MDPI

Article

Simultaneous Burr and Cut Interruption Detection during Laser Cutting with Neural Networks

Benedikt Adelmann * and Ralf Hellmann

Applied Laser and Photonics Group, Faculty of Engineering, University of Applied Sciences Aschaffenburg, Wuerzburger Straße 45, 63739 Aschaffenburg, Germany; ralf.hellmann@th-ab.de

* Correspondence: benedikt.adelmann@th-ab.de; Tel.: +49-6022-81-3693

Abstract: In this contribution, we compare basic neural networks with convolutional neural networks for cut failure classification during fiber laser cutting. The experiments are performed by cutting thin electrical sheets with a 500 W single-mode fiber laser while taking coaxial camera images for the classification. The quality is grouped in the categories good cut, cuts with burr formation and cut interruptions. Indeed, our results reveal that both cut failures can be detected with one system. Independent of the neural network design and size, a minimum classification accuracy of 92.8% is achieved, which could be increased with more complex networks to 95.8%. Thus, convolutional neural networks reveal a slight performance advantage over basic neural networks, which yet is accompanied by a higher calculation time, which nevertheless is still below 2 ms. In a separated examination, cut interruptions can be detected with much higher accuracy as compared to burr formation. Overall, the results reveal the possibility to detect burr formations and cut interruptions during laser cutting simultaneously with high accuracy, as being desirable for industrial applications.

Keywords: laser cutting; quality monitoring; artificial neural network; burr formation; cut interruption; fiber laser

Citation: Adelmann, B.; Hellmann, R. Simultaneous Burr and Cut Interruption Detection during Laser Cutting with Neural Networks. *Sensors* **2021**, *21*, 5831. https://doi.org/10.3390/s21175831

Academic Editor: Bernhard Wilhelm Roth

Received: 8 June 2021
Accepted: 25 August 2021
Published: 30 August 2021

1. Introduction

Laser cutting of thin metal sheets using fiber or disk lasers is now a customary process in the metal industry. The key advantages of laser cutting are high productivity and flexibility, good edge quality and the option for easy process automation. Especially for highly automated unmanned machines, seamlessly combined in line with bending, separation or welding machines, a permanent high cut quality is essential to avoid material waste, downtime or damaging subsequent machine steps in mechanized process chains. As a consequence, besides optimizing the cutting machine in order to reduce the influence of disturbance variables, cut quality monitoring is also of utmost interest.

The most common and disruptive quality defects are cut interruptions and burr formation [1]. To obtain high-quality cuts, process parameters, such as laser power, feed rate, gas pressure, working distance of the nozzle and focus position, respectively, are to be set appropriately. Imprecise process parameters and typical disturbance values like thermal lenses, unclean optics, damaged gas nozzles, gas pressure fluctuations and the variations of material properties may lead to cut poor-quality and, thus, nonconforming products. To ensure a high quality, an online quality monitoring system, which can detect multiple defects, would be the best choice in order to respond quickly and reduce downtime, material waste or cost-extensive rework. Until now, most reviewed sensor systems for monitoring laser cutting focus only on one single fault.

For detecting burr formation during laser cutting, different approaches using cameras, phododiodes or acoustic emission were investigated. In [2,3] burr formation, roughness and striation angle during laser cutting with a 6 kW CO_2 laser are determined by using a NIR camera sampling with 40 Hz. By using two cameras in [4], laser cutting with a CO_2 laser is monitored by observing the spark trajectories underneath the sheet and melt bath geometries and correlate this to the burr formation or overburning defects. A novel approach is used in [5], employing a convolutional neural network to calculate burr formation from camera images with a high accuracy of 92%. By evaluating the thermal radiation of the process zone with photodiodes [6], the burr height during fiber laser cutting can be measured from the standard deviation of a filtered photodiode signal. Results by using photodiode-based sensors integrated in the cutting head [7] showed that the mean photodiode's current increases with lower cut qualities, while similar experiments revealed increasing mean photodiode currents at lower cut surface roughness [8]. An acoustic approach was investigated by monitoring the acoustic emission during laser cutting and deducing burr formation by evaluating the acoustic bursts [9].

Also for cut interruption detection, most approaches are based on photodiode signals or camera images. Photodiode-based methods for cut interruption detection are signal threshold-based [10], done by the comparison of different photodiodes [11] or based on cross-correlations [12]. However, all those methods have the disadvantage of requiring thresholds that vary with the sheet thickness or laser parameters. In addition, an adaptation to other materials or sheet thicknesses requires a large engineering effort to define respective threshold values by extensive investigations. To avoid this problem, [13] uses a convolutional neural network to calculate cut interruptions from camera images during fiber laser cutting of different sheet thicknesses with an accuracy of 99.9%. Another approach is performed by using a regression model based on polynomial logistics [14] to calculate the interruptions from laser machine parameters only.

This literature review reveals that for both burr formation monitoring and cut interruption, individual detection schemes have previously been reported, but a combined and simultaneous detection for both failure patterns has not been reported so far. In addition, many of the previous studies applied CO_2 lasers, which are often replaced nowadays by modern fiber or disk lasers, for which, in turn, fewer reports are available. To detect both failures with the same system, we chose the evaluation of camera images with neural networks, as they are able to achieve a high accuracy in detecting both cut failures [5,13]. The use of neural networks, especially for convolutional neural networks (CNN), has been demonstrated for various image classification purposes, such as face recognition and object detection [15,16], in medicine for cancer detection [17] and electroencephalogram (EEG) evaluations [18] or in geology for earthquake detection [19]. For failure analyses in technical processes, neural networks have also been successfully used for, e.g., concrete crack detection [20], road crack detection [21] or detecting wafer error determinations [22]. In addition, detecting different failure types with the same system has been successfully proven with neural networks, such as detecting various wood veneer surface defects [23] or different welding defects [24] during laser welding.

The objective of this publication is to detect both burr formation and cut interruptions during single-mode laser cutting of electrical sheets from camera images with neural networks. The advantages of our system are, firstly, easy adaption to industrial cutting heads, which often already have a camera interface. Secondly, images are taken coaxially to the laser beam and are therefore independent of the laser cut direction. Thirdly, due to the use of a learning system the engineering effort is low when the system has to be adapted to other materials or sheet thicknesses. Two different neural network types are used, namely a basic neural network and a convolutional neural network. The basic neural network is faster and can detect bright or dark zones but is less able to extract abstractions of 2D features and needs a lot of parameters when the networks get more complex. On the other hand, convolutional neural networks are much better in learning and extracting abstractions of 2D features and usually need fewer parameters. However, they require a higher calculation effort due to many multiplications in the convolution layers [25,26].

The cutting of electrical sheets is chosen because it is an established process in the production and prototyping of electric motors and transformers [27–30], i.e., it is a relevant and contributing process step to foster e-mobility. In order to reduce the electrical losses caused by eddy currents, the rotor is assembled of a stack of thin electrical sheets with electrical isolation layers in between the sheets. The sheet thickness varies typically between 0.35 mm to 0.5 mm, with the eddy currents being lower for thinner sheets. As a result, for an electric motor, a large number of sheets with high quality requirements are necessary. Especially burr formations result in gaps between sheets or the burr can pierce the electrical isolation layer and connect the sheets electrically which both reduce the performance of motors drastically. Therefore, quality monitoring during laser cutting is of great interest for industrial applications.

2. Experimental

2.1. Laser System and Cutting Setup

In this study, a continuous wave 500 W single-mode fiber laser (IPG Photonics, Burbach, Germany) is used to perform the experiments. The laser system is equipped with linear stages (X, Y) for positioning the workpiece (Aerotech, Pittsburgh, PA, USA) and a fine cutting head (Precitec, Gaggenau, Germany) is attached to a third linear drive (Z). The assisted gas nitrogen with purity greater than 99.999% flows coaxially to the laser beam. The gas nozzle has a diameter of 0.8 mm and its distance to the workpiece is positioned by a capacitive closed loop control of the z-linear drive. The emitting wavelength of the laser is specified to be 1070 nm in conjunction with a beam propagation factor of $M^2 < 1{:}1$. The raw beam diameter of 7.25 mm is focused by a lens with a focal length of 50 mm. The according Rayleigh length is calculated to 70 µm and the focus diameter to 10 µm, respectively.

The design of the cutting head with the high-speed camera and a photo of the laser system are illustrated in Figure 1. The dashed lines depict the primary laser radiation from the fiber laser, which is collimated by a collimator and reflected by a dichroic mirror downwards to the processing zone. There the laser radiation is focused by the processing lens through the protective glass onto the work piece, which is placed on the XY stages. The process radiation from the sheet radiates omnidirectional (dash-doted line), thus partly through the nozzle and protective glass and is collimated by the processing lens upwards. The process radiation passes the dichroic mirror and is focused by a lens onto the high-speed camera. The focus of the camera is set to the bottom side of the sheet in order to have a sharp view of possible burr formations.

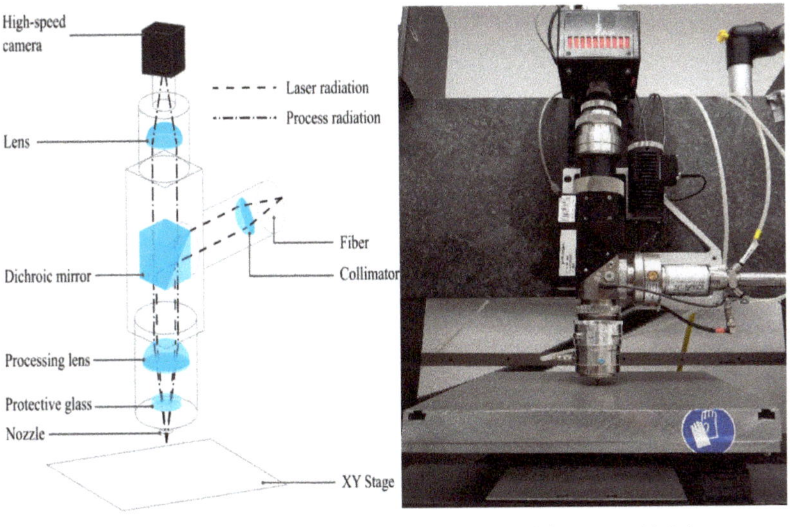

Figure 1. Optical setup of the cutting head (**left**) and image of the system (**right**).

2.2. Laser Cutting

The laser cuts are performed in electrical sheets of the type M270 (according to EN 10106 this denotes a loss of 2.7 W/kg during reversal of magnetism at 50 Hz and 1.5 T) with a sheet thickness of 0.35 mm. This sheet thickness is chosen because it fits well to the laser focus properties e.g., Rayleigh length and it is one of the most often used sheet thicknesses for electrical motors and transformers, because it provides a good compromise between low eddy currents and high productivity. Stacks of thicker sheets are faster to produce because less sheets are required per stack but with increasing sheet thickness also unwanted eddy currents increase. Thinner sheet thicknesses require a higher production effort per stack and are more difficult to cut because they are very flexible, and warp under the gas pressure and thermal influence. In these experiments only one sheet thickness is used, but please note that in previous publications with similar systems an adaptation of the results to other sheet thicknesses was possible with only minor additional expenses [5,13].

As ad-hoc pre-experiments reveal, the parameter combination of a good quality cut is a laser power of 500 W, a feed rate of 400 mm/s and a laser focus position on the bottom side of the metal sheet. The gas nozzle has a diameter of 0.8 mm and is paced 0.5 mm above the sheet surface and the gas pressure is 7 bar. For the experimental design, the parameters are varied to intentionally enforce cut failures. Burr formations are caused by less gas flow into the cut kerf due to higher nozzle to sheet distance, lower gas pressure, an overvalued power to feed rate ratio or damaged nozzles. Cut interruptions are enforced by too high feed rates or too low laser power.

In the experimental design, 39 cuts with different laser parameters are performed for training the neural network and 22 cuts are performed for testing, with the cuts being evenly distributed to the three cut categories (good cut, cuts with burr formation and cut interruptions). A table of all cuts with laser machine parameters, category and use can be found in the Appendix A. The cuts are designed from a straight line including acceleration and deceleration paths of the linear stages. Exemplifying images of the sheets from all three cut categories taken by optical microscope after the cutting process are shown in Figure 2. Firstly, for a good quality cut, both top and bottom side of the cut kerf are characterized by clear edges without damages. Secondly, for a cut with burr, the top side is similar to the good quality cut, however on the bottom side drops of burr formation are clearly visible. Thirdly, the images of the cut interruption reveal a molten line on the sheet top side and only a slightly discolored stripe on the bottom side with both sides of the sheet not being separated.

Good quality cut	Top
	Bottom
Cut with burr	Top
	Bottom
Cut interruption	Top
	Bottom

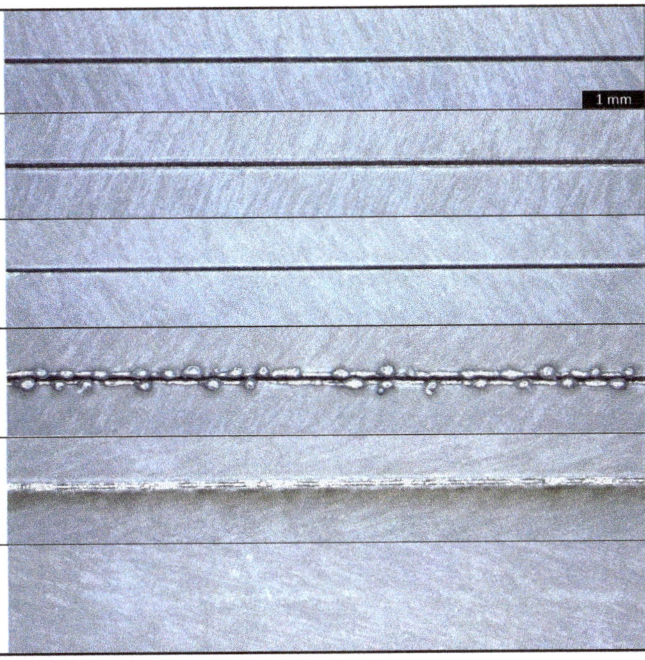

Figure 2. Images of the top and bottom side of laser cuts with and without cut errors taken with an optical microscope after laser cutting.

2.3. Camera and Image Acquisition

For image acquisition during laser cutting, we used a high-speed camera (Fastcam AX50, Photron, Tokyo, Japan) with a maximum frame rate of 170,000 frames per second. The maximum resolution is 1024 × 1024 pixels, with a square pixel size of 20 × 20 μm^2 in combination with a Bayer CFA Color Matrix. For process image acquisition, videos of the laser cutting process are grabbed with a frame rate of 10 kilo frames per second with an exposure time of 2 µs and a resolution of 128 × 64 pixels. Even at this high frame rate, no oversampling occurs and consecutive images are not similar, because the relevant underlying melt flow dynamics are characterized by high melt flow velocities in the range of 10 m/s [31] and vary therefore at estimated frequencies between 100 kHz and 300 kHz [32]. Please note, due to the lack of external illumination in the cutting head, the brightness in the images are caused by the thermal radiation of the process zone.

Two exemplifying images of each cut category are shown in Figure 3 with the cut direction always upwards. The orientation of the images is always the same because the straight lines are cut in the same direction. For complex cuts, images with the same orientation can be transformed from various oriented images by rotation based on the movement direction of the drives. In these images, brightness is caused by the thermal radiation of the hot melt. Good cuts are characterized by a bright circle at the position on the laser focus, and below this, two tapered stripes indicating the flowing melt at the side walls of the cut kerf, because in the middle the melt bath is blown out first. The cuts with burr are similar to the good quality cuts but tapered stripes are formed differently. The cut interruptions are very different to the other categories and are characterized by larger bright areas and a more elliptical shape with no tapered stripes.

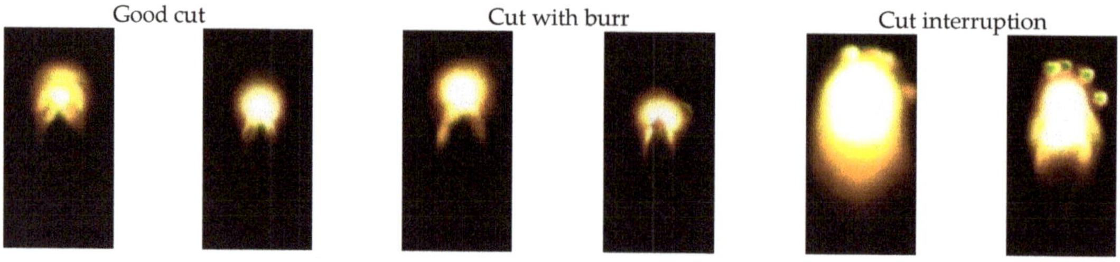

Figure 3. Examples of camera images of the three cut categories taken during laser cutting with the high speed camera.

From the 39 laser cuts, the experimental design delivers the same number of training videos with overall 52 thousand training images, while from the 22 testing cuts 34 thousand test images are provided. It is worth to mention, that the size of several ten thousands of training images is typical for training neural networks [33]. For both training and testing, the images are almost evenly distributed on the three categories with cut interruptions being slightly underrepresented. The reason for this underrepresentation is, that cut interruptions only occur at high feed rates, i.e., images from acceleration and deceleration paths can be used only partially and, in turn, less images per video can be captured.

2.4. *Computer Hardware and Neural Network Design*

For learning and evaluating the neural network, a computer with an Intel Core 7-8700 processor with a 3.2-GHz clock rate in combination with 16-GB DDR4 RAM was used. All calculations are performed with the CPU rather than the GPU to show that the machine learning steps are also possible to run on standard computers, which are usually integrated with laser cutting machines. The used software was TensorFlow version 2.0.0 in combination with Keras version 2.2.4 (Software available from: https://tensorflow.org (accessed on 24 March 2021)).

In most publications about image classification with neural networks, the images have major differences. In contrast, in the images captured in our experiments, the object to analyze always has the same size, orientation and illumination conditions which should simplify the classification when compared to classifying common, moving items like vehicles or animals [34,35]. Furthermore, our images have a rectangular shape with 128 × 64 pixels, while most classification algorithms are optimized on square images sizes having mostly a resolution of 224 × 224 pixels like MobileNet, SqueezeNet or AlexNet [36,37]. Because an enlargement of the image size slows the system drastically, two self-designed and completely different neural network are used with many elements being adapted to other, often used neural networks. The first network, as shown in Figure 4 is a basic network without convolution and only consists of image flattening followed by two fully connected layers with N nodes and ReLU Activation. To classify the three different cut categories, a fully connected layer with 3 nodes and softmax activation completes the network. The second network is a convolutional neural network with four convolution blocks followed by the same three fully connected layers as in the basic network. Each block consists of a convolution layer with a kernel size of 3 × 3 and M filters, which the output of the convolution is added with input of the block. Such bypasses are most common in, e.g., MobileNet [36]. To reduce the number of parameters, a max pooling layer with a common pool size of 2 × 2 is used [26]. In contrast to often neural networks used in the literature, we use a constant instead of an increasing filter number for subsequent convolution layers and we use normal convolutions rather than separable or pointwise convolutions. Because every block halves the image size in 2 dimensions, after 4 blocks the image size is 8 × 4 × M. The fully connected layers after the flattening have the same number of nodes as the number of parameters delivered by the flattened layer. The used model optimizer is Adam, which according to [38], together with SDG (Stochastic Gradient Descent) provides superior optimization results. Furthermore, we use the loss function "categorical crossentropy" to enable categorical outputs (one hot encoding), and the metrics "accuracy".

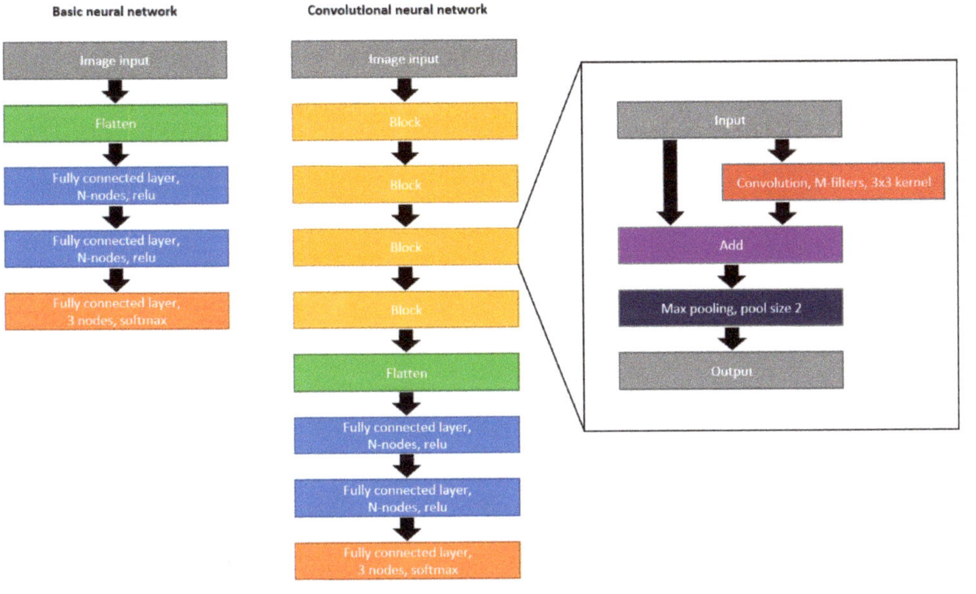

Figure 4. Design of the two neural networks.

2.5. Methodology

The methodology of our experiments is shown in the workflow diagram in Figure 5. In a first step, the laser cuts are performed and during cutting videos are taken from the process zone with the high speed camera, some of these images have been shown in Figure 3. After cutting, the cut kerfs are analyzed by with an optical microscope and categorized manually whether a good cut, burr formation or a cut interruption occurred (examples of these images shown in Figure 2). Based on this classification, the videos taken during laser cutting are labeled with the corresponding class. In case the cut quality changes within one cut, the video is divided, so the quality is constant within a video. Then the videos are separated in training videos and test videos, so the images for testing are not from videos used for training. From the training videos, the single frame is extracted and with these images the neural network is trained. Furthermore, the single frames are extracted from the test videos, and the resulting images are used to test the trained neural network.

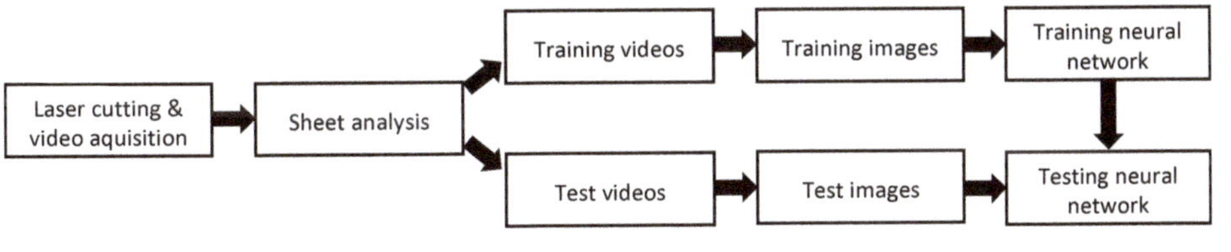

Figure 5. Workflow diagram.

3. Results

3.1. Training Behaviour

The different neural networks are trained on the training dataset and the performance is calculated on the test dataset. Exemplarily, the training behavior of the convolutional neural network with 16 filters in each convolution is shown in Figure 6. Apparently, the training accuracy rises continuously with the training epochs, reaching 99% after 10 epochs and 99.5% after 20 epochs, respectively. On the other hand, the test accuracy reaches 94% after three epochs and fluctuates with further training around this level, which is a typical behavior for neural network training [39]. Even further training, above 20 epochs, results only in a fluctuation of the accuracy rather than a continuous increase. To reduce the deviation of the test results for comparisons between different networks, the mean of the test results between 10 and 20 epochs is used.

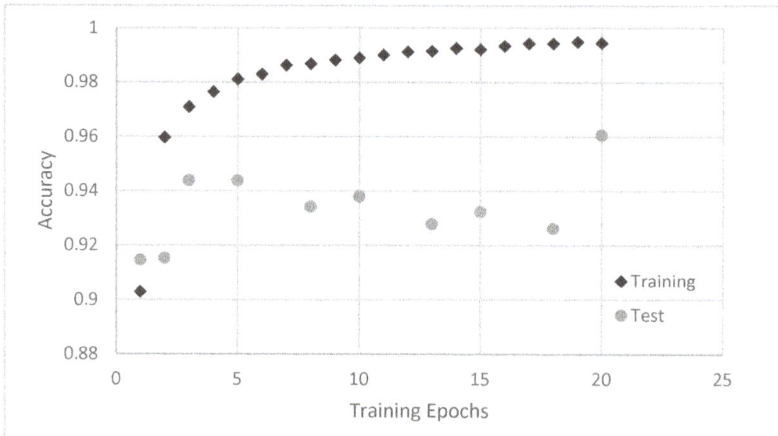

Figure 6. Training accuracy and test accuracy of a convolutional neural network with 16 filters.

3.2. Basic Neural Network

To determine the performance of the basic neural networks, those with node numbers N between 5 and 1000 are trained on the training dataset and tested on the test dataset. The mean test accuracy between 10 and 20 training epochs and the required calculation time per image are shown in Figure 7. It is obvious that the accuracy for a very small network with only five nodes is quite high, being 92.8%, and the calculation time of 0.1 ms per image being very fast. With an increasing number of nodes, the accuracy increases to a maximum of 95.2% at 1000 nodes, which is accompanied by a higher calculation time of 0.32 ms. Parallel to the calculation time, also the trainable parameters increase with the number of nodes starting from 122 thousand parameters for five nodes and reaching 25 million parameters at 1000 nodes. A further increase of the parameters is not considered to be useful, because the training dataset consist of 420 million of pixels (number of images x pixels per image), so the neural network tend to over fit the training dataset rather than developing generalized features. Generally, with the basic neural network accuracies of 94% (mean) are achievable.

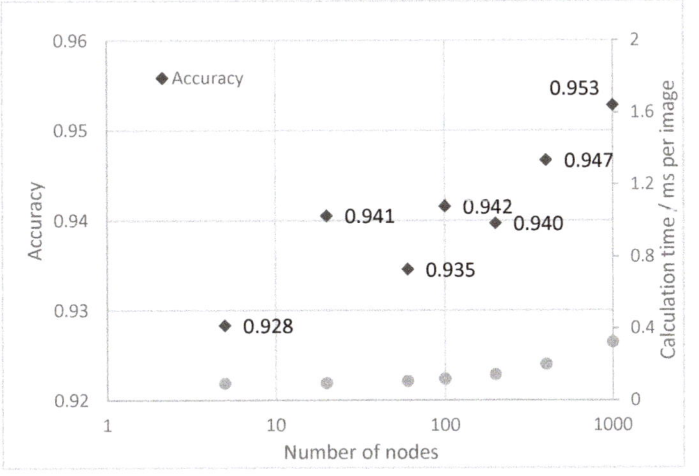

Figure 7. Accuracy of the basic neural network as a function of nodes per fully connected layer.

3.3. Convolutional Neural Network

Under the same conditions as the basic neural network, the convolutional neural network is also trained and tested. The results of the accuracy and calculation time for filter numbers between 4 and 64 are depicted in Figure 8. The accuracy of the neural network is quite high for all filter numbers and fluctuates between 94.6% and 95.8% with no clear trend. In addition, the accuracy also varies for the same network when it is calculated several times. However, the calculation time increases clearly with the number of filters from 0.36 ms per image to 1.77 ms. The number of trainable parameters start with 34 thousand for four filters and increases to 8.4 Million for the 64 filters (details how to calculate the number of parameters are described in [25]). For the mean, the convolutional neural network is able to classify about 95% of the image correctly.

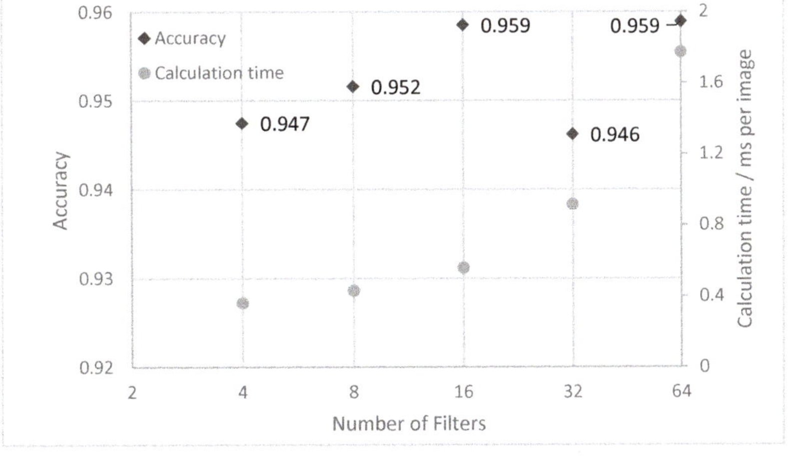

Figure 8. Test accuracy for the convolutional neural network as a function of the number of filters.

3.4. Comparison between Cut Failures

Since the literature is available for both burr detection and cut interruptions during laser cutting, which vary strongly in accuracy, the performance of our neural networks in detecting one cut failure is determined. Therefore, the accuracy in classifying good cuts and cuts with burr as well as good cuts and cut interruptions is calculated separately. For this investigation, the convolutional neural network with 16 filters is chosen, because it provides high accuracy and a comparable moderate calculation time. The results of the separated classification are shown in Figure 9. It is obvious that the detection of cut interruptions is very reliable with the accuracy being 99.5%, as being compared to 93.1% when detecting burr formation. The reason for this can also be seen in Figure 3, where good cuts are much more similar to cuts with burr, while cut interruptions look very different to both of the other failure classes. Both values individually agree with the literature values, which are 99.9% for the cut interruptions [13] and 92% for the burr detection [5], yet for burr detection in the literature a more complex burr definition is chosen. This shows that cut interruptions are much easier to detect from camera images compared to burr formations.

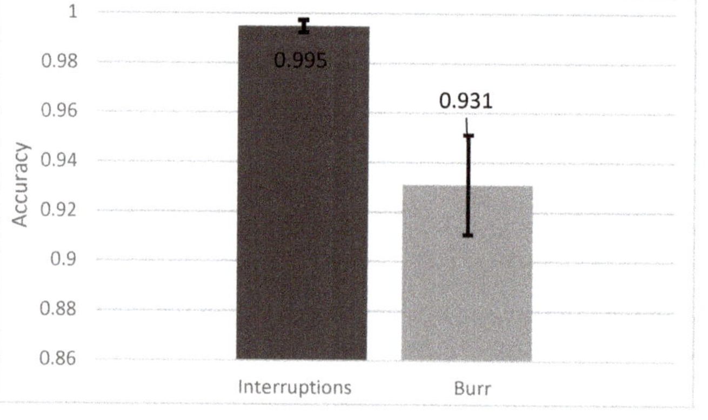

Figure 9. Comparison of the test accuracy between interruptions and burr formations.

3.5. Error Analysis

For the error analysis, single misclassified images and the distribution of misclassifications are analyzed. For the temporal distribution, a video of a cut with different cut qualities is produced. The measured quality obtained by the optical microscope and the prediction of the convolutional neural network with 16 filters is shown in Figure 10. Misclassifications are indicated by red dots that are not placed on the blue line and it can clearly be seen, which misclassifications occur more often than others. The most frequent misclassifications are cuts predicted as burr. Interruptions are rarely misclassified and other images are seldom misclassified as interruptions, which accompanies the results in Section 3.4. The distribution of the misclassifications reveals no concentration on a specific sector but minor accumulations of several misclassifications are observed. In addition, some areas without any misclassification or only single misclassifications can be found. These results reveal that misclassifications do not occur all at once or at a special event but are widely distributed.

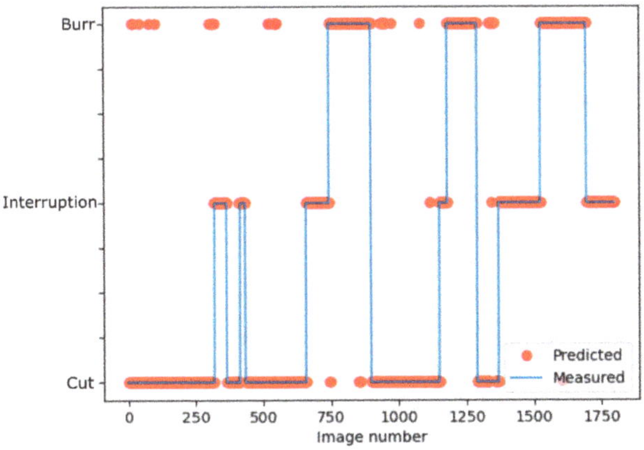

Figure 10. Measured image class and prediction by the neural network.

To analyze the misclassified images, two exemplified images from a good cut classified as cut with burr are shown in Figure 11. In contrast to the images in Figure 3, where the bright area is followed by two tapered stripes, in Figure 11, these stripes are hardly observed. However, these following stripes are important for the classification, because in this area the burr is generated. Therefore, in the case of missing stripes, the classification between cuts with and without burr is difficult and thus characterized by many misclassifications.

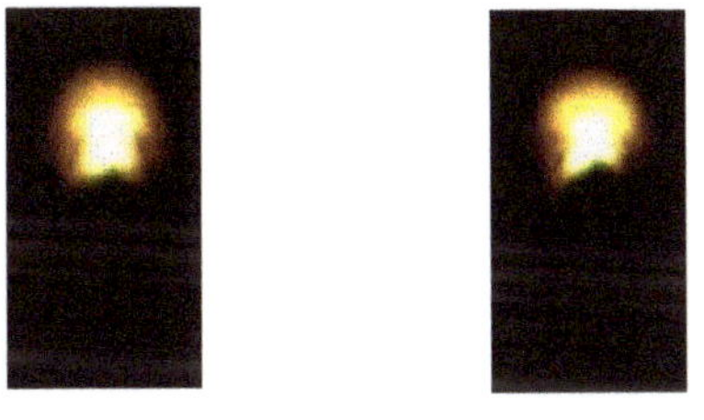

Figure 11. Two examples of cuts missclassified as burr.

4. Discussion

With the classification in three different categories during laser cutting, good cuts can be distinguished from cuts with burr and cut interruptions. The convolutional neural network has, depending on the number of filters, a better classification accuracy by about 1% when compared to the basic networks. The maximum accuracy for the basic neural networks (1000 nodes) is also lower, being 95.2% as compared to a 95.8% accuracy of the convolutional neural network with 16 filters. Nevertheless, the difference between both neural network types is small, which can be explained by the objects in the images always having the same size, orientation and brightness, which is not usually the case for many other classification tasks [34,35]. As a consequence, the basic neural network can classify the images by bright or dark zones and does not necessarily require learning and extracting abstractions of 2D features which is the main advantage of convolutional neural networks [25,26].

For the required accuracy, the size of the cut failure has to be considered. Because of the accuracy being below 100%, a post algorithm is necessary which should report an error only when a certain amount of failures occurs in a sequent number of images. To detect geometrically long failures, which can occur, e.g., by unclean optics, our classification system is adequate. Very short failures, like single burr drops when cutting an edge, are probably not be detectable with our system. It is remarkable for the results with both neural networks, however, that at least 92.8% accuracy (cf. Figure 7) can be achieved with any network configuration independent from network type, number of nodes or filters. This means that about 93% of the images are easy to classify because they differ strongly between the categories. Furthermore, about 2% of the images can be classified by more complex neural networks (cf. Sections 3.2 and 3.3). About 5% of the images, mostly between the categories good cuts and cuts with burr formation, are very difficult to classify because the images are quite similar (cf. Figure 3). For an industrial application, it has to be further considered whether the intentionally enforced cut failures are representative for typical industrial cut failures, e.g., as a result of unclean optics, which are not reproducible in scientific studies.

The main advantage of the basic neural network is the much lower computation time between 0.1 ms and 0.32 ms, while the convolutional neural network requires 0.36 ms to 1.7 ms, respectively. For typical available industrial cameras having maximum frame rates in the range of 1000 Hz, a calculation time for the classification of about 1 ms is sufficient, which is fulfilled by all basic and most of our convolutional neural networks. A similar frame rate was also used by [5] when detecting burr formations during laser cutting. With maximum cutting speeds of modern laser machines in the range of 1000 mm/s still a local resolution of 1 mm is achieved which can clearly be considered as adequate for industrial use.

Following this fundamental and comparative analysis, future investigations have to address field trials of the proposed sensor system and classification scheme in industrial cutting processes. Within such industrial environments additional error sources may appear and further reduce the cut quality, such as damaged gas nozzles or partially unclean optics which in turn are difficult to reproduce under laboratory conditions. The images from these error sources can be added to the training data and improve the detection rate of the classification system. To improve the detection rate it is also possible to classify not a single image but a series of 3 to 10 subsequent images, which reduces the influence of a single misleading image.

5. Conclusions

Overall, with our neural network approach, two cut failures during laser cutting can be detected simultaneously by evaluating camera images with artificial neural networks. With different neural network designs up to 95.8% classification accuracy can be achieved. Generally, convolutional neural networks have only minor classification advantages of about 1% over basic neural networks, while the basic neural networks are considerably faster in calculation. The detection of cut interruptions is remarkably higher when compared to the burr formation, because the images of cut interruptions are more different from the good cuts compared to the images with burr formation. In general, the detection rate is high enough to advance industrial applications.

Author Contributions: Conceptualization, B.A.; methodology, B.A.; software, B.A.; validation, B.A., investigation, B.A.; writing—original draft preparation, B.A. and R.H.; project administration, R.H., All authors have read and agreed to the published version of the manuscript.

Funding: This project was funded by the Bavarian Research Foundation.

Data Availability Statement: Not applicable.

Conflicts of Interest: The authors declare no conflict of interest.

Appendix A

Table of all performed laser cuts with machine parameters, cut category and use for training and test.

Nr:	Laser	Feed Rate	Nozzle Distance	Focus Position	Category	Use
	W	mm/s	mm	mm		
1	500	600	0.5	−1.25	Cut	Training
2	300	400	0.5	−1.25	Cut	Training
3	500	500	0.5	−1.25	Cut	Training
4	500	300	0.5	−1.25	Cut	Training
5	300	600	0.5	−1.25	Interruption	Test
6	200	500	1,0	−1.75	Interruption	Training
7	500	500	0.8	−1.55	Burr	Training
8	500	600	0.5	−1.25	Cut	Test
9	250	500	0.5	−1.25	Interruption	Test
10	500	400	0.5	−1.25	Cut	Test
11	500	600	0.5	−1.25	Interruption	Training
12	500	500	1,0	−1.75	Burr	Training
13	500	500	0.5	−1.25	Cut	Test
14	500	300	0.5	−1.25	Cut	Training
15	500	200	0.5	−1.25	Cut	Test
16	500	500	0.5	−1.25	Cut	Training
17	500	500	0.5	−1.25	Interruption	Test
18	400	500	0.9	−1.65	Burr	Training
19	500	500	0.8	−1.55	Burr	Training
20	200	500	1,0	−1.75	Interruption	Training
21	500	300	0.5	−1.45	Burr	Training
22	150	500	0.5	−1.25	Interruption	Test
23	500	400	0.5	−1.25	Cut	Training
24	500	500	0.5	−1.25	Cut	Test
25	500	400	0.5	−1.25		Training
26	400	500	0.8	−1.55	Burr	Training
27	500	500	0.8	−1.55	Burr	Test
28	150	500	0.5	−1.25	Interruption	Training
29	200	500	1,0	−1.75	Interruption	Test
30	400	500	0.9	−1.65	Burr	Training
31	300	600	0.5	−1.25	Interruption	Training
32	500	500	1,0	−1.75	Burr	Training
33	500	300	0.5	−1.25	Cut	Test

Nr:	Laser	Feed Rate	Nozzle Distance	Focus Position	Category	Use
	W	mm/s	mm	mm		
34	400	500	1,0	−1.75	Burr	Test
35	500	500	0.5	−1.25	Interruption	Training
36	500	400	0.5	−1.25	Cut	Training
37	150	500	0.5	−1.25	Interruption	Training
38	400	500	0.5	−1.45	Burr	Training
39	500	600	0.5	−1.25	Interruption	Training
40	400	400	0.5	−1.25	Cut	Training
41	500	600	0.5	−1.25	Cut	Training
42	500	600	0.5	−1.25	Interruption	Test
43	400	500	0.8	−1.55	Burr	Test
44	400	500	0.5	−1.45	Burr	Test
45	400	400	0.5	−1.25	Cut	Test
46	500	500	0.5	−1.25	Cut	Training
47	500	200	0.5	−1.25	Cut	Training
48	300	400	0.5	−1.25	Interruption	Training
49	400	500	0.5	−1.45	Burr	Training
50	500	400	0.5	−1.25	Cut	Test
51	500	500	0.8	−1.55	Burr	Training
52	400	500	0.9	−1.65	Burr	Training
53	400	500	0.9	−1.65	Burr	Test
54	500	300	0.5	−1.45	Burr	Training
55	300	400	0.5	−1.25	Cut	Training
56	500	300	0.5	−1.45	Burr	Test
57	250	500	0.5	−1.25	Interruption	Training
58	300	400	0.5	−1.25	Cut	Training
59	300	400	0.5	−1.25	Interruption	Training
60	300	600	0.5	−1.25	Interruption	Training
61	300	400	0.5	−1.25	Interruption	Test

References

1. Kratky, A.; Schuöcker, D.; Liedl, G. Processing with kW fibre lasers: Advantages and limits. In Proceedings of the XVII International Symposium on Gas Flow, Chemical Lasers, and High-Power Lasers, Lisboa, Portugal, 15–19 September 2008; p. 71311X.
2. Sichani, E.F.; de Keuster, J.; Kruth, J.; Duflou, J. Real-time monitoring, control and optimization of CO2 laser cutting of mild steel plates. In Proceedings of the 37th International MATADOR Conference, Manchester, UK, 25–27 July 2012; pp. 177–181.
3. Sichani, E.F.; de Keuster, J.; Kruth, J.-P.; Duflou, J.R. Monitoring and adaptive control of CO2 laser flame cutting. *Phys. Procedia* **2010**, *5*, 483–492. [CrossRef]
4. Wen, P.; Zhang, Y.; Chen, W. Quality detection and control during laser cutting progress with coaxial visual monitoring. *J. Laser Appl.* **2012**, *24*, 032006. [CrossRef]
5. Franceschetti, L.; Pacher, M.; Tanelli, M.; Strada, S.C.; Previtali, B.; Savaresi, S.M. Dross attachment estimation in the laser-cutting process via Convolutional Neural Networks (CNN). In Proceedings of the 2020 28th Mediterranean Conference on Control and Automation (MED), Saint-Raphaël, France, 16–18 September 2020; pp. 850–855.
6. Schleier, M.; Adelmann, B.; Neumeier, B.; Hellmann, R. Burr formation detector for fiber laser cutting based on a photodiode sensor system. *Opt. Laser Technol.* **2017**, *96*, 13–17. [CrossRef]

7. Garcia, S.M.; Ramos, J.; Arrizubieta, J.I.; Figueras, J. Analysis of Photodiode Monitoring in Laser Cutting. *Appl. Sci.* **2020**, *10*, 6556. [CrossRef]
8. Levichev, N.; Rodrigues, G.C.; Duflou, J.R. Real-time monitoring of fiber laser cutting of thick plates by means of photodiodes. *Procedia CIRP* **2020**, *94*, 499–504. [CrossRef]
9. Tomaz, K.; Janz, G. Use of AE monitoring in laser cutting and resistance spot welding. In Proceedings of the EWGAE 2010, Vienna, Austria, 8–10 September 2010; pp. 1–7.
10. Adelmann, B.; Schleier, M.; Neumeier, B.; Hellmann, R. Photodiode-based cutting interruption sensor for near-infrared lasers. *Appl. Opt.* **2016**, *55*, 1772–1778. [CrossRef]
11. Adelmann, B.; Schleier, M.; Neumeier, B.; Wilmann, E.; Hellmann, R. Optical Cutting Interruption Sensor for Fiber Lasers. *Appl. Sci.* **2015**, *5*, 544–554. [CrossRef]
12. Schleier, M.; Adelmann, B.; Esen, C.; Hellmann, R. Cross-Correlation-Based Algorithm for Monitoring Laser Cutting with High-Power Fiber Lasers. *IEEE Sens. J.* **2017**, *18*, 1585–1590. [CrossRef]
13. Adelmann, B.; Schleier, M.; Hellmann, R. Laser Cut Interruption Detection from Small Images by Using Convolutional Neural Network. *Sensors* **2021**, *21*, 655. [CrossRef]
14. Tatzel, L.; León, F.P. Prediction of Cutting Interruptions for Laser Cutting Using Logistic Regression. In Proceedings of the Lasers in Manufacturing Conference 2019, Munich, Germany, 24 July 2019; pp. 1–7.
15. Guo, Y.; Liu, Y.; Oerlemans, A.; Lao, S.; Wu, S.; Lew, M.S. Deep learning for visual understanding: A review. *Neurocomputing* **2016**, *187*, 27–48. [CrossRef]
16. Voulodimos, A.; Doulamis, N.; Doulamis, A.; Protopapadakis, E. Deep Learning for Computer Vision: A Brief Review. *Comput. Intell. Neurosci.* **2018**, *2018*, 7068349. [CrossRef] [PubMed]
17. Ting, F.F.; Tan, Y.J.; Sim, K.S. Convolutional neural network improvement for breast cancer classification. *Expert Syst. Appl.* **2018**, *120*, 103–115. [CrossRef]
18. Acharya, U.R.; Oh, S.L.; Hagiwara, Y.; Tan, J.H.; Adeli, H. Deep convolutional neural network for the automated detection and diagnosis of seizure using EEG signals. *Comput. Biol. Med.* **2018**, *100*, 270–278. [CrossRef] [PubMed]
19. Perol, T.; Gharbi, M.; Denolle, M. Convolutional neural network for earthquake detection and location. *Sci. Adv.* **2018**, *4*, e1700578. [CrossRef]
20. Dung, C.V.; Anh, L.D. Autonomous concrete crack detection using deep fully convolutional neural network. *Autom. Constr.* **2019**, *99*, 52–58. [CrossRef]
21. Zhang, L.; Yang, F.; Zhang, Y.D.; Zhu, Y.J. Road crack detection using deep convolutional neural network. In Proceedings of the 2016 IEEE International Conference on Image Processing (ICIP), Phoenix, AZ, USA, 25–28 September 2016; pp. 3708–3712.
22. Nakazawa, T.; Kulkarni, D. Wafer Map Defect Pattern Classification and Image Retrieval Using Convolutional Neural Network. *IEEE Trans. Semicond. Manuf.* **2018**, *31*, 309–314. [CrossRef]
23. Urbonas, A.; Raudonis, V.; Maskeliūnas, R.; Damaševičius, R. Automated Identification of Wood Veneer Surface Defects Using Faster Region-Based Convolutional Neural Network with Data Augmentation and Transfer Learning. *Appl. Sci.* **2019**, *9*, 4898. [CrossRef]
24. Khumaidi, A.; Yuniarno, E.M.; Purnomo, M.H. Welding defect classification based on convolution neural network (CNN) and Gaussian kernel. In Proceedings of the 2017 International Seminar on Intelligent Technology and Its Applications (ISITIA), Surabaya, Indonesia, 28–29 August 2017; pp. 261–265. [CrossRef]
25. Alom, M.Z.; Taha, T.M.; Yakopcic, C.; Westberg, S.; Sidike, P.; Nasrin, M.S.; van Esesn, B.C.; Awwal, A.A.S.; Asari, V.K. The History Began from AlexNet: A Comprehensive Survey on Deep Learning Approaches. March 2018. Available online: http://arxiv.org/pdf/1803.01164v2 (accessed on 26 August 2021).
26. O'Shea, K.; Nash, R. An Introduction to Convolutional Neural Networks Nov 2015. Available online: http://arxiv.org/pdf/1511.08458v2 (accessed on 26 August 2021).
27. Emura, M.; Landgraf, F.J.G.; Ross, W.; Barreta, J.R. The influence of cutting technique on the magnetic properties of electrical steels. *J. Magn. Magn. Mater.* **2002**, *254*, 358–360. [CrossRef]
28. Schoppa, A.; Schneider, J.; Roth, J.-O. Influence of the cutting process on the magnetic properties of non-oriented electrical steels. *J. Magn. Magn. Mater.* **2000**, *215*, 100–102. [CrossRef]
29. Adelmann, B.; Hellmann, R. Process optimization of laser fusion cutting of multilayer stacks of electrical sheets. *Int. J. Adv. Manuf. Technol.* **2013**, *68*, 2693–2701. [CrossRef]
30. Adelmann, B.; Lutz, C.; Hellmann, R. Investigation on shear and tensile strength of laser welded electrical sheet stacks. In Proceedings of the 2018 IEEE 14th International Conference on Automation Science and Engineering (CASE), Munich, Germany, 20–24 August 2018; pp. 565–569.
31. Arntz, D.; Petring, D.; Stoyanov, S.; Quiring, N.; Poprawe, R. Quantitative study of melt flow dynamics inside laser cutting kerfs by in-situ high-speed video-diagnostics. *Procedia CIRP* **2018**, *74*, 640–644. [CrossRef]
32. Tennera, F.; Klämpfla, F.; Schmidta, M. How fast is fast enough in the monitoring and control of laser welding? In Proceedings of the Lasers in Manufacturing Conference, Munich, Germany, 22–25 June 2015.
33. Keshari, R.; Vatsa, M.; Singh, R.; Noore, A. Learning Structure and Strength of CNN Filters for Small Sample Size Training. In Proceedings of the IEEE Conference on Computer Vision and Pattern Recognition (CVPR), Salt Lake City, UT, USA, 18–23 June 2018.

34. Chen, G.; Han, T.X.; He, Z.; Kays, R.; Forrester, T. Deep convolutional neural network based species recognition for wild animal monitoring. In Proceedings of the 2014 IEEE International Conference on Image Processing (ICIP), Paris, France, 27–30 October 2014; pp. 858–862.
35. Dong, Z.; Wu, Y.; Pei, M.; Jia, Y. Vehicle Type Classification Using a Semisupervised Convolutional Neural Network. *IEEE Trans. Intell. Transp. Syst.* **2015**, *16*, 2247–2256. [CrossRef]
36. Zhang, X.; Zhou, X.; Lin, M.; Sun, J. ShuffleNet: An Extremely Efficient Convolutional Neural Network for Mobile Devices. July 2017. Available online: http://arxiv.org/pdf/1707.01083v2 (accessed on 26 August 2021).
37. Iandola, F.N.; Han, S.; Moskewicz, M.W.; Ashraf, K.; Dally, W.J.; Keutzer, K. SqueezeNet: AlexNet-Level Accuracy with 50x Fewer Parameters and <0.5MB Model Size. February 2016. Available online: http://arxiv.org/pdf/1602.07360v4 (accessed on 26 August 2021).
38. Bello, I.; Zoph, B.; Vasudevan, V.; Le, Q.V. Neural Optimizer Search with Reinforcement Learning. September 2017. Available online: http://arxiv.org/pdf/1709.07417v2 (accessed on 26 August 2021).
39. An, S.; Lee, M.; Park, S.; Yang, H.; So, J. An Ensemble of Simple Convolutional Neural Network Models for MNIST Digit Recognition. *arXiv* **2020**, arXiv:2008.10400.

sensors

MDPI

Article

Machine Learning for Sensorless Temperature Estimation of a BLDC Motor

Dariusz Czerwinski [1], Jakub Gęca [2,*] and Krzysztof Kolano [3]

1 Department of Computer Science, Lublin University of Technology, 20-618 Lublin, Poland; d.czerwinski@pollub.pl
2 Doctoral School, Lublin University of Technology, 20-618 Lublin, Poland
3 Department of Electrical Drives and Machines, Lublin University of Technology, 20-618 Lublin, Poland; k.kolano@pollub.pl
* Correspondence: j.geca@pollub.pl

Abstract: In this article, the authors propose two models for BLDC motor winding temperature estimation using machine learning methods. For the purposes of the research, measurements were made for over 160 h of motor operation, and then, they were preprocessed. The algorithms of linear regression, ElasticNet, stochastic gradient descent regressor, support vector machines, decision trees, and AdaBoost were used for predictive modeling. The ability of the models to generalize was achieved by hyperparameter tuning with the use of cross-validation. The conducted research led to promising results of the winding temperature estimation accuracy. In the case of sensorless temperature prediction (model 1), the mean absolute percentage error MAPE was below 4.5% and the coefficient of determination R^2 was above 0.909. In addition, the extension of the model with the temperature measurement on the casing (model 2) allowed reducing the error value to about 1% and increasing R^2 to 0.990. The results obtained for the first proposed model show that the overheating protection of the motor can be ensured without direct temperature measurement. In addition, the introduction of a simple casing temperature measurement system allows for an estimation with accuracy suitable for compensating the motor output torque changes related to temperature.

Keywords: temperature estimation; machine learning; BLDC; electric machine protection

Citation: Czerwinski, D.; Gęca, J.; Kolano, K. Machine Learning for Sensorless Temperature Estimation of a BLDC Motor. *Sensors* **2021**, *21*, 4655. https://doi.org/10.3390/s21144655

Academic Editor: Hyoungsik Nam

Received: 31 May 2021
Accepted: 4 July 2021
Published: 7 July 2021

1. Introduction

Permanent magnet electric motors, in particular permanent magnet synchronous (PMSM) and brushless direct current (BLDC) motors, have gained popularity over the past decade. It is visible both in industrial solutions such as servo drives and actuators, but also in household appliances and traction applications. This is due, inter alia, to the fact that the use of permanent magnets allows the miniaturization of devices (due to the much higher power density in motors), which also increases their reliability and energy efficiency. The elimination of one of the most susceptible to damage elements of DC electric motors, i.e., the mechanical commutator and brushes, made it possible to use BLDC motors in applications requiring increased durability. The authors in [1] presented the percentage share of individual failures in induction motors, which shows that more than 40% of failures are caused by bearing failures, while 38% are problems with the machine stator. Therefore, it can be concluded that in DC brushless motors, damage to these elements will also have a significant impact on the reliability of the device. Diagnostics and detection of BLDC motor failures are devoted to many studies, including methods consisting of monitoring the motor current waveform [2–4], as well as using built-in Hall sensors [5] or additional vibration measurement [6]. A detailed description of faults occurring in brushless motors and the classification of diagnostic methods can be found in [7]. The authors point out that the detection of damage is also possible by measuring the motor temperature.

The analysis of thermal phenomena in electric motors is important for their proper functioning and the possibility of preventing and detecting faults. Excessive temperature rise can destroy the insulation of the stator winding and lead to a short circuit. The increased operating temperature of the motor causes both the aging of the bearings and the degradation of the rotor permanent magnets, which in turn shortens the remaining useful life of the machine. Zhang et al. in [8] also emphasize that the motor life is reduced by 50% for every 10 °C above the maximum temperature limit set by the manufacturer. One of the methods used by manufacturers of electric motors to protect against long-term operation at the upper operating temperature limit is the oversizing of the device or the use of an additional cooling system in the form of a fan placed on the motor shaft. Another cooling method in the case of high-power traction motors is the forced circulation of the coolant in a casing specially designed for this purpose.

According to [9], there are three basic types of thermal losses in permanent magnet motors. The first of these are the losses in the copper winding, the value of which depends on the flowing current. Another is the core iron losses, which mainly depend on the stator voltage. The last type of losses is mechanical, which is influenced by the motor speed. Stator winding insulation is particularly exposed to the effects of temperature and the thermal aging process. Moreover, if the temperature of the permanent magnet motor winding cannot be effectively controlled, the heat will be transferred to the rest of the components through the casing and the air gap, leading to heating of both the bearings and the permanent magnets. The authors in [10] emphasize that the increase in temperature of permanent magnets causes their partial demagnetization, which leads to a drop in motor output torque. If the critical temperature value characterizing a given permanent magnets material is not reached, the process is reversible. Otherwise, the magnets are permanently and irreversibly damaged, resulting in worse motor performance.

The above-mentioned phenomena show that there is a need to monitor the temperature inside DC brushless motors. Therefore, some BLDC motor manufacturers decided to install factory-built winding temperature sensors. This solution entails an increase in the production costs of the device and constitute another element of the machine that may be damaged. On the other hand, the installation of the temperature sensor by the user requires a lot of time and effort, as well as knowledge about the design of the device itself.

In order to optimize costs and eliminate the need for sensors, scientists have made a number of attempts to estimate the temperature of individual internal components of electric motors, such as stator winding [11], rotor [12], or bearings [13]. These efforts are mainly aimed at protecting these components from excessive temperature rise. However, studies show that it is also possible to compensate the torque pulsations on the machine shaft caused by the influence of temperature on the winding resistance and demagnetization of permanent magnets [14–17]. The authors in [18] propose using the BLDC motor thermal model to optimize the trajectory of the industrial robot movements, taking into account thermal constraints. Moreover, based on the research conducted by the authors in [19], it can be concluded that thermal modeling of a traction motor with permanent magnets will have a significant impact on the optimization of V2G (vehicle to grid) systems, because the amount of energy consumed by an electric vehicle is dependent on thermal losses of its motor.

One of the most popular and most effective methods of forecasting the electric motor temperature is lumped parameter thermal networks (LPTN), which simplify the physical model of the motor and allow for temperature estimation based on a set of parameters assigned to network nodes. As described in [20], motor equivalent thermal circuit diagrams can be divided into three basic types depending on the number of nodes in the thermal model. The first is a white box model in which a multi-node network is created that describes the motor based on the theory of heat transfer. In this type of model, there are additional sub-nodes that are designed to even more accurately reflect the actual heat distribution in the machine. Conducting calculations aimed at temperature prediction with such complex network structures requires a lot of computing power and, despite high accuracy, cannot be used for real-time prediction. In addition, creating such accurate models requires knowledge of many parameters and properties of the materials of which the motor is made, as well as expertise in its construction. Therefore, it is obvious that this is difficult to achieve, as manufacturers do not provide complete information about their devices. Light gray box models are another type of thermal equivalent networks. They represent the first degree of simplification compared to the previously described white-box networks and typically have five to fifteen nodes. Thanks to the use of a simpler structure, the complexity of calculations is much lower, although there is still a need for detailed information on the materials and geometry of the motor. In response to the above problems, dark gray box models were created, which have only two to five nodes corresponding to the dominant heat transfer paths and achieve very good prediction accuracy thanks to determining the values of the thermal model parameters based on experimental tests. High efficiency and the possibility of real-time calculations made dark gray box networks popular in the field of thermal modeling of electric motors. However, it is worth noting that they require knowledge of the temperature (application of the sensor) in at least one point in the network, as well as some expertise knowledge of the modeled object. In the literature, a number of publications on temperature estimation based on the created lumped parameter thermal networks of permanent magnet motors can be found [11,21,22].

Another way to predict the temperature of electric motors is to estimate the winding resistance by injecting signals of the appropriate frequency into the stator circuit [12,23–25]. Methods of this type allow for real-time temperature estimation and are resistant to changes in motor cooling conditions (damage to the cooling system), because it is assumed that the relationship between winding resistance and temperature is known and does not change with time. However, the introduction of additional signals causes current and voltage distortions, significantly affecting the electromagnetic compatibility of the device. In addition, the injected signals cause torque pulsations that are unacceptable in some applications.

Thermal modeling of electric motors is also carried out in a purely analytical manner using mathematical [26] and finite element (FEA) methods. It is worth mentioning that there are also hybrid estimation methods such as those described in [27]. An interesting issue concerning motors with permanent magnets is the estimation of the rotor temperature on the basis of the flux measurement with the use of built-in Hall sensors, as described in [5]. However, this method requires the knowledge of the thermal demagnetization constant of the material of which the magnets are made. Moreover, the flux measurement is also affected by the influence of the stator flux, which contributes to erroneous predictions at higher loads.

The modern and very effective methods of estimating the temperature of electric motors include machine learning and deep learning. Their unquestionable advantage is that on the basis of the collected measurement data, a function mapping the relationship between the values of the input features and the output is determined. This means that predictive modeling does not require knowledge of the material properties of a given device or having expertise knowledge about its construction. Both neural networks and other machine learning methods have proven their effectiveness in estimating the temperature of induction motors [28], permanent magnets synchronous motors [9,29–31], as well as brushed DC motors [32,33]. Many of the articles on PMSM temperature prediction using machine learning available in the literature use the motor coolant temperature as an input variable of the algorithm [9,29–31]. Moreover, the authors in [34] emphasize that the stator temperature is strongly correlated with the exponentially weighted moving average of the PMSM motor coolant temperature, and removing this variable from the feature vector results in a significant decrease in the effectiveness of the prediction algorithm.

In this article, we present a comparison of the effectiveness of selected machine learning algorithms in predicting the temperature of BLDC stator winding for a variable load profile and various cooling conditions. An interesting result was achieved even without the knowledge of casing temperature. In addition, the results of the estimation using information about the temperature on the casing and in the absence of additional sensors were compared. The article is organized as follows: Section 2 describes measurement of the BLDC motor winding temperature and the methods used for data preprocessing; Section 3 describes the machine learning algorithms that were used to develop the models; cross-validation and parameter optimization are described in Section 4; Sections 5 and 6 present the results of predictive analysis, while Section 7 presents the summary and conclusions.

2. Measurements and Data Preprocessing

One of the key aspects of predictive modeling with the use of machine learning algorithms is the collection of a sufficiently large set of data, which will be split and then used for training, testing, and validating the results. For this purpose, two mechanically coupled brushless DC machines of the same type (Table 1) were installed on the test stand (Figure 1). One of them worked as a motor during the measurements, while the other one worked as a generator and constituted the adjustable load. The BLDC motor was equipped with two analog LM35DZ temperature sensors supported by the Texas Instruments (Dallas, TX, USA), one of which was mounted to the winding ends with a thermally conductive adhesive, as shown in Figure 2, while the other was placed on the motor casing—Figure 3. The control of the tested motor was carried out with the use of the algorithm described in [35,36] to eliminate the possible impact of Hall sensors' misalignment on the effectiveness of the estimation. The IHM08M1 system, dedicated to work with STM32 series microprocessors supported by the STMicroelectronics N.V. (Amsterdam, The Netherland), was used as the power electronic converter. This converter is equipped with a current measurement system. In addition, the rotational speed of the motor was measured using a preinstalled Hall sensors' system. Data acquisition was carried out using the STM Studio software dedicated to real-time visualization and data acquisition, but due to the fact that thermal processes are slowly changing, it was decided to sample the measurements at a frequency of 4 Hz.

Over 160 h of BLDC motor temperature measurements were carried out on the test stand for a variable load profile and rotational speed, as well as for various cooling conditions. During the first series of measurements lasting more than 80 h, the motor was tested without any additional cooling system. In the second one, a cooling fan was mounted on the motor shaft (Figure 3). In addition, an air duct has been provided to ensure an adequate flow of cooling air.

Table 1. Tested BLDC motor specifications.

Name	Unit	Value
No. of pole	-	8
No. of phase	-	3
Rated voltage	V	48
Rated speed	rpm	3000
Rated torque	Nm	1.4
Max peak torque	Nm	4.2
Torque constant	Nm/A	0.127
Line to line resistance	Ω	0.16
Line to line inductance	mH	0.50
Max peak current	A	33
No-load current	mA	1450
Length	Mm	98
Rotor inertia	g cm^2	1600
Weight	Kg	3.15

Figure 1. Test bench.

Figure 2. Winding temperature sensor placement (**red arrow**).

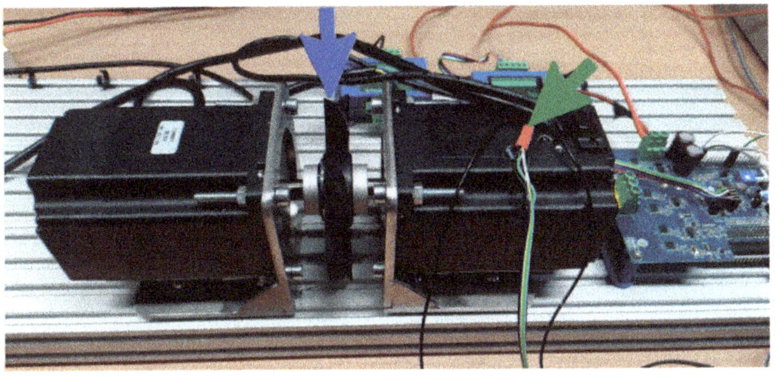

Figure 3. Placement of the cooling fan (**blue arrow**) and the casing temperature sensor (**green arrow**).

The results of the measurements are presented in Figure 4a,b. The load was realized by the machine of the same type coupled on the common shaft. Windings of this auxiliary machine were connected to the variable resistance load. Changes of the rotational speed were achieved by changing the target speed of a closed-loop speed controller. It can be seen that after simultaneous load and rotational speed change, the heating or cooling of the motor is visible for a certain time interval. Therefore, it can be concluded that the important information from the algorithm's point of view will be the time that has elapsed since the last load change. Accurate information about the moment of load change is not available, and its detection is difficult. Therefore, we decided to introduce new features that will constitute a short-term history of the device from previous N_h = 14,400 data records, which corresponds to one hour of measurements (this time was selected on the basis of observations of the obtained waveforms). For this purpose, the mean and standard deviation of the current and rotational speed from the last N_h measurements were calculated for each data record, thus creating additional feature variables. Selection of the optimal length of the short-term history of the device, also known as the so-called look-back parameter, is out of the scope of this work and may form the basis for further research in this field. It is worth emphasizing that for results from Figure 4b, a cooling fan was placed on the common shaft with the motor, so it constitutes the additional load. Moreover, temperature is more dependent on the speed than in the case without a fan. Therefore, when the speed is low, the cooling conditions greatly worsen. If the load is high but the speed is also high, cooling is ensured, and the temperature does not rise too much. As an additional input variable, the winding power losses were also introduced, which were calculated from the formula:

$$P_{wl} = I^2 R_w \tag{1}$$

where I is the winding current and R_w is the stator winding resistance.

An important issue related to the data preparation is the appropriate division into a training set and a test set. In the case of PMSM motor temperature prediction presented by the authors in [34], the 55 h of measurements were enough to train the algorithm properly. Referring to these studies, the dataset was divided into training and test subsets in the proportion of 70:30. As a result, training sets of more than 50 h were obtained for both experiments. It is also worth adding that in the analyzed case, the division had to be carried out with respect to the record occurrence and not in a random manner. As mentioned before, the temperature inside the BLDC motor depends on the time that has elapsed since the last load and/or speed change, which can be seen as the change in the supplying current. Due to the fact that the data are in the form of a timeseries, it is unacceptable to shuffle them.

Measurement data obtained on the test stand and constituting the basis for training and validation of algorithms are expressed in various units and on a different scale (speed is up to 1500 rpm, and current up to 12 A). The use of data in this form could cause a situation where the speed, due to high values, will dominate the cost function values, making it impossible to obtain any useful information from other features, which would significantly worsen the effectiveness of the algorithms. Moreover, some predictive models may generate larger errors if the individual features are not approximately normally distributed. The exception in this case are decision trees algorithms, which are resistant to different scaling of features. In response to the above problem, in this paper, the feature variables have been standardized as follows. First, the mean μ and standard deviation σ of each feature from the subset of training data were calculated. Then, all samples of a given feature were transformed according to the formula:

$$x_{std} = \frac{x - \mu}{\sigma}. \tag{2}$$

The data transformed in this way have approximately zero mean, standard deviation $\sigma = 1$, and are appropriately scaled.

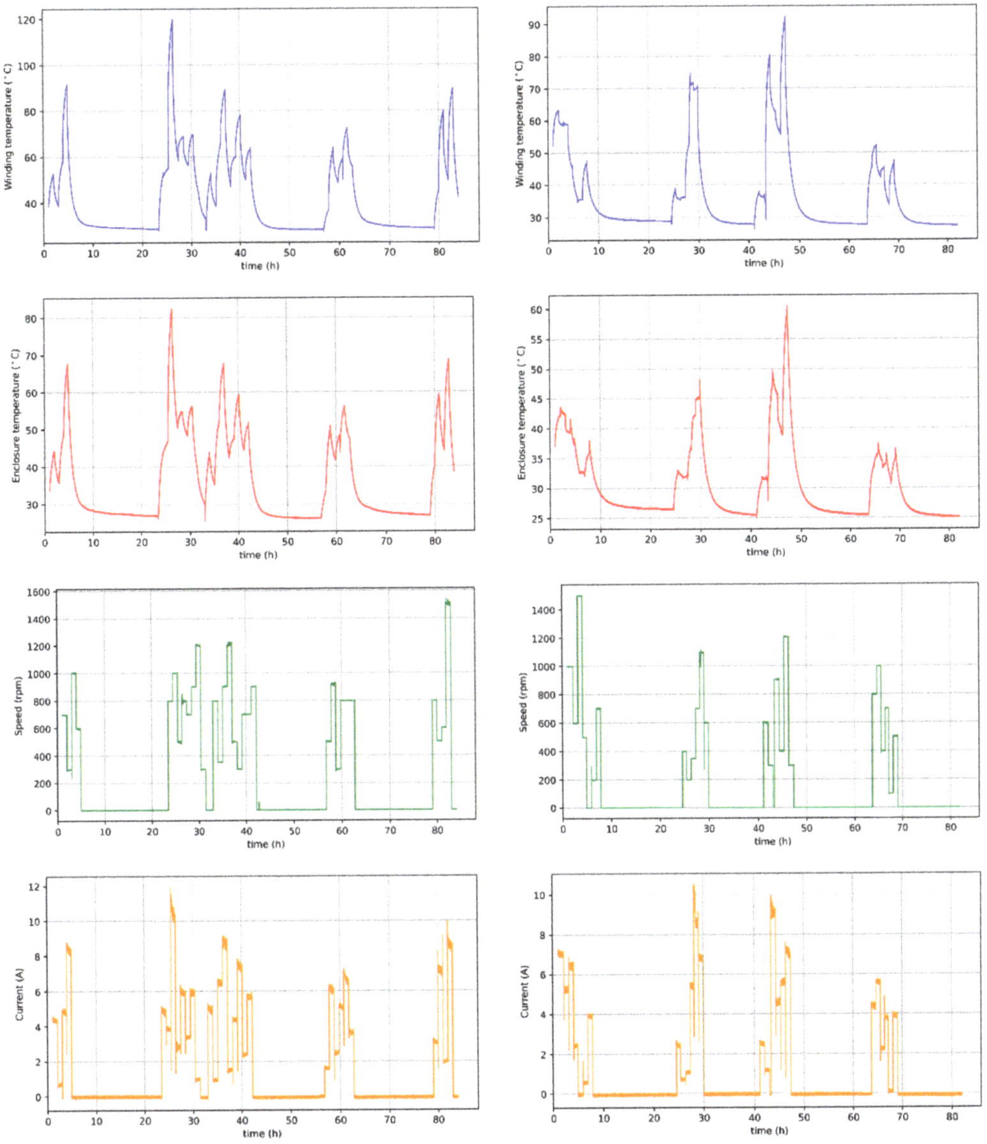

(a) (b)

Figure 4. Measurement results of winding temperature (blue), casing temperature (red), speed (green), and current (orange) of motor: (**a**) without cooling; (**b**) with cooling fan.

3. Machine Learning Algorithms

The purpose of machine learning algorithms for regression is to predict the value of the target variable based on the set of independent variables, which are commonly known as features. The algorithm acquires the ability to forecast in the learning process, which consists of providing examples that allow prediction verification. Then, the algorithm modifies its own structure in such a way as to minimize errors. Linear regression models [37] implement the above statements in the form of an equation:

$$\hat{y} = \mathbf{w}^T \mathbf{x} + b \tag{3}$$

where $\hat{y}$ is the output of the algorithm, i.e., the predicted value, $\mathbf{x} \in \mathbb{R}^n$ is the input features vector, and $\mathbf{w} \in \mathbb{R}^n$ is the vector of the model parameters, which are also called weights. This name results from the fact that they allow defining how strong a given feature should influence the output value. At this point, it is also worth adding that it is the weight values that are optimized in the learning process, thanks to which it is possible to improve the effectiveness of prediction. The component b in (3) is called a bias.

An alternative way of introducing bias in the model is to add the component of value 1 to the feature vector. Then, the weight assigned to this component will act as a bias.

The purpose of modifying the parameters of the model is to find a weight vector that will allow correct prediction of the variable. In other words, the weight update procedure should minimize some prediction error function. One of the most frequently used objective functions in regression is the sum of squared errors (SSE):

$$J(\mathbf{w}) = \sum_{i=1}^{N} (\hat{y} - y)_i^2 \tag{4}$$

where N is the number of samples. The above formula also has its geometric equivalent as the Euclidean distance between the prediction $\hat{y}$ and the real value y of the target variable:

$$J(\mathbf{w}) = \|\hat{y} - y\|_2^2. \tag{5}$$

Since the objective function has been defined as well as the parameter that will be modified, all that remains is to use an optimization algorithm, which can be, for example, the gradient descent. However, note that moving along the gradient decrease may end up reaching a local rather than a global minimum of the cost function. Of course, in many cases, reaching the global minimum is not possible at all, and the solution obtained with this method is sufficient. However, the objective function can have a very complex structure with many local minima, each of which allows for different predictive accuracies. For this reason, scientists have developed many algorithms that are more resistant to getting stuck in small local minima of the cost function and more efficient in terms of computational complexity. Such algorithms include the commonly used optimizer Adam [38].

The main task of machine learning algorithms is to effectively predict specific quantities based on the provided, previously unseen data. It is possible thanks to the previously conducted process of learning the algorithm on the so-called training data. However, from the application point of view, it is most important to generalize well, that is, use the acquired "knowledge" to correctly analyze new data. As previously explained, learning consists in adjusting the parameters of the algorithm based on the determined training error in order to minimize it. The above issue can be treated as an optimization problem, but the test error (also known as the generalization error) should also be as small as possible. Therefore, determining how well a given algorithm will analyze new data is based on the value of the training error and the difference between the training error and the test error. In this way, it is possible to avoid underfitting, which occurs when the algorithm is unable to achieve a sufficiently small value of the training error, and overfitting, the sign of which is a large difference between the training error and the test error. As the algorithm's complexity (also called capacity) increases, the model's variance increases, along with the difference between the value of the training error and the test error, and thus the total error of the system. In the opposite situation, the capacity of the algorithm decreases, the bias increases, and the training error (and thus the total error) increases. This means that a very important element in the design of a machine learning model is to establish an appropriate compromise between bias and variance (underfitting and overfitting). This goal is achieved by adjusting the algorithm's capacity to the true complexity of the problem as well as the amount of training data available. It is a known fact that in practice, there is always some kind of noise and outliers among the real-world data. For this reason, reaching a satisfactory compromise is possible only with the acceptance of training errors resulting from an incorrect classification of outliers. Thanks to this, the prediction of future data will be insensitive to possible noises in the provided test data. In response to the above-presented need to limit or regulate the complexity of the machine learning model, many techniques have been developed that are generally referred to as regularization. Some of them introduce additional constraints on the machine learning model, for example by imposing limits on parameter values. In turn, others extend the objective function with special expressions [39]. Additional restrictions and penalties, if carefully selected, can lead to improved performance on test data. Many methods of regularization are based on limiting the capacity of the model by adding a norm component $\Omega(\mathbf{w})$ to the objective function J. This penalty term is used to limit the values in the parameter matrix $\mathbf{w}$. The regularized objective function $\widetilde{J}$ can be written as a variable:

$$\widetilde{J}(\mathbf{w};\mathbf{x},y) = J(\mathbf{w};\mathbf{x},y) + \alpha\Omega(\mathbf{w}) \tag{6}$$

where $\alpha \in <0,\infty)$ is a hyperparameter that regulates the strength of regularization with respect to the standard cost function [40]. Expression (6) shows that the minimizing cost function $\widetilde{J}$ created in the learning process will result in an optimization of both the original cost function J and some measure of parameter size Ω (or a subset of parameters). A comparison of the regularized objective functions is presented in Table 2.

Table 2. Regularized objective functions.

Regularization	Linear Regressor	Objective Function
L2	Ridge	$\min\|\hat{y} - y\|_2^2 + \alpha\|\mathbf{w}\|_2^2$
L1	Lasso	$\min\frac{1}{2N}\|\hat{y} - y\|_2^2 + \alpha\|\mathbf{w}\|_1$
L1 + L2	ElasticNet	$\min\frac{1}{2N}\|\hat{y} - y\|_2^2 + \alpha\rho\|\mathbf{w}\|_1 + \frac{\alpha(1-\rho)}{2}\|\mathbf{w}\|_2^2$

In this paper, we investigate the effectiveness of temperature prediction by linear regression models as well as those based on decision trees. The following algorithms were tested during the research:

1. Linear regression with the objective function given by Formula (4),
2. Elastic-Net regressor, in which the regularization components are introduced to the objective function (Table 2),
3. Regressor using the stochastic gradient descent optimization algorithm (denoted as SGD),
4. Support vector machine (SVM) with linear kernel,
5. CART (Classification and regression trees) decision trees,
6. AdaBoost—presented in [41], an algorithm that uses boosting to determine the final prediction fitting the sequence of decision trees.

The estimation results using the above-mentioned methods can be found in Sections 5 and 6. In the case of this study, observing the temperature curves of the stator winding and the temperature on the motor casing (Figure 4a,b), one can see an almost perfectly linear relationship between these variables. Therefore, the authors of this article decided to compare the effectiveness of machine learning algorithms in predicting the temperature of the stator winding of a BLDC motor, taking into account the temperature on the casing as well as in the complete absence of additional sensors. It is worth adding that by using the measurement data available from the converter system, Hall sensors, and transformations described in the previous section, the cost of the drive does not increase. This information is available in any drive system that uses current and speed regulators in the control algorithm.

4. Hyperparameters Optimization with Cross-Validation

Achieving the maximum possible predictive accuracy of each model is possible only through hyperparameter tuning, such as the regularization strength described in the previous section. These are parameters that influence the behavior of the algorithm and are not optimized in the learning process—their selection is the programmer's task. For this purpose, it is necessary to carry out a series of tests to identify for which parameter values the temperature prediction error will be the smallest. However, it should be remembered that if during each of the trials, the effectiveness of the algorithms is validated on the same test set, the selected set of hyperparameters will be the best but only for this specific set and may turn out to be inappropriate for predicting new samples. Moreover, the test set has a limited size and will never be able to reflect all the dependencies that occurred in the training data. Therefore, it can be concluded that the model's ability to generalize will be small when its hyperparameters are optimized for a specific case of the test set. One of the solutions to this problem may be to split it into three subsets: training, validation (used to optimize parameters), and test. However, the application of this method requires a significant number of samples, and in addition, the accuracy of the predictions may still depend on some random selection of the dataset split points. In practice, the most frequently used and extremely effective method to avoid problems with generalization is cross-validation.

In this study, an exhaustive gird search method with cross-validation was used to optimize the hyperparameters. For each of the algorithms, a grid consisting of different values of individual parameters was defined. An example of a grid for two parameters is shown in Figure 5. Of course, the dimensionality of the grid depends on the number of unique parameters that can be optimized and is different for each algorithm. In order to determine the effectiveness of the model for a given grid node, cross-validation with respect to the chronology of the data was used. This method consists in dividing the data into k groups in such a way that the algorithm is trained on a k subset and tested on $k+1$. The average of the performance metrics computed in each iteration determines the final score for that grid node. Thanks to this, unlike the traditional k-fold cross-validation, it is possible to prevent the algorithm from being tested on data older than training data. The principle of data division according to the described method for the applied value $k = 5$ is shown in Figure 6. Based on the results for each node of the grid, a set of hyperparameter values is selected for which the compromise between the bias and variance of the model is the most optimal. Then, the algorithm is tested on a set of test data that has been set aside and which it has never actually seen before. This allows obtaining reliable results of the temperature prediction of the BLDC motor winding.

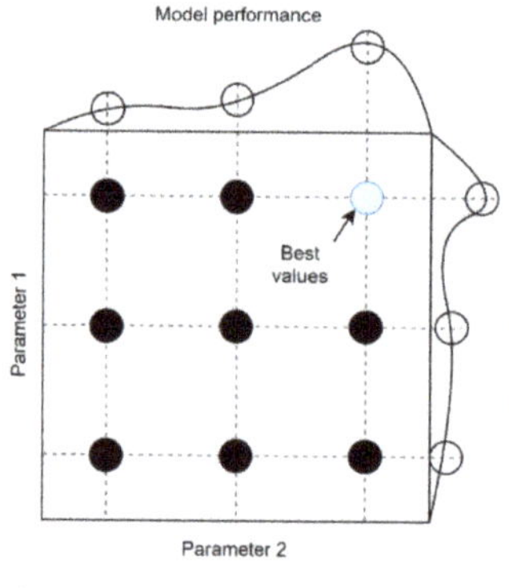

Figure 5. Example of a searching grid for two parameters.

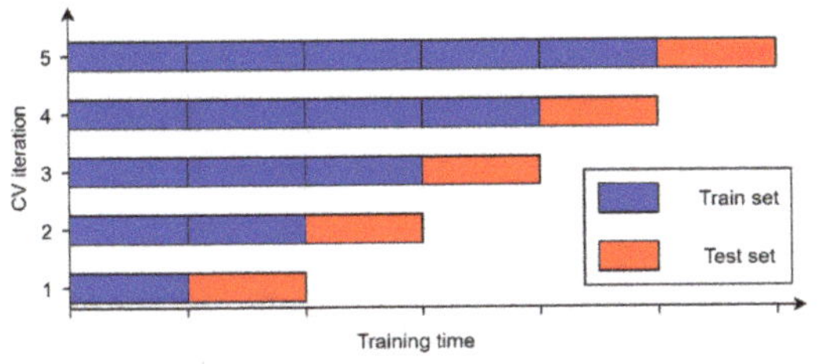

Figure 6. Time-series five-fold cross-validation.

5. Results of the Sensorless Estimation Model

The evaluation of the efficiency of BLDC motor winding temperature estimation was carried out on the basis of the following regression metrics:

- RMSE (root mean squared error) defined as:

$$\mathrm{RMSE} = \sqrt{\frac{1}{N}\sum_{i=1}^{N}(\hat{y}-y)_i^2}. \tag{7}$$

- MAPE (mean absolute percentage error) calculated with the formula:

$$\mathrm{MAPE} = \left(\frac{1}{N}\sum_{i=1}^{N}\left|\frac{\hat{y}-y}{y}\right|_i\right) * 100\%. \tag{8}$$

- Coefficient of determination R^2 (quality of fit) calculated as:

$$\mathrm{R}^2 = 1 - \frac{\sum_{i=1}^{N}(\hat{y}-y)_i^2}{\sum_{i=1}^{N}(y-\overline{y})_i}. \tag{9}$$

In the above equations, N represents the number of samples, $\hat{y}$ is the predicted value, y is the actual value, and the

$$\overline{y} = \frac{1}{N}\sum_{i=1}^{N} y_i \tag{10}$$

is the mean of the actual values.

Figure 7a shows the performance metrics of winding temperature regression for a BLDC motor without additional cooling. As can be seen, all linear models are performing better than decision trees. Only the utilization of many trees and the use of boosting in the AdaBoost algorithm significantly improve the results, allowing for a similar effectiveness as linear models, each of which achieves the value of the determination coefficient above 0.96, the mean absolute percentage error below 5%, and the RMSE error not exceeding 3.2 °C.

Results of the temperature estimation of the motor cooled by the fan placed on the shaft are shown in Figure 7b. One can notice a decrease in the effectiveness of all predictive algorithms, in particular those that use decision trees. The R^2 coefficient of linear models ranges from 0.84 to 0.91, while the MAPE error ranges from 4.3% to 8.5%. On the other hand, the root mean squared errors seem very interesting because they are smaller than in the case of temperature estimation without cooling. This means that the algorithm makes less error on average but has a much bigger problem with fitting, which also causes increased relative errors. The presented difference may result from the fact that the winding temperature of the motor cooled by a fan placed on the shaft depends to a greater extent on the rotational speed. Therefore, the relationship between the features and the target variable, sought by the algorithms, may be of a more complex nature, reducing the effectiveness of the estimation.

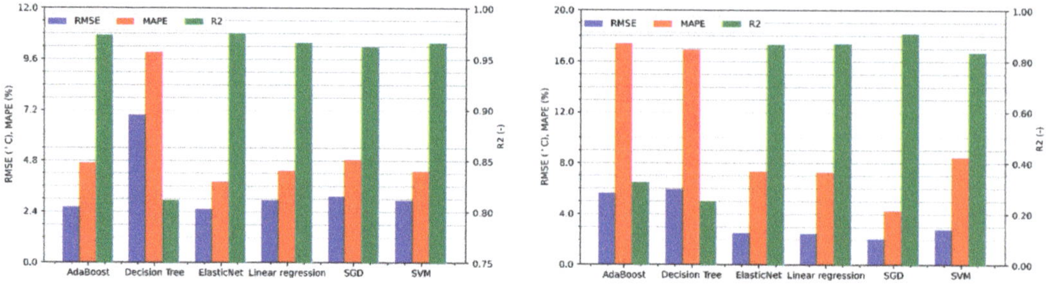

(a) (b)

Figure 7. Regression metrics values for the sensorless temperature estimation of the motor: (**a**) without cooling; (**b**) with a cooling fan.

The highest accuracy of motor temperature prediction without cooling was achieved by the ElasticNet algorithm, for which the regression metrics are respectively RMSE = 2.53 °C, MAPE = 3.82%, and R^2 = 0.975. Analyzing Figure 7b, it can be concluded that the accuracy of the motor temperature prediction with a cooling fan is the highest for the SGD algorithm. It achieved a coefficient of determination equal to 0.909, an RMSE error of 2.07 °C, and a MAPE of 4.3%.

Figure 8a,b show the curves of actual and predicted winding temperatures of a BLDC motor without cooling and with a cooling fan. It is worth noting that some errors are the result of incorrect temperature predictions of an unpowered motor during cooling to the ambient temperature. However, in practice, keeping track of the temperature within this range is not necessary in most applications. On the other hand, there are many important situations in which the motor operates at higher temperatures and should be protected from overheating. The algorithms' behavior in this respect is beneficial, because they more often overestimate the predicted temperature. Thanks to this, the motor protection will be preserved because the information about too high temperature will appear earlier.

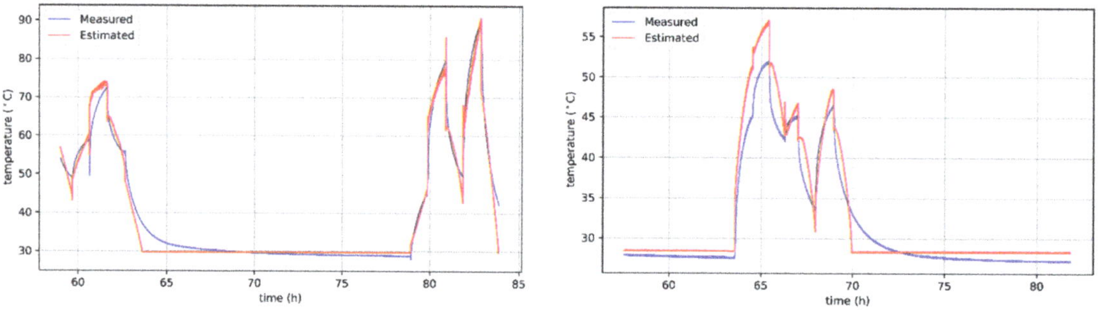

(a) (b)

Figure 8. Results of sensorless temperature estimation for the motor: (**a**) without cooling, obtained with ElaticNet regressor; (**b**) with a cooling fan, obtained with SGD regressor.

6. Results of the Estimation Model with Auxiliary Temperature Sensor

The values of regression accuracy metrics for predicting the winding temperature of a BLDC motor without cooling and taking into account the temperature information on the casing are shown in Figure 9a and with cooling in Figure 9b.

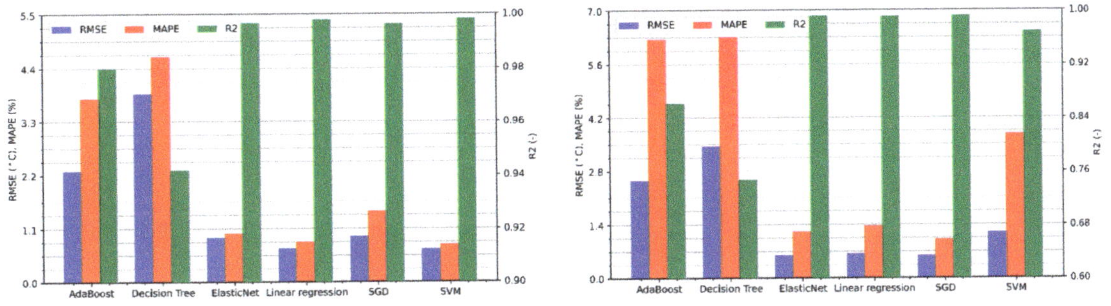

(a) (b)

Figure 9. Regression metrics values for the temperature estimation supported with casing sensor data of the motor: (**a**) without cooling; (**b**) with a cooling fan.

As in the case of sensorless estimation, the results obtained using algorithms based on decision trees are much worse than linear models. In the third section, it was emphasized that on the basis of the winding and the motor casing temperature curves, an almost linear relationship between these variables can be noticed. The estimation results confirm these assumptions because the efficiency of linear models is very good. The RMSE value is less than 1.5 °C and the MAPE is less than 3.8%. Moreover, the coefficient of determination R^2 for each linear algorithm is greater than or equal to 0.97. The best algorithm for estimating the motor temperature without cooling is undoubtedly the linear SVM. Its regression metrics are RMSE = 0.68 °C, MAPE = 0.77%, and R^2 = 0.998, respectively. The curves of the actual and estimated winding temperatures by this algorithm are shown in Figure 10a. Thus, an almost perfect representation of the actual temperature is visible. On the other hand, a slight deterioration in efficiency is visible in all models estimating the motor temperature with an additional cooling fan. The best results during this test were obtained with the linear model optimized with the stochastic gradient descent algorithm, for which RMSE = 0.59 °C, MAPE = 1.02% and R^2 = 0.993. The actual temperature and predicted temperature with the use of SGD are presented in Figure 10b. The decrease in the estimation accuracy for the cooled test is particularly interesting. As noted in the previous section, mounting a cooling fan on the shaft increases the effect of rotational speed on motor temperature and therefore increases the importance of this feature in prediction. Moreover, taking into account the temperature on the motor casing results in a significant improvement in the prediction accuracy. However, it should be remembered that the motor is cooled from the outside, so the rotational speed will have a greater effect on the temperature on the casing than on the inside of the motor. Therefore, it can be inferred that due to changes in rotational speed, the dependence of the winding temperature on the casing temperature will be more non-linear than in the case of a system without cooling. The above phenomenon may cause a significant estimation accuracy decrease.

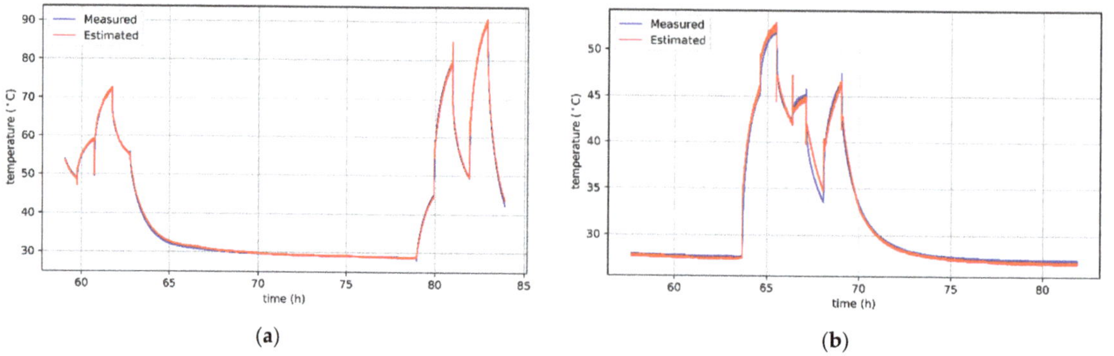

Figure 10. Results of temperature estimation supported with casing sensor data for the motor: (**a**) without cooling, obtained with SVM regressor; (**b**) with a cooling fan, obtained with SGD regressor.

7. Conclusions

This article proposes two models for BLDC motor winding temperature estimation using machine learning methods. The former allows the prediction of temperature without the need for temperature sensors, while the latter involves mounting an auxiliary sensor on the motor casing to improve prediction efficiency. In order to create an appropriate dataset for algorithm training, over 160 h of BLDC motor temperature measurements were carried out for a variable load and speed profile as well as various cooling conditions. Subsequently, data preprocessing was done as follows: additional features containing information about previous states of the device were introduced to create a short-term history of the device, the dataset was divided into training and test data, and then standardization was performed. Selected machine learning algorithms were used to estimate the temperature of the BLDC motor winding, namely: AdaBoost, decision tree, ElasticNet, linear regression, SGD, SVM. For each of the above-mentioned algorithms, a hyperparameter tuning process was performed through the use of a cross-validated grid search mechanism. As a result, it was possible to define a set of parameters that would ensure the appropriate generalization capability.

For the sensorless temperature prediction of BLDC motor without cooling, the greatest effectiveness was achieved by the ElasticNet algorithm reaching MAPE = 3.82%, while the remaining linear models obtained similar, but slightly worse results. In the case of the test with an additional fan on the motor shaft, the SVM turned out to be the best algorithm, for which the mean absolute percentage error was 4.3%. It is worth noting that the obtained error values are comparable with the results of other methods described in the first section of this study. However, they allow completely avoiding the need to use temperature sensors and do not require influencing the current and voltage waveforms of the motor.

The authors of this study also decided to compare the effectiveness of the algorithms taking into account the information about the temperature on the motor casing. In this case, as suspected, the results turned out to be much better. The MAPE error of the best linear models did not exceed 1.5% for each case, while the RMSE was below 0.7 °C. It is worth noting that this good winding temperature prediction results can be used to compensate the temperature effect on the machine output torque.

Comparing the proposed models for motor winding temperature estimation to those described in the literature (Table 3), it can be concluded that the second model gives better [29] or similar [31] results to recurrent and convolutional neural networks. Moreover, this model performs a little worse than the deep neural networks used in [9] and similar to [30] (smaller mean absolute errors, but larger maximum error). However, it is worth mentioning that the algorithms used in model two are much less complex and need fewer resources for training and validation. In addition, model two provides comparable results to those obtained by the authors in [11,20,21] using thermal models of the motor. However, the superiority of the proposed model is that it does not require expertise knowledge of the modeled object. Both models also provide better results than the signal injection method presented in [23], and additionally, it does not require interference with the motor power system. The proposed sensorless model (model 1) gives similar results to the linear models presented in [34], but it does not require any additional temperature sensors if the ambient temperature does not significantly affect the motor temperature. However, it is important to note that the comparison of the results with those available in the literature is indicative, because different engine types and models were tested among the researchers.

Table 3. Comparison of estimation results with methods reported in the literature.

Method		Metric						
		MSE [a]	RMSE	MAPE	R^2	$\max_i \lvert \hat{y}-y \rvert_i$	MAE [b]	MRE [c]
Literature method	[9]	-	0.24 °C	-	0.944	-	0.15 °C	-
	[11]	-	-	-	-	5.2 °C	-	1.50%
	[20]	-	-	-	-	≈8.0 °C	-	-
	[21]	-	-	-	-	8.0 °C	-	-
	[23]	-	-	-	-	-	-	6.14%
	[29]	2.04 K^2	1.43 K	-	-	37.6 K	-	-
	[30]	-	-	-	-	4.5 °C	0.90 °C	-
	[31]	-	-	-	-	10.8 K	-	-
	[34]	6.06 K^2	2.46 K	-	-	11.1 K	-	-
Model 1	Uncooled (ElasticNet)	6.40 $°C^2$	2.53 °C	3.82%	0.975	20.4 °C	1.64 °C	3.82%
	Cooled (SGD)	4.28 $°C^2$	2.07 °C	4.30%	0.909	8.3 °C	1.49 °C	4.30%
Model 2	Uncooled (SVM)	0.46 $°C^2$	0.68 °C	0.77%	0.998	14.0 °C	0.34 °C	0.77%
	Cooled (SGD)	0.35 $°C^2$	0.59 °C	1.02%	0.993	7.3 °C	0.35 °C	1.02%

[a] mean squared error, [b] mean absolute error, [c] mean relative error.

Therefore, it has been proven that overheating protection of the motor can be provided using a trained machine learning algorithm without any additional sensors, thus avoiding the cost of installing additional hardware by the manufacturer or the user. In addition, it has been proven that the use of information from the sensor mounted on the BLDC motor casing allows for very good winding temperature prediction results. This means that using the second described model, it is possible to introduce a compensation mechanism for the temperature impact on the motor output torque. It is worth adding that mounting the sensor on the motor casing is an uncomplicated operation that the user who wants to know the exact temperature inside the device can do by himself and at a low cost. In addition, most motor faults, such as interturn short circuits, bearing damage or magnet degradation, cause the motor temperature to rise significantly. Therefore, it can be anticipated that the described temperature estimation method can be used to detect device components damages.

The increase in ambient temperature may be problematic for the described estimation method. Under laboratory conditions, the ambient temperature was approximately constant and had no effect on the casing temperature and inside the motor. Difficult conditions at the motor site may increase the prediction errors, but a possible solution is to introduce an additional variable informing about the ambient temperature. In addition, future research should verify the effectiveness of temperature estimation for a fast-varying load profile and more sophisticated methods such as neural networks.

Author Contributions: Conceptualization, D.C. and J.G.; methodology, D.C.; software, J.G.; validation, D.C., J.G. and K.K.; formal analysis, D.C.; investigation, J.G.; resources, D.C. and K.K.; data curation, D.C.; writing—original draft preparation, J.G.; writing—review and editing, D.C. and K.K.; visualization, J.G.; supervision, D.C. and K.K.; project administration, D.C.; funding acquisition, D.C. All authors have read and agreed to the published version of the manuscript.

Funding: This work was supported by Lublin University of Technology grant no. FD-20/EE-2/605 and FD-20/IT-3/999.

Institutional Review Board Statement: Not applicable.

Informed Consent Statement: Not applicable.

Data Availability Statement: The database developed for this article [42] is available from IEEE Dataport at https://ieee-dataport.org/documents/bldc-motor-temperature (accessed on 6 July 2021).

Conflicts of Interest: The authors declare no conflict of interest.

References

1. Dambrauskas, K.; Vanagas, J.; Zimnickas, T.; Kalvaitis, A.; Bizimavicius, T. Induction Motor's Bearing Condition Motoring and Diagnosis Applying Cloud Services and Artificial Neural Networks. Advances in Information, Electronic and Electrical Engineering. In Proceedings of the AIEEE 2019-Proceedings of the 7th IEEE Workshop, Liepaja, Latvia, 15–16 November 2019; Volume 2019.
2. Medeiros, R.L.V.; Filho, A.C.L.; Ramos, J.G.G.S.; Nascimento, T.P.; Brito, A.V. A Novel Approach for Speed and Failure Detection in Brushless DC Motors Based on Chaos. *IEEE Trans. Ind. Electron.* **2019**, *66*, 8751–8759. [CrossRef]
3. Park, B.-G.; Lee, K.-J.; Kim, R.-Y.; Hyun, D.S. Low-Cost Fault Diagnosis Algorithm for Switch Open-Damage in BLDC Motor Drives. *J. Power Electron.* **2010**, *10*, 702–708. [CrossRef]
4. Jin, X.; Chow, T.W.S.; Tsui, K.-L. Online Anomaly Detection of Brushless Dc Motor Using Current Monitoring Technique. *Int. J. Perform. Eng.* **2014**, *10*, 263–271.
5. Zandi, O.; Poshtan, J. Fault Diagnosis of Brushless DC Motors Using Built-In Hall Sensors. *IEEE Sens. J.* **2019**, *19*, 8183–8190. [CrossRef]
6. Shifat, T.A.; Hur, J.W. An Effective Stator Fault Diagnosis Framework of BLDC Motor Based on Vibration and Current Signals. *IEEE Access* **2020**, *8*, 106968–106981. [CrossRef]
7. Kudelina, K.; Asad, B.; Vaimann, T.; Rassolkin, A.; Kallaste, A.; Lukichev, D.V. Main Faults and Diagnostic Possibilities of BLDC Motors. In Proceedings of the 2020 27th International Workshop on Electric Drives: MPEI Department of Electric Drives 90th Anniversary, IWED 2020-Proceedings, Moscow, Russia, 27–30 January 2020.
8. Zhang, P.; Lu, B.; Habetler, T.G. An Active Stator Temperature Estimation Technique for Thermal Protection of Inverter-Fed Induction Motors with Considerations of Impaired Cooling Detection. *IEEE Trans. Ind. Appl.* **2010**, *46*, 1873–1881. [CrossRef]
9. Guo, H.; Ding, Q.; Song, Y.; Tang, H.; Wang, L.; Zhao, J. Predicting Temperature of Permanent Magnet Synchronous Motor Based on Deep Neural Network. *Energies* **2020**, *13*, 4782. [CrossRef]
10. Ganchev, M.; Kral, C.; Wolbank, T.M. Compensation of Speed Dependence in Sensorless Rotor Temperature Estimation for Permanent-Magnet Synchronous Motor. *IEEE Trans. Ind. Appl.* **2013**, *49*, 2487–2495. [CrossRef]
11. Sciascera, C.; Giangrande, P.; Papini, L.; Gerada, C.; Galea, M. Analytical Thermal Model for Fast Stator Winding Temperature Prediction. *IEEE Trans. Ind. Electron.* **2017**, *64*, 6116–6126. [CrossRef]
12. Ganchev, M.; Kral, C.; Oberguggenberger, H.; Wolbank, T. Sensorless Rotor Temperature Estimation of Permanent Magnet Synchronous Motor. In Proceedings of the IECON 2011-37th Annual Conference of the IEEE Industrial Electronics Society, Melbourne, Australia, 7–10 November 2011; pp. 2018–2023.
13. Chen, Y.; Zhang, C.; Zhang, N.; Chen, Y.; Wang, H. Multi-Task Learning and Attention Mechanism Based Long Short-Term Memory for Temperature Prediction of EMU Bearing. In Proceedings of the 2019 Prognostics and System Health Management Conference PHM-Qingdao, Qingdao, China, 25–27 October 2019.
14. Wang, Y.; Li, H.; Ren, D. Characteristics of BLDC Motor Drive System at High Temperature. In Proceedings of the 2016 IEEE 8th International Power Electronics and Motion Control Conference, IPEMC-ECCE Asia 2016, Hefei, China, 22–26 May 2016; pp. 856–862.
15. Jayanth, N.; Varunteja, P.V.N.R.; Sreenu, T.T. Direct Torque Control of Sensorless BLDC Motor Using Artificial Neural Networks. *Int. J. Eng. Adv. Technol.* **2019**, *8*, 1228–1231.
16. Ma, Z.; Qi, X. Permanent Magnet Motor Temperature Compensated Constant Torque Control. *IFAC-PapersOnLine* **2018**, *51*, 68–70. [CrossRef]
17. Ma, Z.; Zhang, Q.; Wang, Q.; Liu, T. Temperature Compensation Strategy of Output Torque for Permanent Magnet Synchronous Motor Based on BP Neural Network. In Proceedings of the 2017 12th IEEE Conference on Industrial Electronics and Applications, ICIEA 2017, Siem Reap, Cambodia, 18–20 June 2017; pp. 774–779.

18. Tan, W.X.; Brandão, M.; Hashimoto, K.; Takanishi, A. Trajectory Optimization for High-Power Robots with Motor Temperature Constraints. In *Lecture Notes in Computer Science (Including Subseries Lecture Notes in Artificial Intelligence and Lecture Notes in Bioinformatics)*; Springer: Cham, Switzerland, 2018; pp. 3–14. [CrossRef]
19. Mroczek, B.; Kolodynska, A. The V2G Process with the Predictive Model. *IEEE Access* **2020**, *8*, 86947–86956. [CrossRef]
20. Zhu, Y.; Xiao, M.; Lu, K.; Wu, Z.; Tao, B. A Simplified Thermal Model and Online Temperature Estimation Method of Permanent Magnet Synchronous Motors. *Appl. Sci.* **2019**, *9*, 3158. [CrossRef]
21. Wallscheid, O.; Böcker, J. Global Identification of a Low-Order Lumped-Parameter Thermal Network for Permanent Magnet Synchronous Motors. *IEEE Trans. Energy Convers.* **2016**, *31*, 354–365. [CrossRef]
22. Demetriades, G.D.; De La Parra, H.Z.; Andersson, E.; Olsson, H. A Real-Time Thermal Model of a Permanent-Magnet Synchronous Motor. *IEEE Trans. Power Electron.* **2010**, *25*, 463–474. [CrossRef]
23. Wu, Y.; Gao, H. Induction-Motor Stator and Rotor Winding Temperature Estimation Using Signal Injection Method. *IEEE Trans. Ind. Appl.* **2006**, *42*, 1038–1044. [CrossRef]
24. Jun, B.-S.; Park, J.S.; Choi, J.-H.; Lee, K.-D.; Won, C.-Y. Temperature Estimation of Stator Winding in Permanent Magnet Synchronous Motors Using D-Axis Current Injection. *Energies* **2018**, *11*, 2033. [CrossRef]
25. Sonnaillon, M.O.; Bisheimer, G.; De Angelo, C.; García, G.O. Online Sensorless Induction Motor Temperature Monitoring. *IEEE Trans. Energy Convers.* **2010**, *25*, 273–280. [CrossRef]
26. Breitenbach, E.; Wunsch, E.; Schramm, A. Thermal Monitoring of Electrical Machines by Linear Differential Equations and Numerical Calibration Procedure. In Proceedings of the 2020 International Symposium on Power Electronics, Electrical Drives, Automation and Motion, SPEEDAM 2020, Sorrento, Italy, 24–26 June 2020; pp. 168–172.
27. Erazo, D.E.G.; Wallscheid, O.; Böcker, J. Improved Fusion of Permanent Magnet Temperature Estimation Techniques for Synchronous Motors Using a Kalman Filter. *IEEE Trans. Ind. Electron.* **2020**, *67*, 1708–1717. [CrossRef]
28. Nogay, H.S. Prediction of Internal Temperature in Stator Winding of Three-Phase Induction Motors with Ann. *Eur. Trans. Electr. Power* **2011**, *21*, 120–128. [CrossRef]
29. Kirchgassner, W.; Wallscheid, O.; Bocker, J. Deep Residual Convolutional and Recurrent Neural Networks for Temperature Estimation in Permanent Magnet Synchronous Motors. In Proceedings of the 2019 IEEE International Electric Machines and Drives Conference, IEMDC 2019, San Diego, CA, USA, 12–15 May 2019; pp. 1439–1446.
30. Lee, J.; Ha, J.-I. Temperature Estimation of PMSM Using a Difference-Estimating Feedforward Neural Network. *IEEE Access* **2020**, *8*, 130855–130865. [CrossRef]
31. Wallscheid, O.; Kirchgassner, W.; Bocker, J. Investigation of Long Short-Term Memory Networks to Temperature Prediction for Permanent Magnet Synchronous Motors. In Proceedings of the International Joint Conference on Neural Networks, Anchorage, AK, USA, 4–19 May 2017; pp. 1940–1947.
32. Mellah, H.; Hemsas, K.E.; Taleb, R. CECATI, carlo Estimation of Speed, Armature Temperature and Resistance in Brushed DC Machines Using a CFNN Based on BFGS BP. *Turk. J. Electr. Eng. Comput. Sci.* **2018**, *26*, 3182–3192. [CrossRef]
33. Mellah, H.; Hemsas, K.E.; Taleb, R. Intelligent Sensor Based Bayesian Neural Network for Combined Parameters and States Estimation of a Brushed DC Motor. *Int. J. Adv. Comput. Sci. Appl.* **2016**, *7*. [CrossRef]
34. Kirchgassner, W.; Wallscheid, O.; Bocker, J. Empirical Evaluation of Exponentially Weighted Moving Averages for Simple Linear Thermal Modeling of Permanent Magnet Synchronous Machines. In Proceedings of the IEEE International Symposium on Industrial Electronics, Vancouver, BC, Canada, 12–14 June 2019; pp. 318–323.
35. Kolano, K. Improved Sensor Control Method for BLDC Motors. *IEEE Access* **2019**, *7*, 186158–186166. [CrossRef]
36. Kolano, K. Determining the Position of the Brushless DC Motor Rotor. *Energies* **2020**, *13*, 1607. [CrossRef]
37. Hastie, T.; Tibshirani, R.; Friedman, J. *The Elements of Statistical Learning: Data Mining, Inference, and Prediction*, 2nd ed.; Springer Series in Statistics; Springer: New York, NY, USA, 2009; ISBN 978-0-387-84857-0.
38. Kingma, D.P.; Ba, J. Adam: A Method for Stochastic Optimization. *arXiv* **2017**, arXiv:1412.6980.
39. Friedman, J.; Hastie, T.; Tibshirani, R. Regularization Paths for Generalized Linear Models via Coordinate Descent. *J. Stat. Softw.* **2010**, *33*, 1–22. [CrossRef]
40. Goodfellow, I.; Bengio, Y.; Courville, A. *Deep Learning*; MIT Press: Cambridge, MA, USA, 2016; ISBN 978-0-262-03561-3.
41. Freund, Y.; Schapire, R.E. A Decision-Theoretic Generalization of On-Line Learning and an Application to Boosting. *J. Comput. Syst. Sci.* **1997**, *55*, 119–139. [CrossRef]
42. Czerwinski, D.; Gęca, J.; Kolano, K. BLDC motor temperaturę. *IEEE Dataport* **2021**. [CrossRef]

sensors

MDPI

Article

Constrained Multiple Planar Reconstruction for Automatic Camera Calibration of Intelligent Vehicles

Sang Jun Lee [1], Jae-Woo Lee [2,*], Wonju Lee [2] and Cheolhun Jang [2]

[1] Division of Electronic Engineering, Jeonbuk National University, 567 Baekje-daero, Deokjin-gu, Jeonju-si 54896, Jeollabuk-do, Korea; sj.lee@jbnu.ac.kr
[2] Samsung Advanced Institute of Technology (SAIT), 130 Samsung-ro, Yeongtong-gu, Suwon-si 16678, Gyeonggi-do, Korea; wonjulee@kaist.ac.kr (W.L.); c_h.jang@samsung.com (C.J.)
* Correspondence: magic0ad@gmail.com

Abstract: In intelligent vehicles, extrinsic camera calibration is preferable to be conducted on a regular basis to deal with unpredictable mechanical changes or variations on weight load distribution. Specifically, high-precision extrinsic parameters between the camera coordinate and the world coordinate are essential to implement high-level functions in intelligent vehicles such as distance estimation and lane departure warning. However, conventional calibration methods, which solve a Perspective-n-Point problem, require laborious work to measure the positions of 3D points in the world coordinate. To reduce this inconvenience, this paper proposes an automatic camera calibration method based on 3D reconstruction. The main contribution of this paper is a novel reconstruction method to recover 3D points on planes perpendicular to the ground. The proposed method jointly optimizes reprojection errors of image features projected from multiple planar surfaces, and finally, it significantly reduces errors in camera extrinsic parameters. Experiments were conducted in synthetic simulation and real calibration environments to demonstrate the effectiveness of the proposed method.

Keywords: computer vision; intelligent vehicles; extrinsic camera calibration; structure from motion; convex optimization

Citation: Lee, S.J.; Lee, J.-W.; Lee, W.; Jang, C. Constrained Multiple Planar Reconstruction for Automatic Camera Calibration of Intelligent Vehicles. *Sensors* **2021**, *21*, 4643. https://doi.org/10.3390/s21144643

Academic Editors: Mark Shortis and Gregorij Kurillo

Received: 1 June 2021
Accepted: 4 July 2021
Published: 6 July 2021

1. Introduction

Recovering the positions of 3D points from 2D-2D correspondences is a fundamental building block in geometric computer vision. This is called triangulation, and it is an essential procedure for many applications including structure-from-motion (SfM) [1–3], simultaneous localization and mapping (SLAM) [4–6], and visual odometry [7,8]. Triangulation is conducted based on displacements between image correspondences obtained from stereo cameras or a moving camera. In an ideal case, back-projected rays from an image correspondence intersect at a point in three dimensional space, and it can be simply formulated by a direct linear transformation. However, in practice, the rays do not necessarily intersect due to measurement noise involved in image features, and these features do not in general satisfy the epipolar geometry [9]. Therefore, recovering 3D information is not a trivial problem even in a two-view case.

A standard approach for addressing the problem of noisy measurements is to estimate 2D corrected correspondences which satisfy the epipolar geometry with the minimum geometric cost [9–11]. These 2D corrected points are maximum likelihood estimates under the assumption of zero-mean isotropic Gaussian noise on the measurements [10]. Triangulation is conducted for the corrected correspondences, and it is equivalent to estimate 3D points which minimize the reprojection error. This procedure is called optimal triangulation. In case that all the 3D points are on a plane, their projected points in two views are associated with a projective transformation which is called homography [12]. Chum et al. proposed a method to find optimal 2D correspondences projected from 3D points on a plane, and it is called optimal planar reconstruction [13]. Kanatani et al. further derived an optimal solution for planar scene triangulation in case that plane and camera parameters are unknown [14]. Planar constraint in general reduces a significant amount of reconstruction error by associating multiple image features for correcting individual 2D measurements. This motivates us to associate image features projected from multiple planar surfaces to improve the precision of 3D reconstruction.

In this paper, we propose a multiple planar reconstruction method which can be applicable in a man-made environment: planes of interest are perpendicular to the ground. It is worth noting that this scenario is quite reasonable in environments for end-of-line calibration and indoor camera calibration. This assumption implies that the normal vectors of planes are coplanar. Whereas previous methods reconstruct individual planes, we seek to jointly optimize the structures of multiple planes simultaneously by introducing coplanarity constraints of their normal vectors.

Another main work of this paper is a novel extrinsic camera calibration method. Among various types of extrinsic parameters, our goal is to estimate extrinsic parameters between the camera coordinate and the world coordinate, which are essential prerequisite for high-level functions in intelligent vehicles such as distance estimation and lane departure warning. Extrinsic parameters consist of relative rotation and translation between the camera and world coordinates, and the world coordinate is also called the vehicle coordinate. In recent intelligent vehicles, due to unpredictable mechanical changes or variations on weight distribution, it is desirable to perform extrinsic camera calibration on a regular basis [15], e.g., at the start of every driving. However, camera calibration requires correspondences between 2D image projections and their 3D points [16] to solve a variant of Perspective-n-Point (PnP) problems [17–21], and the procedure for measuring accurate 3D points in the world coordinate is the biggest bottleneck in conventional methods.

Recently, deep learning methods have been utilized in automatic camera calibration for intelligent vehicles [22–24]. However, many original equipment manufacturers (OEMs) and Tier 1 component companies require conventional computer vision methods to guarantee the safety and reliability of the camera calibration function. Moreover, even OEMs, which employ deep learning techniques in recognition and planning algorithms, parallelly utilize conventional methods to acquire a satisfactory level of functional safety for several essential functions such as camera calibration. This paper proposes an extrinsic camera calibration method to reduce manual work in conventional approaches. The proposed method uses 3D cues in the camera coordinate to infer 3D information in the world coordinate by utilizing 3D points on a planar chessboard which is perpendicular to the ground. The main advantage of the method is that it is able to estimate extrinsic parameters without measuring 3D points in the world coordinate.

The contributions of this paper are twofold: (1) multiple planar reconstruction method to jointly recover 3D structures of multiple planar surfaces and (2) extrinsic camera calibration method based on the reconstructed points. This paper is organized as follows. Section 2 presents the multiple planar reconstruction method, and Section 3 explains the extrinsic camera calibration method based on 3D reconstruction. Section 4 presents experimental results in both synthetic and real environments to demonstrate the effectiveness of the proposed methods.

2. Optimal Multiple-Planar Reconstruction

This section presents the proposed method for joint reconstruction of multiple planar surfaces. We assume that planar chessboards are installed perpendicular to the ground, and a vehicle with a camera moves along the ground with a planar motion. Under these assumptions, the plane normal vectors and camera motion vector are orthogonal to the ground normal vector, and therefore, the plane normal and camera motion vectors are coplanar. Although these assumptions are not easy to satisfy in a road situation, it is worth utilizing the proposed method to improve the precision and robustness of end-of-line and indoor calibrations, which are usually conducted in a man-made environment. In

Figure 1, a synthetic configuration containing four planar boards in three dimensional space is projected onto an image plane by using a camera matrix and arbitrary extrinsic parameters. The camera motion vector is indicated by $\mathbf{m}$, and the tetragons filled with a reddish color and the tetragons filled with 2D points depict the projections of the four planar boards from the first and second views, respectively.

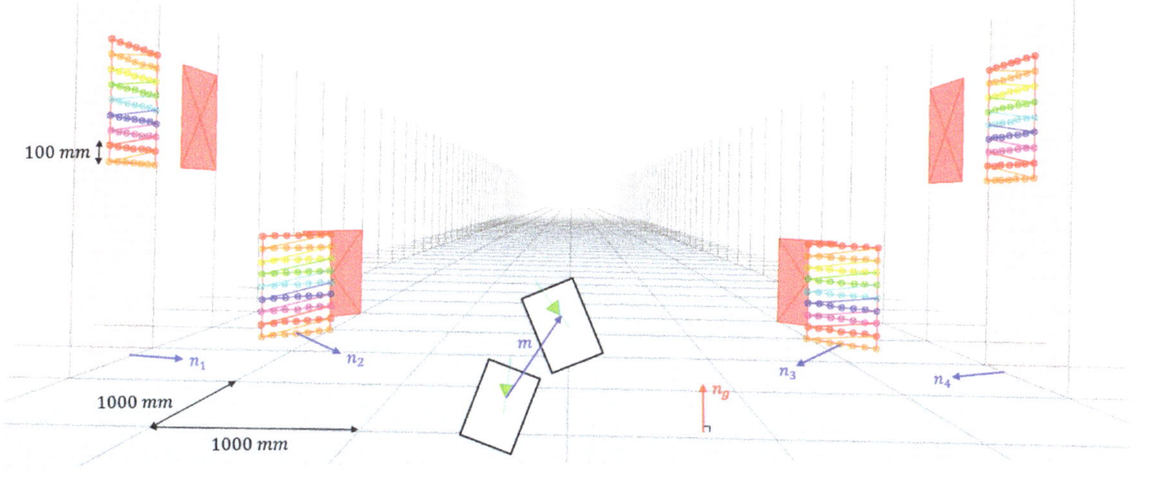

Figure 1. Synthetic simulation environment for multiple planar reconstruction and camera extrinsic calibration.

The outline of the proposed method is as follows. First, normal vectors of the planar chessboards are jointly optimized to minimize the algebraic error of homographies subject to the coplanarity constraints. Then, these homographies are used to correct 2D measurements, and 3D structures of the planar surfaces are recovered via triangulation of the corrected 2D correspondences. Finally, extrinsic parameters are estimated by using the method presented in Section 3.

2.1. Plane Induced Homographies

Suppose that planar surfaces are imaged by a calibrated camera in two views under planar motion of a vehicle. In three dimensional space, the k-th plane can be represented as its scaled normal vector $\mathbf{n}_k$ so that $\mathbf{n}_k^\top \mathbf{X}_C + 1 = 0$, where $\mathbf{X}_C$ is a 3D point in the camera coordinate. Suppose that the essential matrix between the two-view is given by $\mathbf{E} = [\mathbf{t}]_\times \mathbf{R}$, where $\mathbf{R}$ is rotation matrix, $\mathbf{t}$ is translation vector, and $[\cdot]_\times$ is a 3×3 skew-symmetric matrix for representing cross product as a matrix multiplication. Then, it is well-known that a homography induced by the k-th plane can be represented as

$$\mathbf{H}_k = \mathbf{R} - \mathbf{t}\mathbf{n}_k^\top. \quad (1)$$

The essential matrix, $\mathbf{R}$, and $\mathbf{t}$ can be computed via ego-motion estimation. There are several methods to estimate ego-motion of a vehicle, and many previous methods utilize optical flow of background features [25,26].

Suppose that the i-th point on the k-th plane in three dimensional space is projected onto two image planes, and $\mathbf{x}_{i,k}$ and $\mathbf{x}'_{i,k}$ are homogeneous representations of the 2D projections in the first and second views, respectively. Then the 2D correspondence satisfies $\mathbf{x}'_{i,k} \simeq \mathbf{H}_k \mathbf{x}_{i,k} = (\mathbf{R} - \mathbf{t}\mathbf{n}_k^\top)\mathbf{x}_{i,k}$, and it can be reformulated as

$$\mathbf{x}_{i,k}^\top \mathbf{n}_k = \frac{(\mathbf{x}'_{i,k} \times \mathbf{R}\mathbf{x}_{i,k})^\top (\mathbf{x}'_{i,k} \times \mathbf{t})}{(\mathbf{x}'_{i,k} \times \mathbf{t})^\top (\mathbf{x}'_{i,k} \times \mathbf{t})} = b_{i,k}, \quad (2)$$

when $\mathbf{x}'_{i,k}$ and $\mathbf{t}$ are not parallel. The operation $\times$ indicates cross-product between two vectors, and $\simeq$ implies that two vectors are equal up to scale. The scaled normal vector $\mathbf{n}_k$ of the k-th plane can be obtained by solving the unconstrained optimization problem:

$$\underset{\mathbf{n}_k}{\text{minimize}} \quad \|\mathbf{A}_k\mathbf{n}_k - \mathbf{b}_k\|^2, \tag{3}$$

where $\mathbf{b}_k = [b_{1,k}, \cdots, b_{N_k,k}]^\top \in \mathbb{R}^{N_k}$ and $\mathbf{A}_k = [\mathbf{x}_{1,k}, \cdots, \mathbf{x}_{N_k,k}]^\top \in \mathbb{R}^{N_k \times 3}$. Homography optimization in the previous work [14] can be reformulated as a two-step process: optimization of a scaled normal vector by (3) and homography recovery by (1).

2.2. Multiple Planar Reconstruction

This section presents the main idea of the proposed method which introduces coplanarity constraints of plane normal vectors and camera motion vector to jointly reconstruct multiple planar surfaces. In three dimensional space, coplanarity of the normal vectors $\mathbf{n}_i$ $(i = 1, \cdots, K)$ and camera motion $\mathbf{m} = -\mathbf{R}^\top\mathbf{t}$ can be represented as

$$(\mathbf{n}_1 \times \mathbf{n}_k) \cdot \mathbf{m} = 0, \quad k = 2, \cdots, K. \tag{4}$$

To formulate the quadratic constraints in terms of optimization variables, all of the constraints are associated with the camera motion vector $\mathbf{m}$, which is constant in the optimization problem. The number of equations for constraining the coplanarity of K normal vectors and $\mathbf{m}$ is $K(K+1)/2$. However, these equations have redundancy, because, for example, the coplanarities of $(\mathbf{n}_i, \mathbf{n}_j, \mathbf{m})$ and $(\mathbf{n}_i, \mathbf{n}_k, \mathbf{m})$ ensure the coplanarity of $(\mathbf{n}_j, \mathbf{n}_k, \mathbf{m})$. Therefore, the minimum number of equality constraints for the coplanarity of K normal vectors is $K-1$.

Let $\mathbf{w}$ be a concatenated normal vector such that $\mathbf{w} = [\mathbf{n}_1^\top, \cdots, \mathbf{n}_K^\top]^\top$, then the coplanarity (4) can be reformulated as

$$\mathbf{w}^\top\mathbf{C}_k\mathbf{w} = 0, \quad k = 2, \cdots, K, \tag{5}$$

where $\mathbf{C}_k$ is a $3K \times 3K$ symmetric block matrix, of which partitions are 3×3 zero matrices except that $\mathbf{C}_{1k} = [\mathbf{m}]_\times$ and $\mathbf{C}_{k1} = [\mathbf{m}]_\times^\top$; $\mathbf{C}_{ij}$ is a 3×3 submatrix corresponding to the i-th row and j-th column block. By introducing an auxiliary dimension to $\mathbf{w}$ so that $\tilde{\mathbf{w}} = [\mathbf{w}^\top, 1]^\top$, the optimization problem for minimizing the objective function of (3) subject to the coplanarity constraints (4) can be formulated as

$$\begin{aligned} \underset{\tilde{\mathbf{w}}}{\text{minimize}} \quad & \tilde{\mathbf{w}}^\top\mathbf{Q}\tilde{\mathbf{w}} \\ \text{subject to} \quad & \tilde{\mathbf{w}}^\top\tilde{\mathbf{C}}_k\tilde{\mathbf{w}} = 0, \quad k = 2, \cdots, K, \\ & \tilde{\mathbf{w}}^\top\mathbf{C}_0\tilde{\mathbf{w}} = 1, \end{aligned} \tag{6}$$

where $\mathbf{Q} = \begin{bmatrix} \tilde{\mathbf{A}}^\top\tilde{\mathbf{A}} & -\tilde{\mathbf{A}}^\top\mathbf{b} \\ -\mathbf{b}^\top\tilde{\mathbf{A}} & 0 \end{bmatrix}$, $\tilde{\mathbf{C}}_k = \begin{bmatrix} \mathbf{C}_k & \mathbf{0} \\ \mathbf{0}^\top & 0 \end{bmatrix}$, $\mathbf{C}_0 = \begin{bmatrix} \mathbf{0}_{3K} & \mathbf{0} \\ \mathbf{0}^\top & 1 \end{bmatrix}$, $\tilde{\mathbf{A}} \in \mathbb{R}^{(\Sigma_{k=1}^K N_k) \times 3}$ is the block diagonal matrix, of which the k-th diagonal submatrix is $\mathbf{A}_k \in \mathbb{R}^{N_k \times 3}$ and off-diagonal blocks are zero matrices, and $\mathbf{0}_{3K}$ is a $3K \times 3K$ zero matrix. This optimization problem (6) is a quadratically constrained quadratic program (QCQP); the objective is a quadratic function with a positive semidefinite matrix, and the constraints are quadratic with symmetric matrices. Becuase (6) is an NP-hard optimization problem, we reformulate it as a semidefinite program (SDP) by applying the parameterization of $\mathbf{S} = \tilde{\mathbf{w}}\tilde{\mathbf{w}}^\top$ and relaxation of a rank constraint.

2.3. SDP Relaxation

With the parametrization of $\mathbf{S} = \tilde{\mathbf{w}}\tilde{\mathbf{w}}^\top \in S_+$, where S_+ is the set of positive semidefinite matrices, the QCQP (6) can be reformulated in terms of inner products of matrices as

$$\begin{aligned} \underset{\mathbf{S}\in S_+}{\text{minimize}} \quad & \langle \mathbf{Q}, \mathbf{S} \rangle \\ \text{subject to} \quad & \langle \tilde{\mathbf{C}}_k, \mathbf{S} \rangle = 0, \quad k = 2, \cdots, K, \\ & \langle \mathbf{C}_0, \mathbf{S} \rangle = 1, \\ & rank(\mathbf{S}) = 1. \end{aligned} \tag{7}$$

By eliminating the rank constraint in (7), we can obtain the semidefinite relaxation:

$$\begin{aligned} \underset{\mathbf{S}\in S_+}{\text{minimize}} \quad & \langle \mathbf{Q}, \mathbf{S} \rangle \\ \text{subject to} \quad & \langle \tilde{\mathbf{C}}, \mathbf{S} \rangle = 0, \quad k = 2, \cdots, K, \\ & \langle \mathbf{C}_0, \mathbf{S} \rangle = 1. \end{aligned} \tag{8}$$

Becuase the SDP (8) is a convex optimization, we can find the global optimum reliably. Zhao proved the tightness between a primal QCQP and its rank relaxation at noise-free observations, and further showed the stability of rank relaxation at noisy observations [27]. We indeed observe that the solution of rank-relaxed problem (8) always satisfies the rank-1 in both synthetic simulation and real calibration environments.

2.4. Recovering 3D Points

Once the optimal $\mathbf{S}$ of the SDP (8) is obtained, $\tilde{\mathbf{w}}$ can be recovered by computing the eigenvector of $\mathbf{S}$ corresponding to the largest eigenvalue. By utilizing $\tilde{\mathbf{w}}$ and ego-motion of the vehicle, individual homographies are computed by (1). Based on the plane induced homographies, optimal corrections $\hat{\mathbf{x}}_{i,k}$ and $\hat{\mathbf{x}}'_{i,k}$ can be obtained for each 2D measurements $\mathbf{x}_{i,k}$ and $\mathbf{x}'_{i,k}$, by solving a polynomial of degree 8 [13] or by using Sampson's method [28]. The positions of 3D points can be recovered by applying triangulation to the corrected 2D points $\hat{\mathbf{x}}_{i,k}$ and $\hat{\mathbf{x}}'_{i,k}$. Figure 2 shows 3D points on planar surfaces reconstructed by naïve triangulation [28], optimal planar reconstruction [13], and our method.

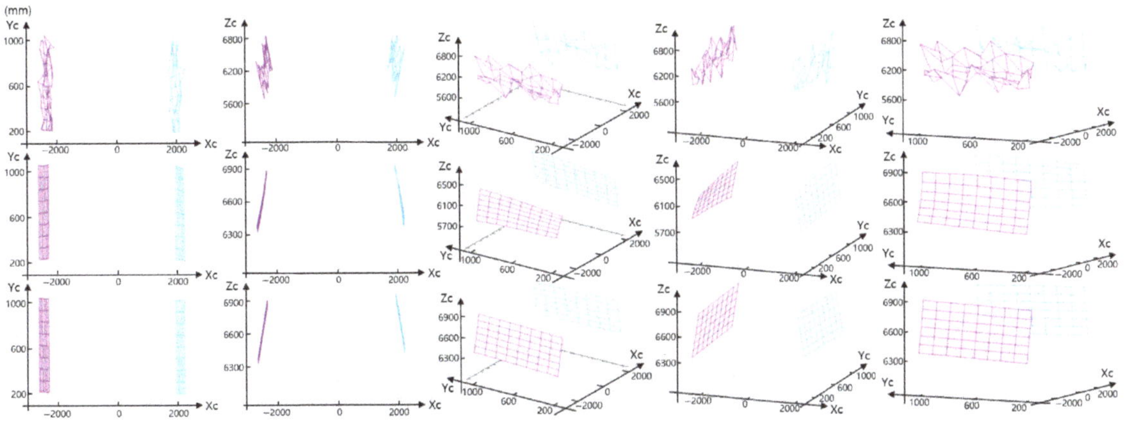

Figure 2. Reconstructed 3D points by using Naïve triangulation (**top** row), optimal planar reconstruction (**middle** row), and proposed reconstruction method (**bottom** row) in different viewpoints.

3. Camera Calibration Based on 3D Reconstruction

3.1. Vehicle Modeling

In this section, we propose a monocular camera calibration method based on 3D reconstruction. Figure 3 shows our vehicle model. Our world coordinate is defined so that its origin is the perpendicular projection of the camera centre to the ground, and the direction of Z_W axis is identical to the normal vector of the world coordinate so that it follows ISO 8855. Under the definition of the world coordinate, fixed values of longitudinal and lateral offsets between the world origin and the foremost point of a vehicle can be compensated at the process of generating signals such as distances to frontal vehicles and time to collision.

The relation between the world and camera coordinates can be formulized in terms of Euler angles (pitch θ, yaw ψ, roll ϕ) and camera height (h) as

$$\mathbf{X}_C = \mathbf{R}(\theta, \phi; \psi)\mathbf{X}_W + \mathbf{t}(\theta, \phi, h; \psi), \tag{9}$$

where $\mathbf{X}_C = [X_C, Y_C, Z_C]^\top$ is a 3D point in the camera coordinate, $\mathbf{X}_W = [X_W, Y_W, Z_W]^\top$ is a 3D point in the world coordinate, and the rotation matrix $\mathbf{R}(\theta, \phi; \psi)$ is defined as (10).

$$R(\theta, \phi; \psi) = \begin{bmatrix} \cos\theta \sin\psi \cos\phi + \sin\theta \sin\phi & -\cos\psi \cos\phi & -\sin\theta \sin\psi \cos\phi + \cos\theta \sin\phi \\ \cos\theta \sin\psi \sin\phi - \sin\theta \cos\phi & -\cos\psi \sin\phi & -\sin\theta \sin\psi \sin\phi - \cos\theta \cos\phi \\ \cos\theta \cos\psi & \sin\psi & -\sin\theta \cos\psi \end{bmatrix}. \tag{10}$$

Since $\mathbf{0} = \mathbf{R}(\theta, \phi; \psi) \cdot [0, 0, h]^\top + \mathbf{t}(\theta, \phi, h; \psi)$, the translation can be represented as

$$\mathbf{t}(\theta, \phi, h; \psi) = -\mathbf{r}_3 h, \tag{11}$$

where $\mathbf{r}_i$ is the i-th column vector of $\mathbf{R}(\theta, \phi; \psi)$.

Suppose that 3D world points of interest are on rectangular planar boards, which are perpendicular to the ground, and their Z_W components (height) are measured beforehand in the world coordinate. Image features projected from these 3D points are detected while a vehicle with a camera moves along the ground, and yaw angle of the camera is estimated by computing a focus of expansion as presented in [29]. The 3D positions corresponding to these image features are recovered in the camera coordinate by using the multiple planar reconstruction method which is explained in Section 2. The objective of the automatic calibration algorithm is to estimate pitch (θ), roll (ϕ), and camera height (h) to recover relative rotation and translation between the camera and world coordinates.

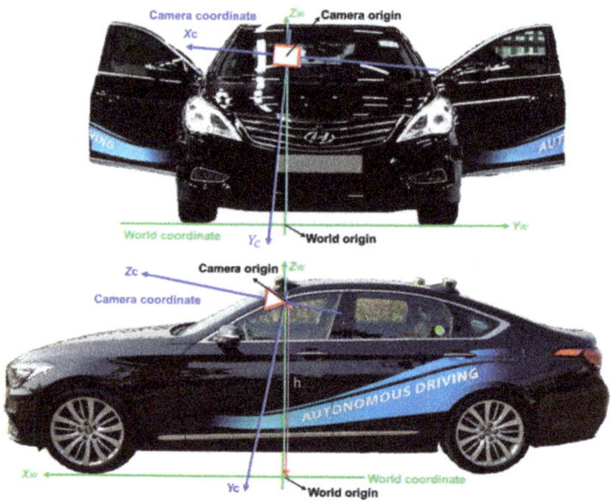

Figure 3. The definition of camera and world coordinates in our vehicle model.

3.2. Estimation of Extrinsic Parameters

Let $\mathbf{X}_W^i$ and $\mathbf{X}_W^j$ be the 3D points on a vertical line which is perpendicular to the ground. Since $X_W^i - X_W^j = 0$ and $Y_W^i - Y_W^j = 0$, component-wise differences between $\mathbf{X}_W^i$ and $\mathbf{X}_W^j$ can be simplified as (12).

$$\begin{aligned} X_C^i - X_C^j &= -(\sin\theta\sin\psi\cos\phi - \cos\theta\sin\phi)(Z_W^i - Z_W^j), \\ Y_C^i - Y_C^j &= -(\sin\theta\sin\psi\sin\phi + \cos\theta\cos\phi)(Z_W^i - Z_W^j), \\ Z_C^i - Z_C^j &= -\sin\theta\cos\psi(Z_W^i - Z_W^j). \end{aligned} \tag{12}$$

Based on (12), pitch angle (θ) can be estimated by

$$\sin\theta = -\frac{Z_C^i - Z_C^j}{\cos\psi(Z_W^i - Z_W^j)}. \tag{13}$$

By solving X_C^i and Y_C^i in terms of $\sin\phi$, roll angle (ϕ) can be estimated by

$$\sin\phi = \frac{\cos\theta(X_C^i - X_C^j) - \sin\theta\sin\psi(Y_C^i - Y_C^j)}{(\sin^2\theta\sin^2\psi + \cos^2\theta)(Z_W^i - Z_W^j)}. \tag{14}$$

After the computation of θ and ϕ, camera height (h) can be obtained by solving the following equation with respect to X_W, Y_W, and h:

$$[\mathbf{r}_1, \mathbf{r}_2, -\mathbf{r}_3][X_W, Y_W, h]^\top = \mathbf{X}_C - \mathbf{r}_3 Z_W. \tag{15}$$

Finally, camera extrinsic parameters can be recovered by using (10) and (11).

4. Experimental Results

The proposed method is composed of constrained multiple planar reconstruction and automatic extrinsic camera calibration. To demonstrate the effectiveness of each method, we synthesized a simulation environment, and the reconstruction and calibration errors were evaluated step by step. In both simulation and real experiments, Naïve triangulation [28] and optimal planar reconstruction method [13] were compared with the proposed method. To analyze the effect of the coplanarity constraint, we evaluate the proposed method with two experimental setups: the coplanarity of two plane normal vectors ($K = 2$) and the coplanarity of four plane normal vectors ($K = 4$). For fairness, we used all of the 3D points on the four planar surfaces in every reconstruction method. For example, in the case of $K = 2$, two SDPs were optimized to use all of the image features projected from the four planar surfaces. The reconstruction and camera height errors were measured in millimetre (mm), and rotation errors were measured in degree.

4.1. Synthetic Environment

To generate a simulation environment, camera extrinsic parameters were randomly sampled under the normal distributions: θ, ψ, $\phi \sim \mathcal{N}(0, 1^2)$ and $h \sim \mathcal{N}(1300, 50^2)$, where $\mathcal{N}$ is normal distribution with a given mean and variance. This synthetic environment reflects the variations of real extrinsic parameters in our vehicle model, and degree and mm units are utilized for representing angles and camera height, respectively. In the simulation environment, known 3D world points on planar surfaces were projected onto two-view images with the size of 1920×1200 by using similar intrinsic parameters to the real case, and Gaussian noise with zero mean and standard deviation of σ was added to the 2D projected image points. To generate the synthetic two-view images, we utilized the vehicle motion when the vehicle moves 1000 mm in forward direction as presented in Figure 4. From the 2D noisy correspondences, reconstruction methods were utilized to recover their 3D points in the camera coordinate, and the proposed calibration method was applied to estimate extrinsic parameters. Each experiment was conducted 100 times, and averaged absolute errors were measured for both reconstructed 3D points and estimated extrinsic parameters.

To evaluate reconstruction accuracy, root-mean-square errors between 3D estimates and their true positions were measured in three dimensional space. Table 1 presents reconstruction errors with respect to various amounts of noise on 2D image projections; the standard deviation σ of the Gaussian pixel-noise was varied from 0.1 to 3.0. We present two cases of simulation results: In one case, virtual planar boards are located at the longitudinal distance of around 8 m from the camera at the first view, and in the other case, those are located around 10 m. With an identical amount of pixel-noise, reconstruction error increases as the distance to the planar boards increases. Although reconstruction error increases as the amount of pixel-noise increases, the proposed reconstruction method consistently shows higher accuracies compared to the other methods. Furthermore, Table 1 demonstrates that joint optimization of one SDP for the four planes is more advantageous than separate optimization of two independent SDPs for upper two planes and lower two planes. It is because normal vectors of planes in upper and lower groups are not associated with a coplanarity constraint in the case of $K = 2$. This result implies that joint reconstruction of entire planar surfaces is effective to reduce the reconstruction error.

Table 1. Reconstruction errors (mm) with respect to various amounts of pixel noise.

Method	Distance	σ 0.1	0.2	0.3	0.5	0.7	1.0	1.5	2.0	3.0
Naïve triangulation	8 m	15.7	30.8	46.6	78.0	110.4	156.0	236.6	318.2	474.8
Optimal triangulation		15.7	30.8	46.7	78.0	110.4	156.0	236.5	317.9	473.8
Planar reconstruction		3.7	7.3	11.4	17.9	26.5	37.7	58.0	77.3	119.5
Propoased method (K = 2)		3.2	6.2	9.8	15.2	22.4	32.4	51.2	70.6	110.5
Propoased method (K = 4)		3.0	5.8	9.1	14.1	20.5	30.1	48.1	67.9	106.3
Naïve triangulation	10 m	30.1	60.7	90.3	152.4	210.6	301.0	455.4	610.2	938.5
Optimal triangulation		30.1	60.7	90.3	152.4	210.5	300.9	455.1	609.1	935.4
Planar reconstruction		6.9	13.9	21.0	34.7	49.3	70.2	108.8	151.9	252.2
Propoased method (K = 2)		6.0	11.6	18.3	30.0	43.0	62.3	99.4	140.8	240.1
Propoased method (K = 4)		5.6	10.9	17.2	28.0	40.1	60.6	95.8	138.0	237.4

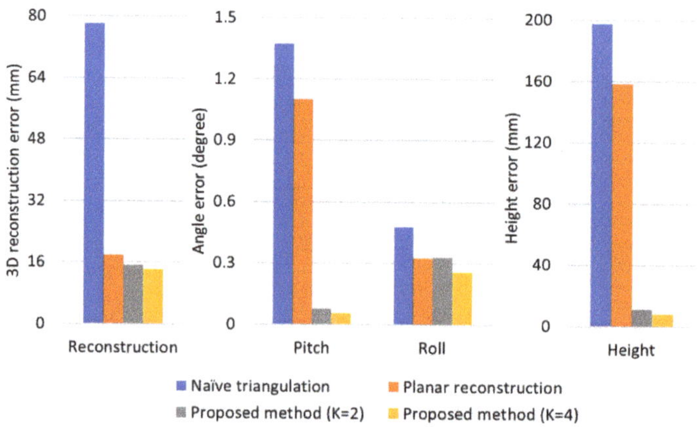

Figure 4. Reconstruction and calibration errors in simulation experiments.

Figure 4 shows the reconstruction and calibration errors in the case that distances to targets were around 8 m and the standard deviation of pixel noise was $\sigma = 0.5$. By reducing the reconstruction error, calibration error was significantly decreased especially for pitch angle and camera height. In the results of the proposed method, the reconstruction and calibration errors of $K = 4$ case were lower than those of $K = 2$ case. It implies that increasing the number of planes was beneficial to reduce the amount of errors. However, it was not practical to setup more than four planes in real experiments, because 2D image features projected from planes which were located far from the vehicle caused a large amount of pixel noise. Therefore, we utilized four planar surfaces for extrinsic calibration in real experiments.

In the proposed reconstruction method, ego-motion was assumed as a planar motion to formulate a coplanarity constraint with plane normal vectors. To analyze the effect of vehicle motion noise to the performance of the proposed method, we conducted simulation experiments with and without vehicle motion noise. The motion noise was modeled as a Gaussian distribution, and we utilized the standard deviation of ego-motions measured in real driving scenarios to generate Gaussian motion noise in the simulation environment. Table 2 presents calibration accuracies with and without vehicle motion noise under various amounts of pixel noise. Although calibration errors were increased by the ego-motion noise, experimental results show that the proposed method was robust compared to previous methods even under the motion noise.

Table 2. Calibration accuracy in simulation environment with respect to various amounts of pixel noise.

Method		$\sigma = 0.3$			$\sigma = 0.5$			$\sigma = 0.7$			$\sigma = 1.0$	
	Pitch	**Roll**	**Height**	**Pitch**	**Roll**	**Height**	**Pitch**	**Roll**	**Height**	**Pitch**	**Roll**	**Height**
					Without camera-motion noise							
Naïve triangulation	0.9829	0.2859	141.3	1.3705	0.4759	197.3	2.4338	0.9065	349.8	3.3727	2.2364	483.3
Planar reconstruction	0.6383	0.2017	91.7	1.0972	0.3225	158.0	1.6317	0.3994	234.8	2.2217	0.7261	319.5
Propoased method (K = 2)	0.0372	0.2048	5.3	0.0751	0.3295	10.9	0.1647	0.4021	23.6	0.3394	0.6990	48.8
Propoased method (K = 4)	0.0312	0.1587	4.5	0.0557	0.2549	8.0	0.0683	0.2890	9.8	0.1330	0.5153	19.1
					With camera-motion noise							
Naïve triangulation	3.0058	0.3214	432.7	3.4568	0.5304	498.9	4.5159	0.9182	646.8	4.9208	1.7227	704.0
Planar reconstruction	2.8668	0.2067	412.8	2.9781	0.3346	429.9	3.8155	0.4555	547.0	4.0049	0.6995	574.8
Propoased method (K = 2)	0.0703	0.2152	10.2	0.1290	0.3486	18.3	0.2741	0.4797	39.4	0.4505	0.7181	65.2
Propoased method (K = 4)	0.0587	0.1501	9.2	0.1076	0.2538	15.1	0.2200	0.3271	31.7	0.3097	0.5023	44.2

4.2. Real Calibration Environment

This section presents experimental results in a real calibration environment to demonstrate the effectiveness of the proposed method. In our garage, chessboards were installed so that they are perpendicular to the ground as shown in Figure 5. While a vehicle moved smoothly, images were collected with the size of 1920 × 1200 by utilizing an in-vehicle frontal camera, FLIR Point Grey Grasshopper 3. Intrinsic parameters of the camera were computed in advance by using the method presented in [30]. Background features were extracted and tracked by grid-based feature detection and Lucas–Kanade method [31], and the essential matrix was computed by the five-point method [32] with RANSAC [33] to estimate camera motion. The correspondences of chessboard features were independently detected, and yaw angle of the camera with respect to moving direction was calculated based on focus of expansion, which was computed from the chessboard features. The multiple planar reconstruction method was applied to recover 3D structures of chessboard features, and finally, camera extrinsic parameters were estimated by using the proposed calibration method. This calibration process was performed multiple times while a vehicle was passing the chessboards, and these estimates were averaged to compute a final calibration parameters. In our experiment, the vehicle moved about 5 km/h to obtain enough number of image pairs, and the averaged values of 10 estimates were utilized as final extrinsic parameters. The number of calibration trials could be affected by vehicle speed, field of view of the camera, and distances between chessboards.

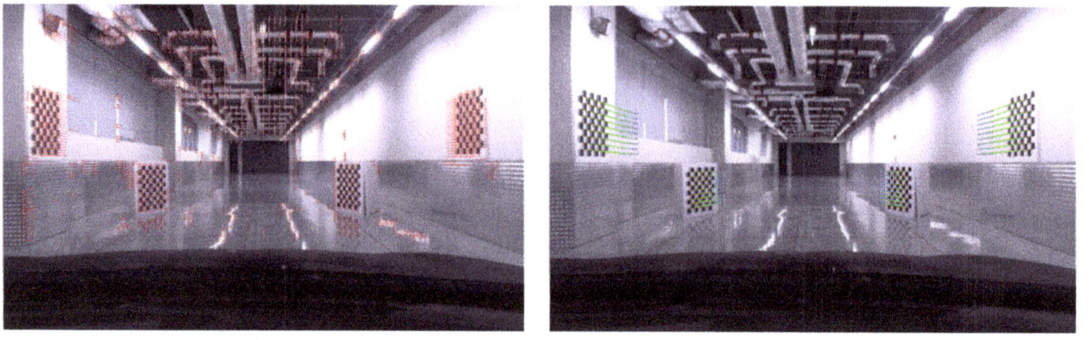

(a) (b)

Figure 5. Real calibration environment: (**a**) optical flow of background features for estimating ego-motion of the vehicle. (**b**) Optical flow of chessboard features for 3d reconstruction.

To evaluate the accuracy of the proposed method, we collected reference values of extrinsic parameters from an identical experiment environment. In the procedure for generating reference parameters, we manually measured 3D locations of multiple feature points with respect to the world coordinate using a laser range finder, and corresponding 2D projections in the image domain were labelled. 2D-3D correspondences were used to solve a Perspective-n-Point (PnP) problem to compute extrinsic parameters. All the procedures took around 30 min, and it was repeated eight times to obtain averaged extrinsic parameters; the reference values for camera height, pitch angle, roll angle are 1195.48 mm, 0.2413 degree, 0.3663 degree, respectively. In the real experiment, we measured absolute errors between the reference parameters and estimated extrinsic parameters.

To demonstrate the effectiveness of the proposed method, we conducted experiments with four different reconstruction methods: Naïve triangulation [28] and optimal planar reconstruction method [13], and the proposed constrained multiple planar reconstructions ($K = 2$ and $K = 4$). Calibration errors in the real calibration scenario are presented in Figure 6, and the proposed reconstruction method gives much lower calibration errors compared to the conventional methods. Similar to experimental results in synthetic simulation, calibration accuracy was improved by utilizing a greater number of planar surfaces in a SDP. Compared to the previous planar reconstruction method [13], height error of the proposed method ($K = 4$) was reduced from 110.1 mm to 23.9 mm, and pitch angle and roll angle errors were reduced from 0.2764 degree to 0.0470 degree and from 1.1098 degree to 0.0859 degree, respectively; about 78% and 87% of height and angle errors were reduced by using the coplanarity constraint. Because angle errors less than 0.1 degree and height error less than 30 mm were not significant to perform high-level functions such as distance estimation and lane departure warning, the proposed method was able to be utilized in intelligent vehicle industries for computing extrinsic parameters between the camera coordinate and the world coordinate.

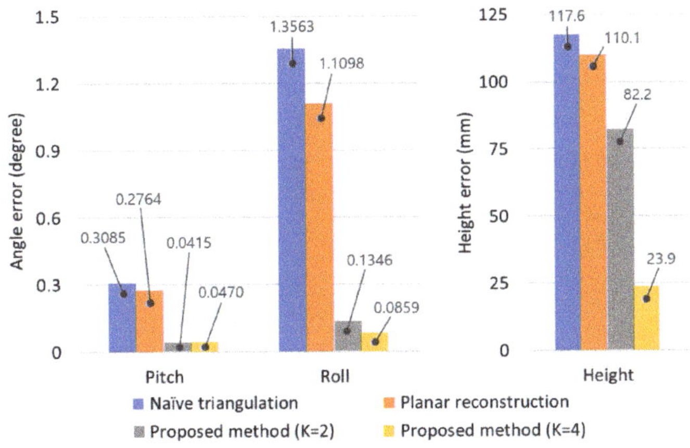

Figure 6. Calibration errors in real calibration scenario.

5. Conclusions

In this paper, we propose a method for automatic camera calibration of intelligent vehicles. The proposed method is based on 3D reconstruction of a man-made environment, and the key contribution of this paper is novel multiple planar reconstruction method to reduce errors in camera extrinsic parameters. We first formulate a QCQP with the coplanarity constraints between plane normal vectors and camera motion vector. The QCQP is reformulated into an SDP, and the optimal solution is obtained using rank-1 relaxation. From the optimal solution of the relaxed SDP, normal vectors are computed for 3D reconstruction of planar surfaces.

We also propose a method to compute camera extrinsic parameters by utilizing planar surfaces which are perpendicular to the ground. This man-made environment is quite reasonable for end-of-line calibration and indoor camera calibration. Main benefit of the proposed method is that it does not require 3D measurements of image features, and thus, extrinsic calibration can be conducted automatically at the start of every driving. In both synthetic simulation and real calibration environment, the proposed reconstruction method significantly outperformed the previous 3D reconstruction methods, and thus errors in extrinsic parameters were dramatically reduced.

Author Contributions: Conceptualization, S.J.L.; methodology, S.J.L., J.-W.L. and W.L.; software, S.J.L. and J.-W.L.; validation, J.-W.L.; formal analysis, S.J.L. and W.L.; data curation, J.-W.L.; writing—original draft preparation, S.J.L.; writing—review and editing, J.-W.L. and W.L.; visualization, C.J. All authors have read and agreed to the published version of the manuscript.

Funding: This work was supported by the National Research Foundation of Korea (NRF) grant funded by the Korea government (MSIT) (No. 2021R1G1A1009792). This paper was supported by research funds for newly appointed professors of Jeonbuk National University in 2020.

Institutional Review Board Statement: Not applicable.

Informed Consent Statement: Not applicable.

Data Availability Statement: Not applicable.

Acknowledgments: This work was supported by the National Research Foundation of Korea (NRF) grant funded by the Korea government (MSIT) (No. 2021R1G1A1009792). This paper was supported by research funds for newly appointed professors of Jeonbuk National University in 2020. We deeply appreciate Daehyun Ji (SAIT) and Jahoo Koo (SAIT) for their technical support and project management.

Conflicts of Interest: The authors declare no conflict of interest.

References

1. Song, S.; Chandraker, M. Robust scale estimation in real-time monocular SFM for autonomous driving. In Proceedings of the IEEE Conference on Computer Vision and Pattern Recognition, Columbus, OH, USA, 23–28 June 2014; pp. 1566–1573. [CrossRef]
2. Schonberger, J.L.; Frahm, J.M. Structure-from-motion revisited. In Proceedings of the IEEE Conference on Computer Vision and Pattern Recognition, Las Vegas, NV, USA, 27–30 June 2016; pp. 4104–4113. [CrossRef]
3. Zhou, Z.; Jin, H.; Ma, Y. Robust plane-based structure from motion. In Proceedings of the IEEE Conference on Computer Vision and Pattern Recognition, Providence, RI, USA, 16–21 June 2012; pp. 1482–1489. [CrossRef]
4. Davison, A.J.; Reid, I.D.; Molton, N.D.; Stasse, O. MonoSLAM: Real-time single camera SLAM. *IEEE Trans. Pattern Anal. Mach. Intell.* **2007**, *29*, 1052–1067. [CrossRef] [PubMed]
5. Mur-Artal, R.; Montiel, J.M.M.; Tardos, J.D. ORB-SLAM: A versatile and accurate monocular SLAM system. *IEEE Trans. Robot.* **2015**, *31*, 1147–1163. [CrossRef]
6. Bustos, A.P.; Chin, T.J.; Eriksson, A.; Reid, I. Visual SLAM: Why Bundle Adjust? In Proceedings of the 2019 International Conference on Robotics and Automation (ICRA), Montreal, QC, Canada, 20–24 May 2019; pp. 2385–2391. [CrossRef]
7. Mouragnon, E.; Lhuillier, M.; Dhome, M.; Dekeyser, F.; Sayd, P. Real time localization and 3d reconstruction. In Proceedings of the IEEE Conference on Computer Vision and Pattern Recognition, New York, NY, USA, 17–22 June 2006; pp. 363–370. [CrossRef]
8. Li, Y.; Ushiku, Y.; Harada, T. Pose Graph Optimization for Unsupervised Monocular Visual Odometry. In Proceedings of the 2019 International Conference on Robotics and Automation (ICRA), Montreal, QC, Canada, 20–24 May 2019; pp. 5439–5445. [CrossRef]
9. Hartley, R.I.; Sturm, P. Triangulation. *Comput. Vis. Image Underst.* **1997**, *68*, 146–157. [CrossRef]
10. Kanatani, K.; Sugaya, Y.; Niitsuma, H. Triangulation from two views revisited: Hartley-Sturm vs. optimal correction. In Proceedings of the British Machine Vision Conference, Leeds, UK, 1–4 September 2008; pp. 18.1–18.10. [CrossRef]
11. Lindstrom, P. Triangulation made easy. In Proceedings of the IEEE Conference on Computer Vision and Pattern Recognition, San Francisco, CA, USA, 13–18 June 2010; pp. 1554–1561. [CrossRef]
12. Kanatani, K.; Sugaya, Y.; Kanazawa, Y. *Guide to 3D Vision Computation*; Springer: Berlin/Heidelberg, Germany, 2016.
13. Chum, O.; Pajdla, T.; Sturm, P. The geometric error for homographies. *Comput. Vis. Image Underst.* **2005**, *97*, 86–102. [CrossRef]
14. Kanatani, K.; Niitsuma, H. Optimal two-view planar scene triangulation. *IPSJ Trans. Comput. Vis. Appl.* **2011**, *3*, 67–79. [CrossRef]
15. Miksch, M.; Yang, B.; Zimmermann, K. Automatic extrinsic camera self-calibration based on homography and epipolar geometry. In Proceedings of the 2010 IEEE Intelligent Vehicles Symposium, La Jolla, CA, USA, 21–24 June 2010; pp. 832–839. [CrossRef]
16. Bazargani, H.; Laganière, R. Camera calibration and pose estimation from planes. *IEEE Instrum. Meas. Mag.* **2015**, *18*, 20–27. [CrossRef]
17. Haralick, R.M.; Lee, C.N.; Ottenburg, K.; Nölle, M. Analysis and solutions of the three point perspective pose estimation problem. In Proceedings of the IEEE Conference on Computer Vision and Pattern Recognition, Maui, HI, USA, 3–6 June 1991; pp. 592–598. [CrossRef]
18. Horaud, R.; Conio, B.; Leboulleux, O.; Lacolle, B. An analytic solution for the perspective 4-point problem. *Comput. Vis. Graph. Image Process.* **1989**, *47*, 33–44. [CrossRef]
19. Quan, L.; Lan, Z. Linear n-point camera pose determination. *IEEE Trans. Pattern Anal. Mach. Intell.* **1999**, *21*, 774–780. [CrossRef]
20. Gao, X.S.; Hou, X.R.; Tang, J.; Cheng, H.F. Complete solution classification for the perspective-three-point problem. *IEEE Trans. Pattern Anal. Mach. Intell.* **2003**, *25*, 930–943. [CrossRef]
21. Lepetit, V.; Moreno-Noguer, F.; Fua, P. Epnp: An accurate o(n) solution to the pnp problem. *Int. J. Comput. Vis.* **2009**, *81*, 155–160. [CrossRef]
22. Lee, J.; Sung, M.; Lee, H.; Kim, J. Neural Geometric Parser for Single Image Camera Calibration. In Proceedings of the European Conference on Computer Vision, Glasgow, UK, 23–28 August 2020; pp. 541–557. [CrossRef]
23. Itu, R.; Borza, D.; Danescu, R. Automatic extrinsic camera parameters calibration using Convolutional Neural Networks. In Proceedings of the 2017 13th IEEE International Conference on Intelligent Computer Communication and Processing (ICCP), Cluj-Napoca, Romania, 7–9 September 2017; pp. 273–278. [CrossRef]
24. Antunes, M.; Barreto, J.P.; Aouada, D.; Ottersten, B. Unsupervised vanishing point detection and camera calibration from a single manhattan image with radial distortion. In Proceedings of the IEEE Conference on Computer Vision and Pattern Recognition, Honolulu, HI, USA, 21–26 July 2017; pp. 4288–4296. [CrossRef]
25. Wang, X.; Ban, K.; Ishii, K. Estimation of mobile robot ego-motion and obstacle depth detection by using optical flow. In Proceedings of the IEEE International Conference on Systems, Man and Cybernetics, San Antonio, TX, USA, 11–14 October 2009; pp. 1770–1775. [CrossRef]
26. Yamaguchi, K.; Kato, T.; Ninomiya, Y. Vehicle ego-motion estimation and moving object detection using a monocular camera. In Proceedings of the 18th International Conference on Pattern Recognition (ICPR), Hong Kong, China, 20–24 August 2006; pp. 610–613. [CrossRef]
27. Zhao, J. An efficient solution to non-minimal case essential matrix estimation. *IEEE Trans. Pattern Anal. Mach. Intell.* **2020**. [CrossRef] [PubMed]
28. Hartley, R.; Zisserman, A. *Multiple View Geometry in Computer Vision*; Cambridge Univertity Press: Cambridge, UK, 2003.
29. Kolupaev, A.V.; Metelyov, A.P.; Prozorov, D.E.; Kurbatova, E.E.; Kharina, N.L. Speed increasing of FOE calculation in autonomous vehicle control systems. In Proceedings of the 2018 IEEE East-West Design & Test Symposium (EWDTS), Kazan, Russia, 14–17 September 2018; pp. 1–4. [CrossRef]

30. Zhang, Z. A flexible new technique for camera calibration. *IEEE Trans. Pattern Anal. Mach. Intell.* **2000**, *22*, 1330–1334. [CrossRef]
31. Lucas, B.D.; Kanade, T. An iterative image registration technique with an application to stereo vision. In Proceedings of the 7th International Joint Conference on Artificial Intelligence, Vancouver, BC, Canada, 24–28 August 1981; pp. 121–130. Available online: https://www.researchgate.net/publication/215458777 (accessed on 6 July 2021).
32. Nistér, D. An efficient solution to the five-point relative pose problem. *IEEE Trans. Pattern Anal. Mach. Intell.* **2004**, *26*, 756–770. [CrossRef] [PubMed]
33. Fischler, M.A.; Bolles, R.C. Random sample consensus: A paradigm for model fitting with applications to image analysis and automated cartography. *Commun. ACM* **1981**, *24*, 381–395. [CrossRef]

sensors

MDPI

Article

Piston Error Measurement for Segmented Telescopes with an Artificial Neural Network

Dan Yue *, Yihao He and Yushuang Li

College of Science, Changchun University of Science and Technology, Changchun 130022, China; 2019100133@mails.cust.edu.cn (Y.H.); 2020100186@mails.cust.edu.cn (Y.L.)
* Correspondence: yuedan@cust.edu.cn; Tel.:+86-158-4314-9856

Abstract: A piston error detection method is proposed based on the broadband intensity distribution on the image plane using a back-propagation (BP) artificial neural network. By setting a mask with a sparse circular clear multi-subaperture configuration in the exit pupil plane of a segmented telescope to fragment the pupil, the relation between the piston error of segments and amplitude of the modulation transfer function (MTF) sidelobes is strictly derived according to the Fourier optics principle. Then the BP artificial neural network is utilized to establish the mapping relation between them, where the amplitudes of the MTF sidelobes directly calculated from theoretical relationship and the introduced piston errors are used as inputs and outputs respectively to train the network. With the well trained-network, the piston errors are measured to a good precision using one in-focused broadband image without defocus division as input, and the capture range achieving the coherence length of the broadband light is available. Adequate simulations demonstrate the effectiveness and accuracy of the proposed method; the results show that the trained network has high measurement accuracy, wide detection range, quite good noise immunity and generalization ability. This method provides a feasible and easily implemented way to measure piston error and can simultaneously detect the multiple piston errors of the entire aperture of the segmented telescope.

Keywords: piston error detection; segmented telescope; BP artificial neural network; modulation transfer function

Citation: Yue, D.; He, Y.; Li, Y. Piston Error Measurement for Segmented Telescopes with an Artificial Neural Network. *Sensors* **2021**, *21*, 3364. https://doi.org/10.3390/s21103364

Academic Editor: Hyoungsik Nam

Received: 19 March 2021
Accepted: 7 May 2021
Published: 12 May 2021

1. Introduction

To fulfill the demands of increasing space exploration, the aperture diameter of the telescope is getting larger and larger [1]. However, the diameter is limited by current technology to about 10 m for ground-based telescopes and a few meters for space-based telescopes because of volume and mass considerations. The segmented and deployable primary mirror was proposed to address this problem, which can effectively reduce the mass of the main mirror and the difficulty of manufacturing and transportation [2]. Ground-based telescopes that have adopted the segmented primary mirror include the already existing 10 m aperture Keck telescopes [3,4] made of 36 segments, the 42 m diameter European Extremely Large Telescope (E-ELT) [5] made of 984 segments and the Thirty Meter Telescope (TMT) made of 492 segments with a 30 m diameter [6], which are planned in the near future. Space segmented telescope projects aim at astronomical or earth observations, such as the 6.5 m James Webb Space Telescope (JWST) [7] composed of 18 segments, to be launched very soon, and the 9.2 m Advanced Technology Large Aperture Space Telescope (ATLAST) [8] composed of 36 segments for a next-generation flagship astrophysics mission to study the universe.

However, segmented-mirror technology introduced new problems. The optical path difference (OPD) between segments, which can be described by the first three Zernike polynomials (piston and tip-tilt), must be reduced to a small fraction of the wavelength to achieve a high-resolution equivalent to that of a monolithic mirror [9]. The crucial point is correcting the piston errors, which cannot be directly detected by traditional wavefront sensors like Shack–Hartmann. The research shows that, when the piston error between segments is reduced to about $\lambda/20$, the angle resolution is $1.22\lambda/D$, where D is the diameter of the primary mirror and λ is the observation wavelength. If the piston error is larger than λ, the angle resolution decreases to $1.22\lambda/d$, where d is the diameter of the sub-mirror. For the JWST, which is composed of 18 1.32 m sub-mirrors, the angular resolution difference can reach five times. Thus, in order to guarantee the imaging quality of the segmented telescope, the piston error between the segments must be accurately measured, and the measurement accuracy should be 30~40 nm at least. To achieve this, the detection method of co-phasing the piston error should meet the requirements of large range and high precision.

Currently, many piston error detection methods have been successfully applied to the segmented telescope. The modified Shack–Hartmann wavefront detection method has been successfully used on Keck telescopes. It measures the piston error by installing a Hartmann micro-lens array at the exit pupil. The reliable capture range of the narrow Shack–Hartmann method [10] is $\pm\lambda/4$ while the measurement accuracy is as high as 6 nm. The broadband Shack–Hartmann algorithm [11] has a wider measurement range and generally reaches 30 μm, while the measurement accuracy is around 30 nm. The quadrilateral pyramid detection method [12,13], which utilizes a quadrangular pyramid mirror, relay lens and CCD camera, can achieve several tens of nanometers' detection accuracy in the $\pm\lambda/2$ detection range. The dispersed fringe sensor (DFS) [14,15] is a new piston error detection technique proposed for the next-generation space telescope. The DFS is composed of a prism, a micro-lens array and a CCD camera, which can quickly and accurately detect the piston error from a few microns to tens of microns. All above methods need optical devices which add to the detection hardware complexity. Focal-plane wavefront sensing is an elegant solution to measure the OPD since this wavefront sensor is included in the main imaging detector, simplifying the hardware and minimizing differential paths. The classic focal-plane detection methods, such as phase diversity [16–18] and phase retrieval [19,20], have high measurement accuracy but narrow detection range and are very time-consuming. In 2015, Simar [21] found out that the modulation transfer function (MTF) amplitude part of the optical transfer function (OTF) has a relationship with the piston error based on the coherence measurement of a star image when using broadband input light, using Gaussian function to fit the relationship between the MTF and the piston error. The capture range of this method is close to the coherence length of the input light, but its accuracy decreases sharply when approaching the extremes of capture range. Junlun Jiang et al. [22] utilized piecewise quartic polynomial function to fit the relationship; the accuracy of this method is higher than that of Gaussian fitting but the complexity of curve fitting is increased since different mathematical expressions are needed in different intervals of the capture range.

This paper proposes a different method to detect the piston error based on analyzing the broadband intensity distribution on the image plane of a star image with a back-propagation (BP) artificial neural network. By attaching a mask with a sparse multi-subaperture configuration in the exit pupil plane of the segmented telescope, we can get the non-redundant MTF sidelobes distribution. Then, based on Fourier optics principle [23], the theoretical relation between peak heights of MTF sidelobes and piston error is rigorously derived, and we have obtained a more universally applicable and accurate theoretical conclusion than the work presented in paper [22], which clearly indicates that the peak height of the MTF sidelobe is only related to the number of sub-pupils, input wavelength and the piston error between segments. Instead of fitting the relation by Gaussian function or two piecewise quartic polynomials, we utilize the BP artificial neural network to establish the nonlinear mapping relationship between the piston error and the amplitudes of the MTF's surrounding peaks. The BP artificial neural network [24,25] is simple in structure and very convenient to train. Here, the peak heights of MTF sidelobes directly calculated from the obtained theoretical formula served as the inputs and the introduced piston errors served as outputs, respectively, to train the artificial neural network. Once well trained, it can estimate the piston error to a good precision with high efficiency and robustness using the PSF image collected from the optics imaging system as input. The detection accuracy is both higher than Gaussian function fitting and piecewise quartic polynomial function fitting, and the implementation complexity is lower than piecewise quartic polynomial function fitting since there is no need to build different computation models in different intervals of the capture range. The capture range is close to the coherence length of the input light; thus the piston error detection no longer needs to be divided into coarse and fine regimes which involve separate dedicated hardware solutions. The influence of the CCD camera noise and the existence of tip-tilt aberrations on the accuracy of piston error detection are discussed. The results indicate that the proposed piston error detection algorithm has very good noise immunity, but the existence of tip-tilt aberrations has a great influence on the detection accuracy. Besides, the generalization ability of the network is also discussed by changing the F-number ($F^{\#}$) of the imaging system to generate different intensity images as testing sets to examine the trained network. The results show that our network has quite good generalization ability, since the datasets used to train the network are the same when the optical systems have the same number of sub-pupils and working wavelength. Moreover, by matching up the MTF sidelobes with their associated sub-pupils, multiple piston error measurements of the whole aperture can be implemented simultaneously by one detection of a single broadband image.

This paper is structured as follows. In Section 2, we derive the theoretical relation between the peak heights of MTF sidelobes and piston error, and verify the correctness of the theoretical relation by MATLAB simulation. Section 3 describes how to use the BP artificial neural network to establish the nonlinear mapping relationship between the amplitudes of the MTF sidelobes and the piston error, and how to implement the established network to detect piston errors. Adequate simulation validations and discussions on the proposed approach are presented in Section 4. In Section 5, we conclude the paper.

2. Theoretical Relation between the Piston Error and Amplitudes of MTF Sidelobes

In this part, we mainly focus on establishing the theoretical relationship between the amplitudes of the MTF sidelobes and the piston error based on the Fourier optics principle at first. Then, we verify the correctness of the established theoretical relation through MATLAB simulation.

2.1. Establishing the Theoretical Relationship Based on Fourier Optics Principle

In order to simplify the theoretical derivation process, we used a primary mirror composed of two hexagonal segments as the optics system model shown in Figure 1. In order to separate the sidelobes of the MTF from the main peak, a mask with two circles was set on the exit-pupil plane of the primary mirror to fragment the pupil. If p is the piston error between the two segments, the generalized pupil function (GPF) can be written as:

$$G(x,y) = A(x,y)\left[circ\left(\frac{x-b/2,y}{d/2}\right)\cdot e^{i\phi_1} + circ\left(\frac{x+b/2,y}{d/2}\right)\cdot e^{i\phi_2}\right], \tag{1}$$

where (x,y) is the coordinate vector of the pupil plane, b is the distance between the center of the two circle pupils on the mask, d is the diameter of the circle pupil, λ is the observation wavelength, $circ()$ stands for circle function, the phase difference between the two segments is $\Delta\phi = \phi_1 - \phi_2 = \frac{2\pi}{\lambda}2p$, here we set $\phi_1 = \frac{2\pi}{\lambda}p$ and $\phi_2 = -\frac{2\pi}{\lambda}p$ to facilitate the following calculations. $A(x,y)$ is the binary shape function of the hexagon segment which is shown as:

$$A(x,y) = \begin{cases} 1 & inside\ \ the\ \ pupil \\ 0 & others \end{cases}. \tag{2}$$

We performed Fourier transform for the GPF, and based on the properties of Fourier transform we can derive the following equation:

$$\Im\{G(x,y)\} = \frac{d}{2}\left[\frac{J_1(\pi d\sqrt{u^2+v^2})}{\sqrt{u^2+v^2}}\right] \cdot \left(e^{i(\frac{\pi}{\lambda}2p-\pi ub)} + e^{-i(\frac{\pi}{\lambda}2p-\pi ub)}\right), \tag{3}$$

where (u,v) is the coordinate vector of the image plane, $\Im\{\}$ stands for Fourier transform operation, $J_1\{\}$ is first order Bessel function.

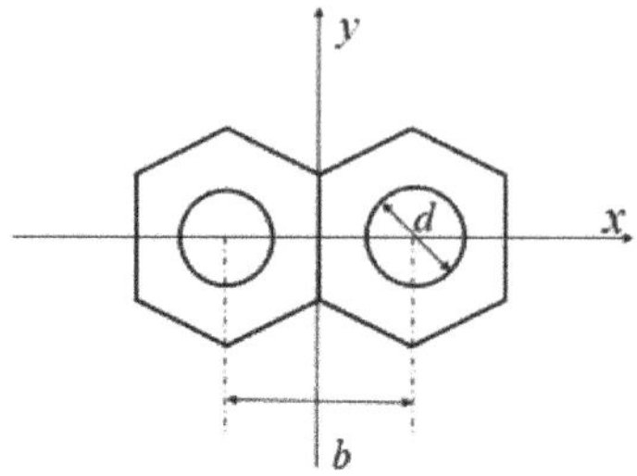

Figure 1. The schematic diagram of the telescope composed of two hexagonal segments with sparse circles configuration.

Based on the Fourier optics, the point spread function (PSF) of the system is the squared modulus of the Fourier transform of the GPF which is given by:

$$\begin{aligned} PSF(u,v) = |\Im\{G(x,y)\}|^2 = (d/2)^2\left|\left[\frac{J_1(\pi d\sqrt{u^2+v^2})}{\sqrt{u^2+v^2}}\right] \cdot \left(e^{i(\frac{\pi}{\lambda}2p-\pi ub)} + e^{-i(\frac{\pi}{\lambda}2p-\pi ub)}\right)\right|^2 \\ = 2\frac{(d/2)^2 {J_1}^2(\pi d\sqrt{u^2+v^2})}{u^2+v^2}\left[1+\cos\left(\frac{2\pi}{\lambda}2p - 2\pi ub\right)\right]. \end{aligned} \tag{4}$$

From Equation (4), we can see that the PSF includes two parts: the diffraction part and the interference part. $\frac{(d/2)^2 {J_1}^2(\pi d\sqrt{u^2+v^2})}{u^2+v^2}$ is the diffraction intensity of the single circle aperture; the coefficient 2 means the diffraction part is simple superposition formed by the two single circle apertures' diffraction. $\left[1+\cos\left(\frac{2\pi}{\lambda}2p - 2\pi ub\right)\right]$ is the interference part which is coherent superposition formed by the sub-waves sampled by the two single circle apertures. As we can see, the piston error is included in the interference part. When the observed target is a point source, the image obtained on the focal plane can be simplified as the PSF of the system, thus the piston error between segments is included in the CCD captured image. According to the interference principle, the interference factor will disappear when the piston error exceeds the coherence length of the light used in the optical imaging system, then the intensity distribution on the image plane becomes the simple superposition of the two sub-pupils' diffraction. Hence, the capture range of the piston error is limited by the coherence length of the input light.

Based on Fourier optics, the OTF is the 2D Fourier transform of the PSF shown as:

$$OTF(f_x, f_y) = \Im\{PSF(u,v)\}, \tag{5}$$

where (f_x, f_y) is the spatial frequency in the x and y directions, respectively. According to the OTF calculation method of the diffraction limited system, the expression for the OTF of the segmented system can be given by:

$$OTF(f_x, f_y) = 2OTF_{sub}(f_x, f_y) + OTF_{sub}(f_x + \frac{b}{\lambda f}, f_y)e^{-i\frac{2\pi}{\lambda}2p} + OTF_{sub}(f_x - \frac{b}{\lambda f}, f_y)e^{i\frac{2\pi}{\lambda}2p}, \tag{6}$$

where $OTF_{sub}(f_x, f_y)$ is the OTF of a single circle aperture diffraction system, which is given by:

$$OTF_{sub}(f_x, f_y) = \begin{cases} \frac{2}{\pi}\left[\arccos\left(\frac{\rho}{2\rho_0}\right) - \frac{\rho}{2\rho_0}\sqrt{1-\left(\frac{\rho}{2\rho_0}\right)^2}\right], & \rho \leq 2\rho_0 \\ 0 & , others \end{cases} \quad (7)$$

where $\rho = \sqrt{f_x^2 + f_y^2}$ is the radial coordinate on the frequency plane, $\rho_0 = \frac{d}{2\lambda f}$ is the system cut-off frequency, f is focal length of the imaging lens.

Equation (6) shows that the OTF of the segmented telescope composed of two hexagonal segments with mask includes three parts: the central part and two sidelobes. The central part's spatial frequency is $(0,0)$ and its value is the maximum modules of the sidelobe without piston error multiplied by the number of sub-pupils. Since the OTF is normalized, the value of the central part is always 1, so the maximum height of the sidelobe part without piston error is $1/N$, where N is the number of sub-pupils, and $N = 2$ is the telescope composed of two segments. The spatial frequencies of the two sidelobes are $(\pm b/\lambda f, 0)$ respectively, and they are symmetrically distributed about the central part. Based on the Fourier optics, we can get the MTF of the system by doing a modulus operation for Equation (6), which is shown as:

$$\begin{aligned} & MTF(f_x, f_y) = |OTF(f_x, f_y)| \\ = & |2OTF_{sub}(f_x, f_y)| + \left|OTF_{sub}(f_x + \tfrac{b}{\lambda f}, f_y)e^{-i\frac{2\pi}{\lambda}2p}\right| + \left|OTF_{sub}(f_x - \tfrac{b}{\lambda f}, f_y)e^{i\frac{2\pi}{\lambda}2p}\right|. \end{aligned} \quad (8)$$

As we can see, only the value of the MTF sidelobe is modulated by the piston error. We take one sidelobe out marked as $MTF_{sidelobe}$, which is shown as:

$$MTF_{sidelobe}(f_x, f_y) = \left|OTF_{sub}(f_x + \frac{b}{\lambda f}, f_y)e^{-i\frac{2\pi}{\lambda}2p}\right|. \quad (9)$$

Then the peak height value of the sidelobe with piston error marked as $MTF_{sidelobe-peak}$ is obtained by:

$$MTF_{sidelobe-peak} = \frac{1}{N}\left|e^{-i\frac{2\pi}{\lambda}2p}\right|. \quad (10)$$

From Equation (10), we can see that the peak height of the MTF sidelobe is only related to the number of sub-pupils, input wavelength and piston error. The effective detection range of the piston error is limited for a single wavelength due to the 2π ambiguity. Therefore, we need a broadband input light value to expand the piston error detection range.

For a broadband input light centered at λ_0 with the bandwidth $\Delta\lambda$, the PSF of the system is the sum of all PSFs at different monochromatic wavelengths which is given by:

$$\begin{aligned} PSF_{broad}(u, v, \lambda) &= \int_{\lambda_0-\frac{\Delta\lambda}{2}}^{\lambda_0+\frac{\Delta\lambda}{2}} PSF(u, v, \lambda)S(\lambda)d\lambda \\ &= \int_{\lambda_0-\frac{\Delta\lambda}{2}}^{\lambda_0+\frac{\Delta\lambda}{2}} 2\left(\frac{d}{2}\right)^2 \frac{J_1{}^2(\pi d\sqrt{u^2+v^2})}{u^2+v^2}\left[1+\cos\left(\frac{2\pi}{\lambda}2p - 2\pi ub\right)\right]S(\lambda)d\lambda, \end{aligned} \quad (11)$$

where $S(\lambda)$ is PSF weight of different wavelengths, assuming $S(\lambda) = 1$.

Since the integral is difficult to calculate, a differential summation approximation is used. The $\Delta\lambda$ is divided into n intervals equally, then the PSF can be rewritten as:

$$PSF_{broad}(u, v, \lambda) = \frac{\Delta\lambda}{n}\sum_{t=1}^{n} 2\left(\frac{d}{2}\right)^2 \frac{J_1{}^2(\pi d\sqrt{u^2+v^2})}{u^2+v^2}\left[1+\cos\left(\frac{2\pi}{\lambda_t}2p - 2\pi ub\right)\right]. \quad (12)$$

By performing Fourier transform for the PSF_{broad}, the corresponding OTF of the system in broadband light marked as OTF_{broad} is given by:

$$OTF_{broad}(f_x, f_y, \lambda) = \frac{\Delta\lambda}{n}\sum_{t=1}^{n}\begin{bmatrix} 2OTF_{sub}(f_x, f_y) + OTF_{sub}(f_x + \frac{b}{\lambda_t f}, f_y)e^{-i\frac{2\pi}{\lambda_t}2p} + \ldots \\ OTF_{sub}(f_x - \frac{b}{\lambda_t f}, f_y)e^{i\frac{2\pi}{\lambda_t}2p} \end{bmatrix}. \quad (13)$$

As we know, performing a modulus operation for the OTF, we can get the MTF of the system and the piston error is only related to the MTF sidelobes, thus the MTF sidelobe marked as $MTF_{sidelobe-broad}$ is extracted out from the modulus of the OTF_{broad} and shown as:

$$MTF_{sidelobe-broad}(f_x, f_y, \lambda) = \frac{\Delta\lambda}{n}\left|\sum_{t=1}^{n}\left[OTF_{sub}(f_x + \frac{b}{\lambda_t f}, f_y)e^{-i\frac{2\pi}{\lambda_t}2p}\right]\right|. \quad (14)$$

In fact, for different wavelengths, the position coordinates of the MTF sidelobes are different. However, the coordinate difference introduced by different wavelengths is very small when the bandwidth is much smaller than the central wavelength [26], so the position coordinate $(f_x + \frac{b}{\lambda_t f}, f_y)$ with different wavelengths can be approximated as $(f_x + \frac{b}{\lambda_0 f}, f_y)$. The effect of different wavelengths is mainly reflected in the e-index part. Thus Equation (14) can be rewritten as:

$$MTF_{sidelobe-broad}(f_x, f_y, \lambda) = \frac{\Delta\lambda}{n}\left|OTF_{sub}(f_x + \frac{b}{\lambda_0 f}, f_y)\right|\left|\sum_{t=1}^{n} e^{-i\frac{2\pi}{\lambda_t}2p}\right|. \quad (15)$$

For the broadband input light value, when the OTF_{broad} is normalized, the value of the central peak is 1. From Equation (13) we can see that the peak height of the MTF sidelobe without piston error, namely $|OTF_{sub}(f_x, f_y)|$, is equal to $\frac{1}{N\Delta\lambda}$ ($N = 2$ for Equation (13)), thus the peak height value of the MTF sidelobe with piston error is

$$MTF_{sidelobe-broad-peak} = \frac{1}{nN}\left|\sum_{t=1}^{n} e^{-i\frac{2\pi}{\lambda_t}2p}\right|. \quad (16)$$

Based on the above derivation, we obtained a more universally applicable and precise theoretical conclusion shown in Equation (16). We can see that the peak height of the MTF sidelobe is only related to the number of sub-pupils of the segmented telescope, input wavelength and piston error between segments based on the coherence measurement of a star image when using broadband input light. So the amplitudes of the MTF sidelobe can be easily calculated when the piston error is known. However, on the other hand, due to the complexity and high nonlinearity of Equation (16), the piston error cannot be directly solved when the peak height of the MTF sidelobe is measured.

2.2. Verification of the Correctness of the Theoretical Relation by MATLAB Simulation

In order to verify the correctness of the mathematical derivation for Equation (16), we set up an optical system in MATLAB and simulated the corresponding MTF using the MATLAB program. The simulation segmented telescope system we used here is composed of two sub-pupils with mask, as shown as Figure 1. The left sub-mirror is the reference pupil; the piston error was introduced on the right pupil. The sampling grid of the exit pupil plane was set as 256 × 256 pixels, the pixel size of the CCD was 3.5 µm and the F# of the optical system was 8. Thus the circumscribed circle diameter of the single hexagonal sub-mirror was 59 pixels, the diameter of the circle on the mask was 18 pixels and the distance between the centers of the two circles was 52 pixels to satisfy the Nyquist sampling criterion. The central wavelength of the input broad light was 632.8 nm and its bandwidth was 1 nm. The effective detection range of the proposed method is half of the coherent length:

$$L = \frac{L_c}{2} = \frac{{\lambda_0}^2}{2\Delta\lambda} = \frac{(632.8\ nm)^2}{2 \times 1\ nm} \approx 200\ \mu m, \quad (17)$$

where L_c is the coherent length of the input light, the factor 1/2 is due to the input light reflection on the segmented primary mirror giving an OPD equal to twice the step between segments.

Based on the set system model, different piston errors were introduced to the right sub-pupil. We first obtained the system MTF using the MATLAB program, which directly performs 2D discrete Fourier transform for the simulated CCD captured image of the point-source object. Then by taking the corresponding piston errors to Equation (16), we got the theoretical values of $MTF_{sidelobe-broad-peak}$. By comparing the theoretical calculated values from Equation (16) with the MATLAB simulation results, we verified the correctness of the theoretical formula established based on the Fourier optics principle.

Figure 2 shows four sets of MTF obtained by the MTALAB program and the peak height value of the MTF sidelobe is marked in the corresponding figure, where the introduced piston errors were 0 μm, 50 μm, 100 μm and 200 μm, respectively. We took the four sets of piston errors into Equation (16), the n was set to 100 in order to make the differential summation closer to the integral, and $N = 2$. Then the calculated corresponding peak heights of the MTF sidelobe were 0.5, 0.4503, 0.3187 and 6.0708×10^{-4}, respectively. We can see that the surrounding peaks of the MTF calculated by Equation (16) were the same as the results directly obtained from the MATLAB program.

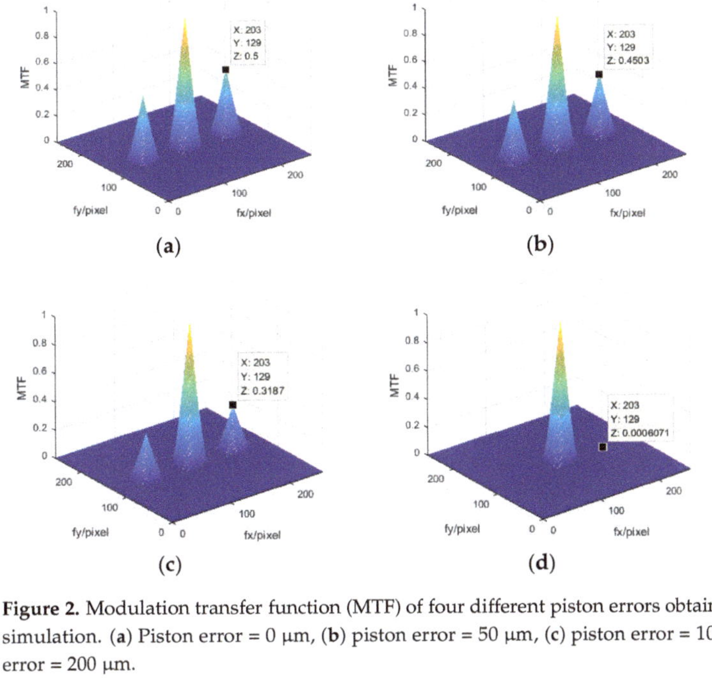

Figure 2. Modulation transfer function (MTF) of four different piston errors obtained by MATLAB simulation. (**a**) Piston error = 0 μm, (**b**) piston error = 50 μm, (**c**) piston error = 100 μm, (**d**) piston error = 200 μm.

In order to further verify the correctness of Equation (16), 201 sets of piston errors were introduced in 1 μm steps from 0 to 200 μm. The peak heights of the MTF sidelobes calculated from Equation (16) and the MATLAB program are shown in Figure 3a,b. Figure 3 verifies the correctness of the theoretical relationship between the peak heights of the MTF sidelobes and the piston error shown as Equation (16). When the MTF was obtained, the piston error should have been solved by Equation (16) theoretically. However, due to the nonlinearity and complexity of Equation (16), it is difficult to calculate the piston error in a realistic segmented telescope directly.

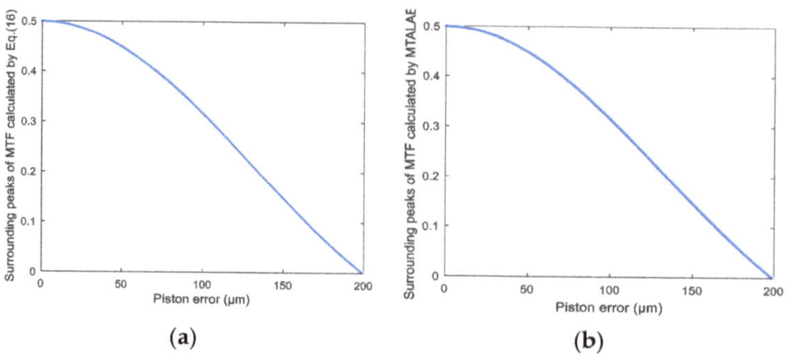

Figure 3. Surrounding peak heights of MTF varying with the piston error (**a**) calculated by Equation (16) and (**b**) obtained by the MATLAB program.

Note that Figure 3 only shows the positive side of the piston error. If two piston errors have the same absolute value but opposite signs, they will have the same peak heights of MTF sidelobes due to the fact that Equation (16) is a symmetrical function. Namely, we need to know the relative spatial position between the sub-mirror to be measured and the reference sub-mirror in advance in order to obtain the correct piston error. This is one of the inconveniences when using our proposed algorithm.

3. Piston Error Detection Method Using BP Artificial Neural Network

3.1. Brief Introduction of the BP Artificial Neural Network

Equation (16) directly gives the theoretical relationship between the piston error and the peak heights of the MTF sidelobes. However, due to the high nonlinearity and complexity of Equation (16), it is difficult to calculate the piston error in a realistic segmented telescope when the MTF is obtained. Thus, we turned to machine learning and neural networks [27] to establish the mapping relationship between the MTF's surrounding peak heights and the piston error rather than fitting Equation (16) with precise mathematical expressions.

Artificial neural networks [28–30], which belong to machine learning, are input–output information processors composed of parallel layers of elements or neurons, loosely modeled on biological neurons, which possess local memory and are capable of elementary arithmetic. They can be used to learn and store a great deal of nonlinear mapping relations from the input–output model. The artificial neural network, composed of many neurons, has a parallel distribution structure. Each neuron has a single output and can be connected to other neurons; there are multiple output connection methods, and each connection method corresponds to a connection weight coefficient. The artificial neural network can be regarded as a directed graph formed by connecting weighted directed chains with processing elements as nodes. For the kth neuron, if the signal from another neuron is x_i, their interaction weight is w_{ik} and the internal threshold of the processing unit is θ_k, then the input of the kth neuron is $\sum_{i=1}^{m} x_i w_{ik}$, where m is the number of the input neurons connected to the kth neuron. Thus, the output of the kth neuron y_k is:

$$y_k = f\left(\sum_{i=1}^{m} x_i w_{ik} - \theta_k\right), \tag{18}$$

where $f()$ is called the activation (or transfer) function. The purposes of the activation function are introducing nonlinearity to neural networks and bounding the value of the neuron so that the neural network is not paralyzed by divergent neurons. A common example of activation function is the sigmoid (or logistic) function, which is shown as:

$$\phi(z) = \frac{1}{1 + \exp(-z)}. \tag{19}$$

When the neural network structure is designed and has inputs and outputs, the network needs to be trained so it can learn the appropriate mapping relationships between the inputs and the outputs. The BP algorithm is one of the most widely used algorithms for training neural networks, and utilizes a gradient descent method to minimize the square of the error between the network output value and the target value to adjust the weights of each neuron. The BP algorithm can be divided into two steps: forward propagation and back propagation. The forward propagation process is propagating the input signals from the input layer, via the hidden layer, to the output layer to get the final output results. The back-propagation process is propagating the error signal, which is the difference between the real output and the expected output of the network, from the output end to the input layer in a layer-by-layer manner to adjust the weights and thresholds. Through one forward propagation and one back propagation, we can update the parameters of the network once. The network training process is to constantly reciprocate the forward and back propagation and continuously update the network parameters, and finally make the network precisely map the real relationships between the inputs and outputs. A sketch graph of a BP artificial neural network with two hidden layers is shown in Figure 4.

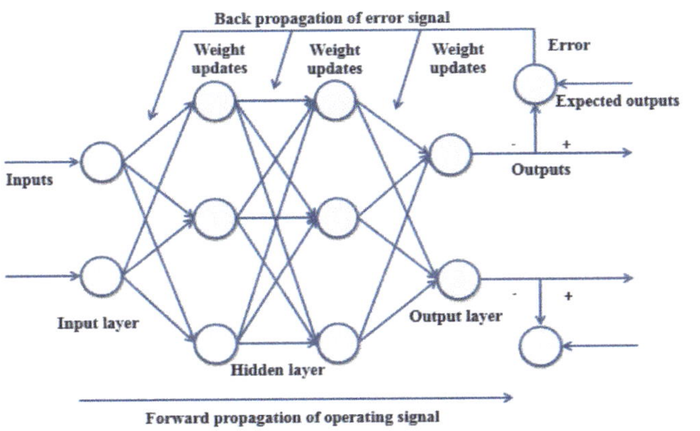

Figure 4. Sketch graph of a back-propagation (BP) artificial neural network with two hidden layers.

3.2. Piston Error Detection Approach with BP Artificial Neural Network

The piston error detection for multiple sub-mirrors of a segmented telescope with a BP artificial neural network can be divided into three steps:

(1) Determine the system parameters and generate the datasets for training the neural network under the specified system parameters. From Equation (16), we can see that the peak heights of the MTF sidelobes are only related to the input broadband light, the number of sub-pupils and the piston errors between the multiple sub-mirrors. When the segmented telescope and its working wavelength are determined, within the half of the coherent length of input light, a set of piston errors between segments is randomly introduced, and by taking the corresponding piston errors into Equation (16), the peak heights of MTF sidelobes can be obtained. The peak heights of MTF sidelobes directly calculated from Equation (16) served as one column of input matrix and the corresponding piston errors served as one column of the output matrix. Thus the input dataset and output dataset of the network could be generated.

(2) Establish the neural network and train it with the input dataset and the corresponding output dataset. Here we utilized the neural network fitting tool in MATLAB, and by properly setting the number of neurons in each layer, the neural network was established. In the training process, the dataset is separated into three parts including training set, validation set and test set, then a specific training algorithm is set to train the network.
(3) Once the network is well trained, we can apply the trained neural network to determine the piston errors with the PSF image collected from the optical system. Note that the image should be Fourier transformed first to get the peak heights of the MTF sidelobes before they can be handled with the neural network. By establishing the correspondence of the MTF sidelobes with their associated sub-pupils, multiple piston error measurements of the whole aperture can be implemented simultaneously by one detection of a CCD broadband image.

The application procedure of the piston error detection approach with BP artificial neural network is summarized in Figure 5.

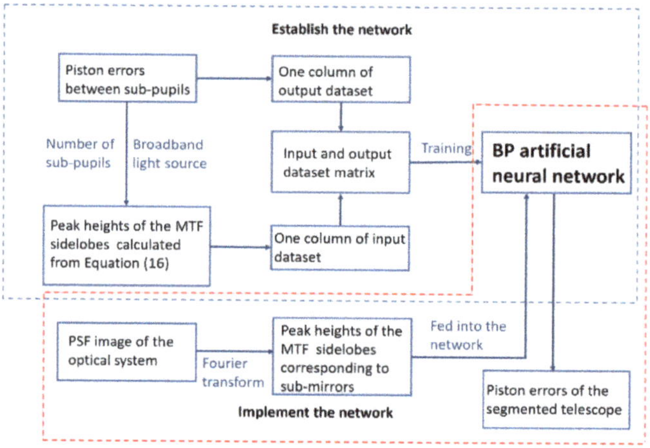

Figure 5. The detailed piston error detection approach with a BP artificial neural network.

4. Simulation

4.1. Piston Error Detection for the Telescope Composed of Two Segments

Here we first utilized the telescope model composed of two hexagonal segments in Figure 1 to test the proposed method. According to Equation (16), the peak height of the MTF sidelobe is only related to the number of sub-pupils, input wavelength and piston error. Here the number of sub-pupils was $N = 2$, the light source was broadband light centered at 632.8 nm with bandwidth 1 nm, the left sub-mirror was set as the reference pupil, and a series of piston errors were introduced to the right pupil to generate the corresponding peak heights of MTF sidelobes. Considering the coherence length, 100,000 sets of piston errors were randomly generated between 0 μm and 200 μm to produce a corresponding 100,000 sets of surrounding peak heights of the MTF based on Equation (16). Thus, we could obtain the dataset for training the network. Then the neural network could easily be trained using the neural network fitting tool in MATLAB (the number of nodes in the hidden layer was set as 50), the peak heights of the MTF sidelobes were fed into the network and the corresponding piston errors constituted the output dataset. In this process, the dataset was separated into three parts; i.e., training set, validation set and test set. The training set was used for learning, which was to fit the weights of the network; the validation set was used for tuning the final architecture of the network; the test set was only used for assessing the performance of the network. The ratio between them in this work was 70%: 15%: 15% (i.e., 70,000 sets for training, 15,000 sets for validation and 15,000 sets for testing).

The training results are shown in Figure 6, which provides the distribution of the error between the targets and the actual outputs of the network in the form of a histogram. From the results, we can see that for the vast majority of the cases the errors between the targets and outputs were within 0.01 μm, which equals 1.5% of the input central wavelength. More specifically, the root mean square errors (RMSE) between the targets and outputs of the network in the training set, validation set and testing set were 2.531×10^{-5} μm, 1.914×10^{-5} μm and 2.465×10^{-5} μm, respectively. This demonstrates the high accuracy of the neural network with the peak heights of the MTF sidelobes serving as the input to solve the piston error.

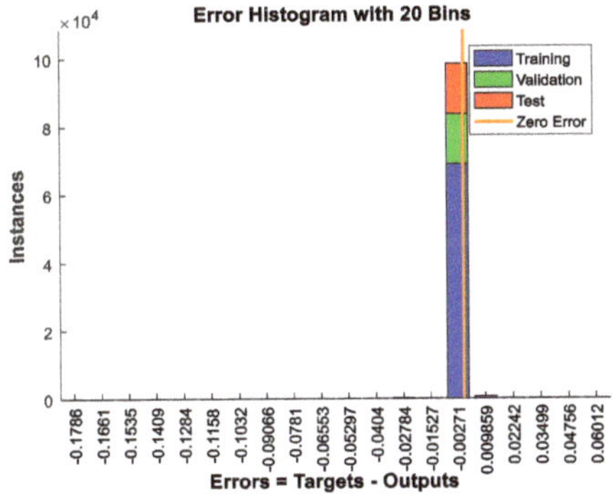

Figure 6. Distribution of the errors between the targets and the actual outputs of the network in the form of a histogram.

Then we apply the obtained neural network to the PSF images collected from the established optical system composed of two hexagonal segments in MATLAB described in Section 3.2. To approximate the practical imaging environment, a zero mean and 0.05 variance Gaussian distribution noise were introduced in the simulated PSF images. A Fourier transform was performed for the noised PSF image to obtain the OTF, which was composed of the MTF and phase transfer function (PTF), where the peak heights of the MTF sidelobes served as the input. The well-trained neural network could directly output the piston error introduced on the right sub-mirror corresponding to each PSF image; we named the actual output of the network as the measured piston error. The difference between the measured piston errors and the settled piston errors is shown in Figure 7. Seventeen sets of experimental results are listed in this figure. From the error analysis, the RMSE of all the difference was 1.3 nm. Here the RMSE was calculated from the following formula:

$$RMSE = \sqrt{\frac{1}{N}\sum_{n=1}^{N} Difference_n^2}, \tag{20}$$

where N is the total number of the datasets.

From the simulation results, we can see that the piston error detection method based on a BP artificial neural network had a quite high accuracy and large capture range. However, when approaching the two ends of the capture range, the detection accuracy decreased slightly. This was due to the peak heights of the MTF sidelobes approaching the edges of the capture range not changing with piston error very obviously; namely the gradients of the curve were small in these parts, and we can see this directly from the theoretical relationship curve between the piston errors and peak heights of MTF sidelobes shown in Figure 3a. This results from the insensitiveness of the network to data approaching the ends of the capture range, so the detection accuracy decreased a little bit, but the accuracy at the two ends of the detection range was still high enough to meet the requirement of co-phasing the piston error.

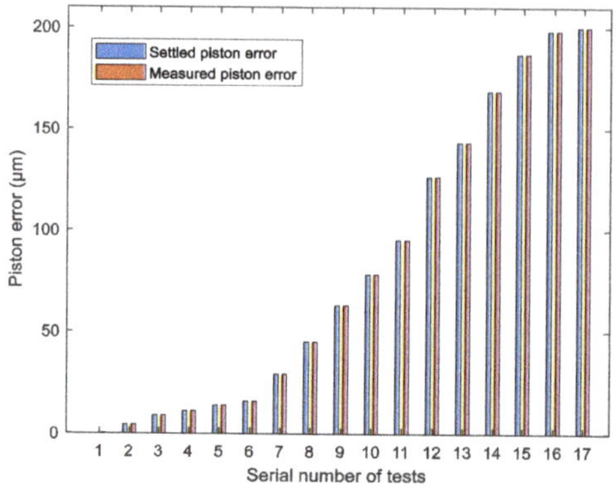

Figure 7. The piston error detection results in bar chart form for two segments system with a BP artificial neural network.

We also compared our proposed method with the Gaussian fitting and two-piece quartic polynomial function fitting methods presented in [21,22], respectively, and the comparison results on detection accuracy are shown in Figure 8. From Figure 8a, we can see that the detection accuracy of our proposed method is much higher than that of the Gaussian fitting method. Figure 8b shows the detection accuracy curve of our method compared with that of the two-piece quartic polynomial function fitting method. Our method produced larger differences (~5 nm) for small piston errors, but with the increase of the piston error, especially after 125 μm, the two-piece quartic polynomial function fitting method generated much larger differences. From the error analysis, the difference of our method was 1.3 nm RMS, which is smaller than the 1.8 nm RMS presented in [22]. This demonstrates that our method also has higher detection accuracy than the two-piece quartic polynomial function fitting method, while the implementation complexity is lower since there is no need to build different computation models in different intervals of the capture range.

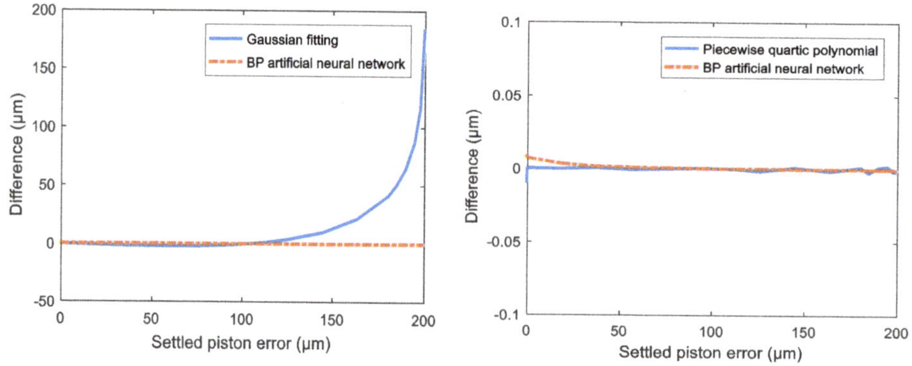

(**a**) (**b**)

Figure 8. The comparison results on detection accuracy of a BP artificial neural network approach with (**a**) the Gaussian fitting method (**b**) and the two-piece quartic polynomial function fitting method.

To further evaluate the ability of the trained network, we designed some other simulations to discuss the effect of image noise and the tip-tilt aberrations of the segments on the detection accuracy of the proposed method. To approximate the practical imaging environment, different intensity Gaussian distribution noises were introduced in the simulated PSF images. Figure 9a gives the detection accuracy under three different image noise intensities by the trained network, and the simulation result showed that PSF image noise had little effect on the piston error detection accuracy. This was due to the fact that the Gaussian distribution noise contributed little to the peak heights of MTF sidelobes, so the proposed piston error detection algorithm had a very good noise immunity. Additionally, since the tip-tilt aberrations could not be corrected entirely before measuring the piston error, slight tip-tilt distortions were introduced to the optics system while generating the simulated PSF images. Here we introduced two different tip-tilt aberrations: the RMS errors were 0.01λ and 0.05λ ($\lambda = 632.8\ nm$), respectively. Since the introduced tip-tilt aberrations were quite small, we assumed that the aberration distribution for each wavelength was the same. The effect on the piston error detection accuracy is shown in Figure 9b, which confirms that the existence of the tip-tilt error had a great influence on the detection accuracy. This was because that the tip-tilt error greatly changed the amplitudes of MTF sidelobes. So, in order to guarantee the detection accuracy, the tip-tilt aberrations should be corrected well enough (at least less than 0.01λ RMS) before piston error measurement.

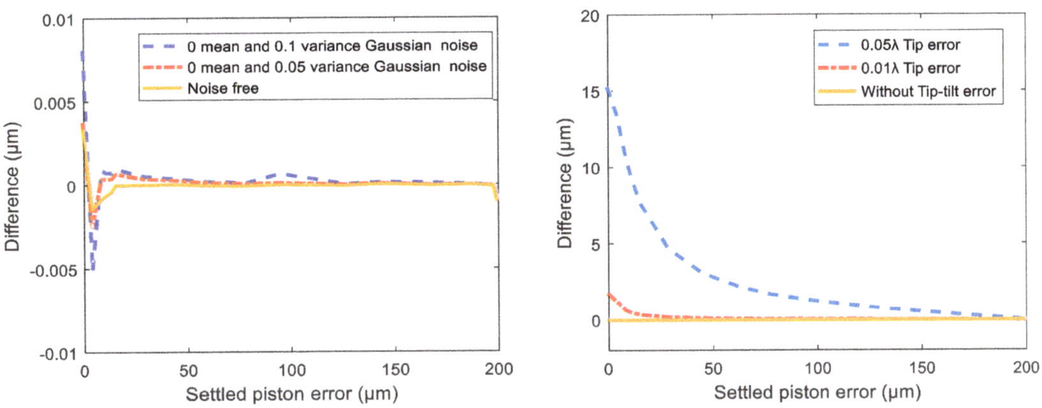

(a) (b)

Figure 9. The piston error detection accuracy under (**a**) different CCD image noise values and (**b**) different tip-tilt aberrations.

From Equation (16), we can conclude that the datasets used to train the network are only related to the numbers of telescope sub-pupils and input light sources. When the numbers of the sub-pupils and working wavelength were determined, the network could be trained to map the relation between the peak heights of MTF sidelobes and piston error, since learning mapping relationships is a data-driven process. Thus, even for different imaging systems, if they have the same numbers of sub-pupils and work under the same operating light, the trained network should have quite good generalization ability. In order to verify this, we designed another simulation based on the previous one. In this simulation, we changed the $F^{\#}$ of the imaging system and generated different intensity images from these systems as testing sets to examine the generalization ability of the trained network (original imaging system parameters are given in Section 3.2 with $F^{\#} = 8$) for the new systems. The $F^{\#}$ of the imaging systems used for simulations were 10, 20, 30, 40, 50 respectively, while the piston error measurement results of all the new systems using the trained network had the same accuracy as those of the original imaging system with $F^{\#} = 8$. This demonstrates that our network has quite good generalization ability.

4.2. Simultaneous Multi-Piston Measurement

In the part, we utilized the proposed method to simultaneously detect multi-piston errors of the whole aperture. Based on the theoretical deduction in Section 3.1, we can see that the MTF model of the N sub-pupils segmented telescope consisted of N^2 sub-MTFs. In the spatial frequency domain, the N sub-MTFs overlapped at the position where the center spatial frequency was zero to form the central peak, while the other N $(N - 1)$ sub-MTFs distributed around the central peak to form the sidelobes. Every pair of sub-pupils produced a pair of MTF sidelobes, the sidelobes symmetrically distributed on both sides of the central peak. If all of the sidelobes did not overlap, their amplitudes could be obtained at the same time by one CCD image, hence the piston errors of all sub-mirrors were retrieved at the same time by inputting the peak height of sub-MTF corresponding to each sub-mirror into the trained network.

In order to verify the feasibility of simultaneous multi-piston measurement with the proposed method, we took a primary mirror composed of four hexagonal sub-pupils as an example. We modeled the four-segment imaging system using MATLAB, results of which are shown in Figure 10a. The No. 1 sub-mirror was set as reference pupil, and the piston errors were introduced on the No. 2, No. 3 and No.4 sub-pupils, respectively. The corresponding system MTF without piston errors is shown in Figure 10b: there was one MTF central peak and 12 MTF sidelobes (N $(N - 1) = 4 \times (4 - 1) = 12$). Four sub-MTFs ($N = 4$) overlapped at the position where the center spatial frequency was zero to form this central peak, and every pair of sub-pupils produced a pair of MTF sidelobes. Note that a colored MTF was used here in order to describe which peaks were actually used to solve the piston error of the corresponding sub-mirror. When the No. 1 sub-mirror was set as reference pupil, No. 2 sub-mirror produced the two red sub-MTFs, No. 3 sub-mirror produced the two green sub-MTFs and No. 4 sub-mirror produced the two yellow sub-MTFs. The six light blue sub-MTFs at the outermost periphery were produced either by No. 2 and No. 3 sub-mirrors together, or by No. 2 and No. 4 sub-mirrors together, or by No. 3 and No. 4 sub-mirrors together, which could not be used to measure piston errors. Thus, we could use any one of the red sub-MTFs to measure the piston error of No. 2 sub-mirror, any one of the green sub-MTFs to measure the piston error of No. 3 sub-mirror and any one of the yellow sub-MTFs to measure the piston error of No. 4 sub-mirror, respectively. For the detailed generation principle between the sub-MTFs and sub-mirrors, refer to paper [31].

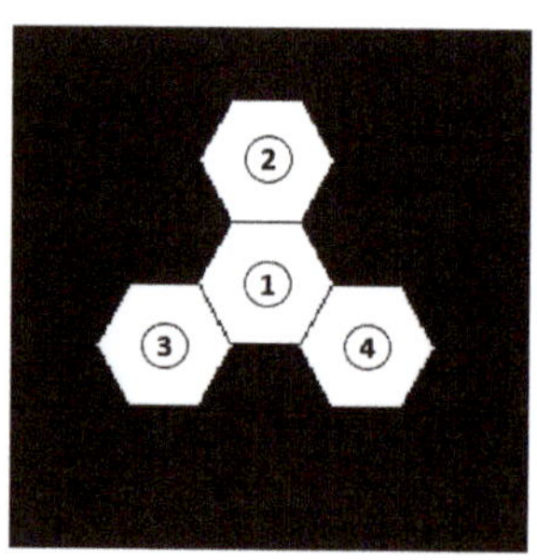

(a)

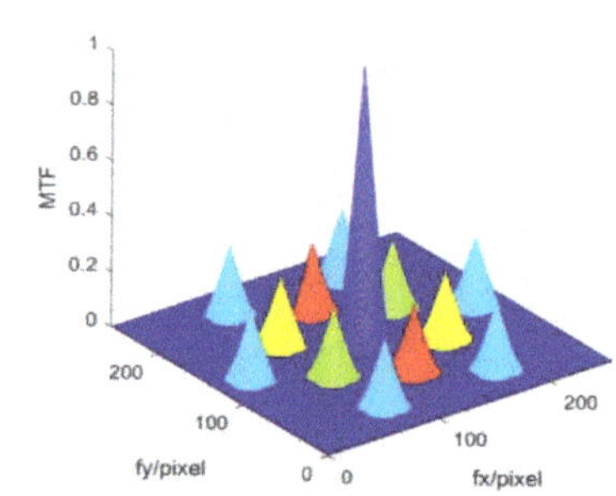

(b)

Figure 10. The 4 hexagonal segmented telescope showing (**a**) the position arrangement of the 4 sub-mirrors and (**b**) the corresponding system MTF colored without piston error.

In order to simultaneously detect piston errors of the three segments, we first had to establish the neural network and train it with the generated dataset. Here we still used the broadband light source centered at 632.8 nm with bandwidth 1 nm, since $N = 4$, and according to Equation (16) we could directly obtain the datasets of the peak heights of the MTF sidelobes varying with the piston errors to train the network. The curve of the relation between the peak heights of the sub-MTF and the piston errors is shown in Figure 11. One hundred thousand sets of data were fed into the network to train it, and the network settings were the same as those of the two sub-pupils' telescope except the ratio between the training set, validation set and test set was 65%: 20%: 15%. We added more data to the validation set to avoid the overfitting problem. The training results are shown in Figure 12.

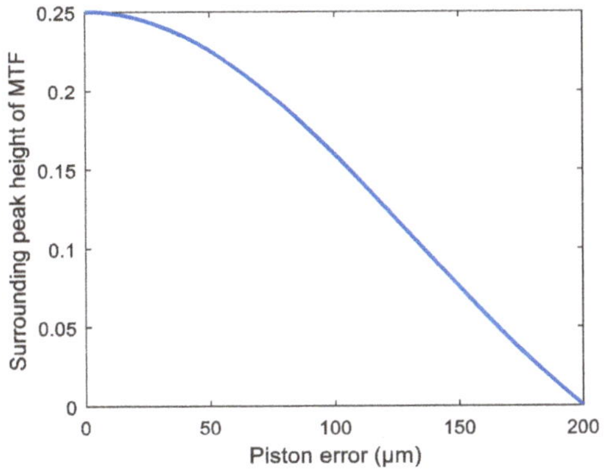

Figure 11. Peak heights of the MTF sidelobes varying with the piston errors of the optics system composed of 4 sub-mirrors.

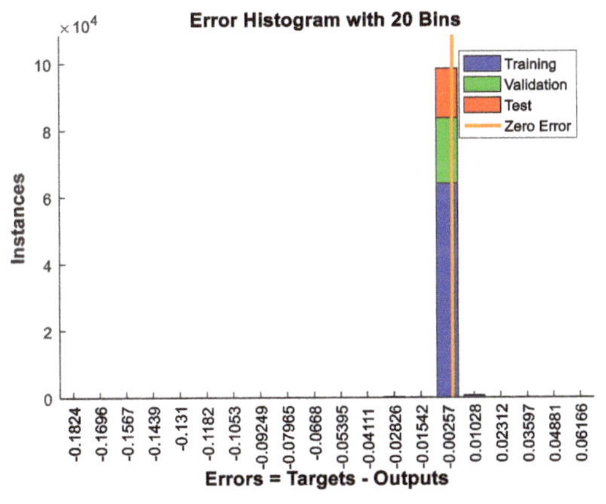

Figure 12. The network training result for the optics system composed of 4 sub-mirrors.

Then we used the PSF images directly collected from the established optical system composed of 4 hexagonal segments shown in Figure 9a to simultaneously measure piston errors for all sub-mirrors based on the trained network. We randomly introduced different piston errors between 0 μm and 200 μm to No. 2, No. 3 and No. 4 sub-mirrors at the same time, then performed Fourier transform for the obtained PSF image to get the system MTF. Here we also added zero mean and 0.05 variance Gaussian distribution noise to the simulated PSF images. According to the correspondence relation between the MTF sidelobes and sub-mirrors (here No. 4 sub-MTF corresponded to No. 2 sub-mirror, No. 6 sub-MTF to the No. 3 sub-mirror and No. 5 sub-MTF to No. 4 sub-mirror, respectively), by inputting the peak height of sub-MTF corresponding to each sub-mirror into the trained network, the piston errors of all the sub-mirrors could be measured by one CCD image at one time. A part of the simulation results is listed in Figure 13. The piston error measurement results of No. 2, No. 3 and No.4 sub-mirrors are given by Figure 13a–c, respectively. Figure 13d shows the total RMSE of the three sub-mirrors' piston error measurement accuracy during each test.

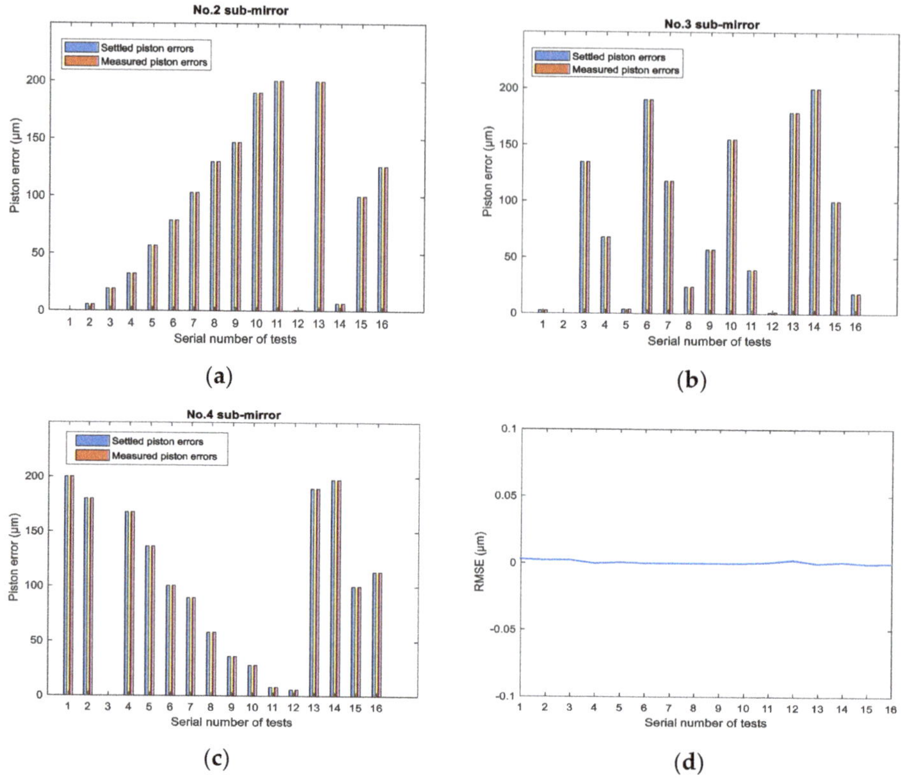

Figure 13. Piston error measurement results for the four-segment system: (**a–c**) piston error measurement results of No. 2, No. 3 and No.4 sub-mirrors, respectively, and (**d**) the total RMSE of the three sub-mirrors' piston error measurement accuracy during each test.

From the simulation results we can see that the multi-piston errors between the segments of the entire aperture could be simultaneously detected with high accuracy: the average value of the RMS errors over the testing samples was about 1.4 nm. Since the peak height of sub-MTF corresponding to each sub-mirror was fed into the trained network to measure the piston errors, the most important issue was to avoid the MTF sidelobes respectively produced by the reference sub-pupil and all of the measured sub-pupils overlapping. For the detailed arrangement rules, refer to paper [31] to confirm the MTF sidelobe distribution was non-redundant.

Finally, we tried to compare our work with Ma's work presented in paper [32], since this also used a single broadband image to sense the piston errors between sub-pupils. Rather than establishing the theoretical relationship between the MTF sidelobes of PSF images and the piston errors like us, this research directly used in-focused broadband images as the input, and established one deep convolutional neural network (DCNN) to learn to sense pistons with a single broadband focal image. The simulation results for the two-pupil imaging system and four-pupil imaging system are shown in Figure 14.

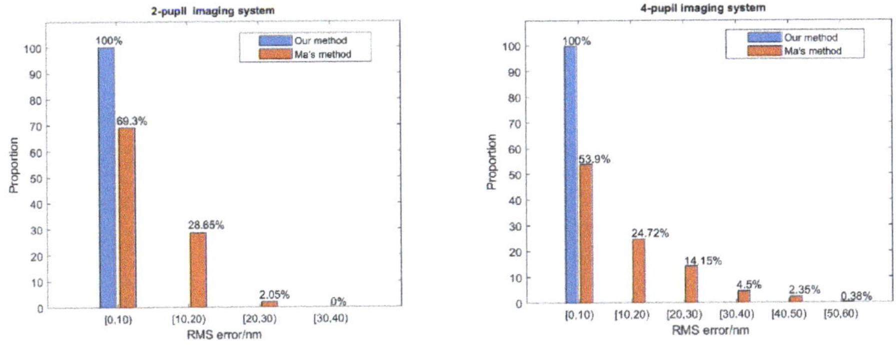

(a) (b)

Figure 14. The piston error detection accuracy comparison between our method and Ma's method (**a**) for the 2-pupil imaging system and (**b**) the 4-pupil imaging system.

From the simulation results, we can see that our method has higher piston error measurement accuracy than Ma's method presented in paper [32]. This improvement of the detection accuracy is at the expense of extensive complex optical calculations in advance in our work, since we used the MTF sidelobes of PSF images as inputs for the network based on the established theoretical relationship between the MTF sidelobes and the piston errors, while Ma directly used in-focused broadband images as the input.

5. Conclusions

In this paper, we put forward a method to simultaneously detect the multi-piston errors between the segments based on the broadband intensity distribution on the image plane by a BP artificial neural network. A mask with a sparse sub-pupil configuration was set on the exit-pupil plane to sample the wave from the segments. Based on the Fourier optics principle, the accurate theoretical relation between peak heights of MTF sidelobes and piston errors was obtained. Instead of fitting the relation by Gaussian function or two piecewise quartic polynomials, we utilized a BP artificial neural network to establish the nonlinear mapping relationship between the piston errors and the amplitudes of the MTF's surrounding peaks. By introducing different piston errors into the segmented optics system, the corresponding peak heights of MTF sidelobes could be directly calculated by the established theoretical formula, while the corresponding amplitudes of MTF sidelobes and the introduced piston errors were used as the inputs and outputs, respectively, to train the network. Once well trained, it could estimate the piston error to a good precision with high efficiency and robustness using the PSF images collected from the optics imaging system as inputs. Since the MTF sidelobes of all the sub-pupils can be simultaneously obtained by one detection of a CCD image, multiple piston errors of the entire aperture can be retrieved at one time.

Adequate simulation experiments were implemented to demonstrate the effectiveness and accuracy of the proposed approach. We established the models of a segmented telescope composed of two hexagonal segments and four hexagonal segments successively in MATLAB and implemented the PSF images collected from these simulated optical systems to test the performance of the corresponding trained networks, respectively. The piston sensing simulation results showed that the average values of RMS errors on the two-segment imaging system and the four-segment imaging system could achieve 1.3 nm and 1.4 nm, respectively, and the method's capture range was the operating light's coherence length. Compared to Gaussian fitting and two-piece quartic polynomial function fitting methods, our method has higher detection accuracy and is easier to implement. The influence of the CCD camera noise and the existence of tip-tilt aberrations on the accuracy of piston error detection were also discussed. The results indicated that the proposed piston error detection algorithm has a very good noise immunity, but the tip-tilt aberration should be corrected well enough before the piston error measurements. Besides, different intensity images from the imaging system with different $F^{\#}$ values were generated and fed into the original trained network to test the network generalization ability. The piston error detection accuracy was the same as that of the original imaging system, which demonstrates that our network has quite good generalization ability.

The hardware cost of our method is quite small; only a mask with a sparse multi-subaperture configuration is needed to attach in the exit pupil plane of the segmented telescope, and this mask should ensure the MTF sidelobes' non-redundant distribution. Thus, with this method, multi-piston measurements of the whole aperture can be implemented simultaneously, and piston detection no longer need be divided into coarse and fine regimes, which involves separate dedicated hardware solutions. In view of the efficiency and superiority, it is expected that the piston sensing method based on the BP artificial neural network proposed in this paper can be adapted to any segmented and deployable primary mirror telescope, no matter the shape of the segmented mirror and the number of the segments.

Author Contributions: Methodology, D.Y.; Software, D.Y.; Writing—original draft, D.Y.; Data curation, Y.H.; Writing—review & editing, Y.L. All authors have read and agreed to the published version of the manuscript.

Funding: National Natural Science Foundation of China (NSFC) (61905023); The Fourth Lifting Project of Young Science and Technology Talents in Jilin Province (QT202026); Youth Fund Project of Changchun University of Science and Technology (XQNJJ-2018-01); National Natural Science Foundation of China (NSFC) (62004016).

Institutional Review Board Statement: Not applicable.

Informed Consent Statement: Not applicable.

Data Availability Statement: Not applicable.

Conflicts of Interest: The authors declare no conflict of interest.

References

1. Bely, P.Y. *The Design and Construction of Large Optical Telescopes*; Springer Science & Business Media: Berlin/Heidelberg, Germany, 2003.
2. Noethe, L. History of mirror casting, figuring, segmentation and active optics. *Exp. Astron.* **2009**, *26*, 1–18. [CrossRef]
3. Nelson, J.E.; Gillingham, P.R. Overview of the performance of the W. M. Keck Observatory. *Proc. SPIE* **1994**, *2199*, 82–93.
4. Chanan, G.A.; Mast, T.S.; Nelson, J.E. Keck telescope primary mirror segments: Initial alignment and active control. In *Proceedings of European Southern Observatory Conference on Very Large Telescopes and their Instrumentation*; Ulrich, M.-H., Ed.; ESO: Garching, Germany, 1988; pp. 421–428.
5. Gilmozzi, R.; Spyromilio, J. The 42m European ELT: Status, SPIE Astronomical Telescopes Instrumentation. *Int. Soc. Opt. Photonics* **2008**, *7012*, 701219.
6. Nelson, J.; Sanders, G.H. The status of the Thirty Meter Telescope project. *Proc. SPIE* **2008**, *7012*, 70121A.
7. Clampin, M. Status of the James Webb Space Telescope observatory. *Proc. SPIE* **2012**, *8442*, 84422A.
8. Oegerle, W.R.; Feinberg, L.D.; Purves, L.R.; Hyde, T.T.; Thronson, H.A.; Townsend, J.A.; Postman, M.; Bolcar, M.R.; Budinoff, J.G.; Dean, B.H.; et al. ATLAST-9.2m: A large-aperture deployable space telescope. *Proc. SPIE* **2010**, *7731*, 52–56.
9. Acton, D.S.; Atcheson, P.; Cermak, M. James Webb Space Telescope Wavefront Sensing and Control Algorithms, SPIE Astronomical Telescopes Instrumentation. *Int. Soc. Opt. Photonics* **2004**, 887–896. [CrossRef]
10. Chanan, G.; Ohara, C.; Troy, M. Phasing the mirror segments of the Keck telescopes II: The narrow-band phasing algorithm. *Appl. Opt.* **2000**, *39*, 4706–4714. [CrossRef]
11. Chanan, G.; Troy, M.; Dekens, F. Phasing the mirror segments of the Keck telescopes: The broadband phasing algorithm. *Appl. Opt.* **1998**, *37*, 140–155. [CrossRef] [PubMed]
12. Esposito, S.; Puglisi, A.; Tozzi, A. Co-phasing of segmented mirror using pyramid sensor. *SPIE* **2003**, *5169*. [CrossRef]
13. Esposito, S.; Puglisi, A.; Tozzi, A. Pyramid sensor for segmented mirror alignment. *Opt. Lett.* **2005**, *30*, 2572–2574. [CrossRef]
14. Shi, F.; Redding, D.C.; Lowman, A.E.; Bowers, C.W.; Burns, L.A.; Petrone, P., III; Ohara, C.M.; Basinger, S.A. Segmented mirror coarse phasing with a dispersed fringe sensor: Experiment on NGST's wavefront control testbed. *Proc. SPIE* **2003**, *4850*, 318–328.
15. Van Dam, M.A.; Mcleod, B.A.; Bouchez, A.H. Dispersed fringe sensor for the Giant Magellan Telescope. *Appl. Opt.* **2016**, *55*, 539–547. [CrossRef]

16. Yue, D.; Xu, S.Y.; Nie, H.T. Co-phasing of the segmented mirror and image retrieval based on phase diversity using a modified algorithm. *Appl. Opt.* **2015**, *54*, 7917–7924. [CrossRef]
17. Zhang, N.; Zhang, X.; Xu, S.; Liu, N.; Zhao, L. Simplified Phase Diversity algorithm based on a first-order Taylor expansion. *Appl. Opt.* **2016**, *55*, 7872.
18. Paxman, R.G.; Schulz, T.J.; Fienup, J.R. Joint estimation of object and aberrations by using phase diversity. *J. Opt. Soc. Am. A* **1992**, *9*, 1072–1085. [CrossRef]
19. Baron, F.; Mocoeur, I.; Cassaing, F.; Mugnier, L.M. Unambiguous phase retrieval as a co-phasing sensor for phased array telescopes. *J. Opt. Soc. Am. A* **2008**, *25*, 1000–1015. [CrossRef]
20. Ju, G.; Yan, C.; Dan, Y.; Gu, Z. Field diversity phase retrieval method for wavefront sensing in monolithic mirror space telescopes. *Appl. Opt.* **2017**, *56*, 4224. [CrossRef]
21. Simar, J.F.; Stockman, Y.; Surdej, J. Single-wavelength coarse phasing in segmented telescopes. *Appl. Opt.* **2015**, *54*, 1118–1123. [CrossRef]
22. Jiang, J.; Zhao, W. Phasing piston error in segmented telescopes. *Opt. Express* **2016**, *24*, 19123–19137. [CrossRef]
23. Lauterborn, W.; Kurz, R.; Wiesenfeldt, D. *Fourier Optics*; Springer Science & Business Media: Berlin/Heidelberg, Germany, 1995.
24. Li, Z.; Zhao, X. BP artificial neural network-based wave front correction for sensor-less free space optics communication. *Opt. Commun.* **2017**, *385*, 219–228. [CrossRef]
25. Guo, H.; Xin, Q.; Ma, H.; Yan, C. Feature-based phase retrieval wavefront sensing approach using machine learning. *Opt. Express* **2018**, *26*, 31767–31783.
26. Paykin, I.; Yacobi, L.; Adler, J.; Ribak, E.N. Phasing a segmented telescope. *Phys. Rev. E Stat. Nonlinear Soft Matter Phys.* **2015**, *91*, 023302. [CrossRef] [PubMed]
27. Hagan, M.T.; Beale, M. *Neural Network Design*; China Machine Press: Bejing, China, 2002.
28. Guo, H.; Xu, Y.; Li, Q.; Du, S.; He, D.; Wang, Q.; Huang, Y. Improved Machine Learning Approach for Wavefront Sensing. *Sensors* **2019**, *19*, 3533. [CrossRef] [PubMed]
29. Hu, S.; Hu, L.; Zhang, B.; Gong, W.; Si, K. Simplifying the detection of optical distortions by machine learning. *J. Innov. Opt. Health Sci.* **2019**, *13*. [CrossRef]
30. Tian, Q.; Lu, C.; Liu, B.; Zhu, L.; Pan, X.; Zhang, Q.; Yang, L.; Tian, F.; Xin, X. DNN-based aberration correction in a wavefront sensorless adaptive optics system. *Opt. Express* **2019**, *27*, 10765–10776. [CrossRef] [PubMed]
31. Zhao, W.; Zeng, Q. Simultaneous multi-piston measurement method in segmented telescopes. *Opt. Express* **2017**, *25*, 24540–24552. [CrossRef]
32. Ma, X.; Xie, Z.; Ma, H.; Xu, Y.; Ren, G.; Liu, Y. Piston sensing of sparse aperture systems with a single broadband image via deep learning. *Opt. Express* **2019**, *27*, 16058. [CrossRef]

sensors

MDPI

Article

Object Detection Combining CNN and Adaptive Color Prior Features

Peng Gu [1,2], Xiaosong Lan [2] and Shuxiao Li [1,2,*]

1 School of Artificial Intelligence, University of Chinese Academy of Sciences, Beijing 100049, China; gupeng2018@ia.ac.cn
2 Institute of Automation, Chinese Academy of Sciences, Beijing 100190, China; lanxiaosong2012@ia.ac.cn
* Correspondence: shuxiao.li@ia.ac.cn

Abstract: When compared with the traditional manual design method, the convolutional neural network has the advantages of strong expressive ability and it is insensitive to scale, light, and deformation, so it has become the mainstream method in the object detection field. In order to further improve the accuracy of existing object detection methods based on convolutional neural networks, this paper draws on the characteristics of the attention mechanism to model color priors. Firstly, it proposes a cognitive-driven color prior model to obtain the color prior features for the known types of target samples and the overall scene, respectively. Subsequently, the acquired color prior features and test image color features are adaptively weighted and competed to obtain prior-based saliency images. Finally, the obtained saliency images are treated as features maps and they are further fused with those extracted by the convolutional neural network to complete the subsequent object detection task. The proposed algorithm does not need training parameters, has strong generalization ability, and it is directly fused with convolutional neural network features at the feature extraction stage, thus has strong versatility. Experiments on the VOC2007 and VOC2012 benchmark data sets show that the utilization of cognitive-drive color priors can further improve the performance of existing object detection algorithms.

Keywords: convolutional neural network; color prior model; object detection

Citation: Gu, P.; Lan, X.; Li, S. Object Detection Combining CNN and Adaptive Color Prior Features. *Sensors* **2021**, *21*, 2796. https://doi.org/10.3390/s21082796

Academic Editor: Hyoungsik Nam

Received: 19 February 2021
Accepted: 12 April 2021
Published: 15 April 2021

1. Introduction

Over the past few decades, the amount of visual image information has grown at an explosive rate. As a relatively accurate and vivid description of the objective world, visual image information is one of the mainstream forms for humans for understanding the world and receiving external information. Visual image information is difficult for computers to understand. How to process these image data and use the visual image information to improve people's lives to the greatest extent is a very important subject, which is also the main problem in the field of computer vision [1]. Object detection is a very challenging research subject in the field of computer vision. Its main task is to output the bounding box position and classification confidence score of the target of interest in the test image [2]. Object detection algorithms can bring great convenience to our lives, such as the use of face detection algorithms to assist the camera's autofocus, the use of medical image detection algorithms to assist doctors in accurately identifying disease features, and the use of object detection and tracking algorithms to realize the following flight of the UAV to the interested target, etc.

The main difficulties faced by the object detection algorithm are, as follows: (1) there are factors, such as uneven illumination, different postures, large scales, mutual occlusion, etc., which lead to problems, such as missed detection and false detection; (2) under the interference of complex background, the accuracy of the algorithm will drop sharply. In some environments with simple backgrounds and fixed scenes, traditional object detection algorithms can usually achieve satisfactory results. However, with the increasing demand of detection accuracy and speed in the fields of target tracking and automatic driving, traditional detection algorithms become unqualified. Therefore, how to detect targets in complex scenes and improve the detection speed is the focus of research in the field of object detection [3–6].

In recent years, deep learning technology has swept the field of artificial intelligence. Deep learning technology has made significant progress in computer vision [7], natural language processing [8], speech recognition [9], and other fields. Among them, the convolutional neural network technology relies on its advantages, such as scale invariance and translation invariance, and it has set off a storm of deep learning in the field of computer vision. Therefore, the convolutional neural network technology has become the mainstream algorithm in the current object detection field [10–14] .

Visual attention is the cognitive process by which people obtain the most critical area information when observing natural scenes. Different parts of the human retina have different information processing capabilities. Therefore, this mechanism is very important for the efficient operation of the human visual system [15,16]. The human visual system will not consider equally for all areas in a natural scene in order to make reasonable use of the limited visual processing resources. Actually, the human visual system will selectively pay attention to the target area of interest, and ignore or suppress some less important background areas. There are two different methods of visual attention: one is bottom-up attention; the other is top-down driven attention that is based on specific tasks.

Being inspired by the top-down visual attention mechanism, we introduced the concept of color prior features [17]. Because different types of objects usually have different color features, this can help us to improve the ability to discern objects. Therefore, we propose a cognitive-driven color prior model for object detection. Two main contributions are included in our method. One aspect is that we propose an adaptive color prior model that comprehensively considers the fusion and competition among category pattern distribution (the category pattern distribution represents the probability of each pattern occurring in the target image block in labeled dataset for the class), scene pattern distribution (scene pattern distribution represents the probability of each pattern in all visual scene images), and test image pattern distribution (test image pattern distribution represents the probability of each pattern in the test image). It uses test image features to dynamically and adaptively adjust the memory prior to obtain more robust color prior information. The other aspect is that saliency images that are acquired by the proposed color prior model are regarded as feature maps and they are fused with those by convolutional neural networks, regardless of backbone network types, thus having strong versatility. The experiments show that the utilization of cognitive-drive color priors can further improve the performance of existing object detection algorithms.

The rest of the paper is arranged, as follows: Section 2 introduces related work, and Section 3 details the proposed methods. Section 4 analyzes the ablation experiments of the color prior model in Faster R-CNN and the contrast experiments in different types of object detection algorithms. Section 5 gives the main conclusions.

2. Related Work

This section mainly introduces related work from two aspects: object detection algorithm and saliency detection algorithm.

2.1. Object Detection Algorithm

According to different feature extraction methods, object detection algorithms can be roughly divided into traditional object detection algorithms and object detection algorithms that are based on deep neural networks.

Traditional object detection algorithms rely on carefully designed features to detect targets [18]. This type of method usually first designs a specific algorithm to screen out the most likely target areas, calculates the feature description of the candidate region according to the manually designed feature extraction rules, and then finally uses the pattern classification model to determine whether these candidate regions are real targets. The representative work of candidate region selection is sliding window and selective search algorithm [19]. The candidate region selection algorithm that is based on sliding window requires a large number of sizes and shapes of sliding windows, which will lead to huge computational redundancy. The selective search algorithm first divides the image into multiple small regions, and then further merges to obtain a larger candidate region, which improves the computational efficiency. Feature extraction is mainly designed for different tasks such as face detection and pedestrian detection. It includes the cognitive thinking wisdom of human experts. The most influential feature description methods include Scale-Invariant Feature Transform (SIFT) [20], Local Binary Pattern (LBP) [21], Histogram of Oriented Gradient (HOG) [22], etc. The main goal of pattern classification is to remove invalid candidate regions, which can be achieved by training effective classifiers. Commonly used algorithms are Support Vector Machines (SVM) [23], AdaBoost [24], etc. Although traditional object detection algorithms that are based on manual feature extraction have been developed to some extent since the end of the last century, there are still many shortcomings in practice. For example, HOG, SIFT, etc. are basically low-level features, such as contours and textures, which are less robust and cannot cope with object detection requirements in a diversified and complex environment.

The object detection algorithm that is based on deep neural network is a new type of method that has emerged in recent years. It can automatically form more abstract high-level features by combining low-level features from samples. These features have powerful expression and generalization capabilities. This is the current mainstream method of object detection [12,25,26]. In 2012, the AlexNet network [27] proposed by Alex Krizhevsky et al. achieved results far surpassing traditional object detection algorithms in the large-scale visual recognition challenge, which made deep neural network technology attract people's attention in the field of image recognition and object detection. After several years of development, deep neural networks have been widely used in object detection tasks [28–33]. These algorithms are mainly divided into object detection algorithms that are based on candidate boxes and object detection algorithms based on regression. The object detection algorithm based on the candidate boxes is also called a two-stage type algorithm. It first extracts the region proposal, and then performs candidate boxes recognition and boxes regression. R-CNN series is the representative work [30,31,34]. R-CNN [34] uses a selective search algorithm to extract candidate frames, and it starts to use deep neural networks to extract features, and finally uses support vector machines to complete the classification of the target. Fast R-CNN [30] performs feature pooling for each candidate frame, and uses the softmax classifier to replace the support vector machine. It only needs to extract the features of the image once, which improves the training and reference speed. Faster R-CNN [31] uses neural networks to generate candidate boxes, instead of selective search algorithms, so that the entire object detection truly realizes end-to-end calculation. At the same time, region proposal, classification, and regression share convolution features, greatly improving the accuracy and computational efficiency of the algorithm. The object detection algorithm that is based on the regression idea is also called a single-stage type algorithm, which skips candidate box extraction stage and directly regards the object detection algorithm as a regression task, such as YOLO [32] and SSD [33]. For an input image, the regression-based object detection model directly outputs the categories and positions of all targets for each image block, and all of the detection steps are completed in a neural network. The advantage is that the algorithm is efficient, but the accuracy is usually slightly lower than that of the two-stage algorithm.

2.2. Saliency Detection Algorithm

In human cognitive science, different parts of the retina have different information processing capabilities. In order to make rational use of limited visual information processing resources, the human visual system usually selectively focuses on specific parts of the visual scene. This phenomenon is called the visual attention mechanism, which is the theoretical basis of saliency detection. The current saliency detection methods can be roughly divided into two types, one is a bottom-up saliency detection algorithm, which is data-driven, and the other is a top-down saliency detection algorithm, which is task-driven.

The bottom-up saliency detection algorithm directly extracts the underlying information of the image for detection, and it does not need to specify the type of target of interest or provide training samples in advance, so it has a wide range of adaptations. Itti and Koch et al. proposed a saliency detection algorithm earlier by learning from neuron structure and animal vision [35,36]; Xie et al. [37] proposed using the Bayesian model to construct saliency maps; and, Cheng et al. [38] proposed a saliency detection method based on global contrast.

The top-down saliency detection method is task-driven, which is, it is guided by human subjective consciousness or target tasks. It first learns the basic information of the target object at the training stage with supervision, and then uses the learned information to increase the significance of the expected region or the target of interest. Jing et al. [39] proposed the use of supervision to obtain prior information from the data set. Liu et al. [40] proposed using conditional random fields to generate a saliency map based on the extracted features, such as the center surround histogram, multi-scale contrast, and color space distribution. Zhang et al. [41] proposed not using a specific image set, but to obtain statistical information directly from the natural image set, and then build a saliency detection model that is based on the Bayesian method.

3. Proposed Method

3.1. Overall Structure

Figure 1 shows the overall flow of the algorithm. Figure 1a is the off-line memory stage. Based on training images, color priors are calculated through probability statistics and saved in the form of look-up tables for on-line mapping. Figure 1b is the on-line mapping stage. Adaptive color prior weights are first calculated through pattern competition and fusion. Subsequently, the acquired color prior saliency features of the test image are obtained through indexing the lookup table. Finally, the image pyramid technique is utilized to obtain the corresponding levels of color prior features $\{C4, C5, C6\}$, which are fused with deep neural network features $\{P4, P5, P6\}$ to perform the subsequent detection tasks. The ideas behind the two stages are detailed below.

(1) "Off-line memory" stage: drawing on the characteristics of human memory, the category pattern distribution of typical targets and the category-independent scene pattern distribution are separately established based on object detection databases and stored in a table. In order to facilitate memory and calculations, patterns are used to represent the local characteristics of the target (the patterns are essentially limited one-dimensional discrete numbers, such as grayscale, serialized colors, local binary patterns, etc., this article focuses on YUV color features). For a given object detection dataset, the category pattern distribution is established by aggregating the pattern frequencies in the target labeling boxes of the same category, and the scene pattern distribution is calculated by aggregating the pattern frequencies of all training images.

(2) "On-line mapping" stage: the object of interest in object detection usually only occupies a small part of the image, and it is "significant" relative to the test image. Therefore, only colors that are "significant" in the category pattern distribution and "rare" in both the scene pattern distribution and the test image pattern distribution should be enhanced. Thus, the saliency value of the color prior needs to be adaptively adjusted according to the test image. The saliency value of the color prior is obtained through the fusion and competition between the category pattern, the scene pattern, and the test image pattern, thereby establishing a dynamic adaptive color prior model that is represented by $\{pattern, saliency\}$. Borrowing the characteristics of human memory, the color saliency features of the test image are then obtained in the form of index lookup table [17], and are finally fused with the convolutional neural network features to work together on the object detection task.

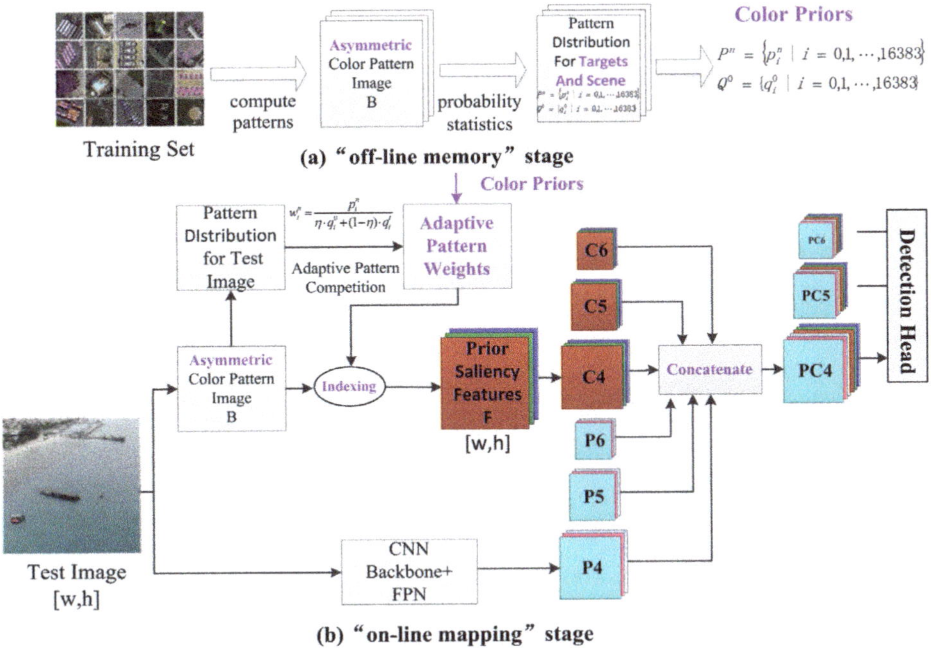

Figure 1. Overall architecture of object detection method combined with color prior model.

3.2. "Off-Line Memory" Stage

This stage first introduces the representation method of the asymmetric color pattern image, and then gives the calculation process of the category pattern distribution and the scene pattern distribution, respectively, which can be saved offline.

3.2.1. Asymmetric Color Pattern Image

In this paper, patterns are used to represent the local features of images, so that memory and mapping can be achieved through probability statistics and indexing, respectively. Let $I \in R^{w \times h \times 3}$ be the input image, where w and h are the image width and image height, respectively. We use the pattern operation Φ to transform the input image to a pattern image $B \in Z^{w \times h}$ (discrete integer space), which can be expressed as:

$$B = \Phi(I) \tag{1}$$

Pattern operation Φ can be the serialization of color features, or texture feature operators, such as local binary pattern and ordinal features. In fact, when humans perform color matching, they pay more attention to the difference of color components, and allow a certain change in the brightness component to achieve illumination invariance to some extent. In order to express this characteristic, the RGB format image is converted to the YUV format image, and the asymmetric color pattern image is obtained. As we know, the YUV color model is closer to the human visual mechanism than RGB color model, thus it is more suitable for establishing the color prior model. In this paper, we quantize and splice the values of the three channels of Y, U, and V to obtain an asymmetric color pattern image. Quantizing the color channels can reduce the total number of patterns. In this paper, the brightness component Y is quantized to 16 levels, and the color components U and V are quantized to 32 levels, so as to allow certain brightness changes under the premise of better color distinctness. Thus, each pattern occupies 14 bits, and the value range is [0,16383]. Let

I_Y, I_U and I_V be the values of the Y, U, and V channels of the input image I, the calculation process of the asymmetric color pattern image can be expressed as:

$$\begin{aligned} B_c &= Con\{I_Y \gg 4, I_U \gg 3, I_V \gg 3\} \\ &= (I_Y \gg 4) \ll 10 + (I_U \gg 3) \ll 5 + (I_V \gg 3) \end{aligned} \tag{2}$$

Converting YUV images to pattern images is beneficial for memory and mapping. Memory can be simply done by summing up the pattern occurrence frequencies in target labeling images, from which color prior saliency for different types of targets can be learned and stored in advance. When recognizing the world, the memorized priors can be picked up by indexing, which is similar to human conditioned reflex.

3.2.2. Category Pattern Distribution and Scene Pattern Distribution

There are many ways to express color features. This article adopts the form of pattern histogram. The pattern histogram expresses the probability distribution of different color patterns in typical category targets and scene images. It discards the spatial location information of color patterns in different images, thus it is translation and rotation invariant. For the object detection data set, each image may contain multiple targets of different categories, so we perform category pattern statistics that are based on the target block image I_o in the target labeling box. Denote $I_o^{n,k}$ as the k-th target block image of the n-th category target, then its pattern probability distribution can be obtained by statistical histogram and normalization, which is recorded as:

$$\begin{aligned} P^{n,k} &= Normalize\{Hist\lfloor \Phi_c(I_o^{n,k}) \rfloor\} \\ &= \{p_i^{n,k}, i = 0, 1, \cdots, 16383\} \end{aligned} \tag{3}$$

The category pattern distribution can be obtained by continuously "memorizing" the pattern probability distribution of all the target block images of the same category. For the training data set without time label, we take the mean value of all target block images of the same category as the final probability pattern distribution:

$$P^n = \left\{ p_i^n = \frac{1}{N^n} \sum_{k=1}^{N^n} p_i^{n,k}, i = 0, 1, \cdots, 16383 \right\} \tag{4}$$

where N^n is the total number of labeled samples of the n-th category.

There are always many different objects in natural scenes. The visual system usually weights key areas via a top-down manner based on experience or interest in order to make rational use of visual resources. This can be achieved through a visual competition mechanism. If the occurrence frequency of pattern i of the k-th image is higher than that of natural scenes or other types of targets, it means that the pattern i of the k-th image is "competitive" and it should be enhanced, otherwise it will be suppressed. Denote $P^m = \{p_i^m, i = 0, 1, ..., 16383\}$ as the pattern distribution of the m-th training image; we approximate the scene pattern distribution Q^0 with the mean of all training images:

$$Q^0 = \left\{ q_i^0 = \frac{1}{M} \sum_{k=1}^{M} p_i^k, i = 0, 1, \cdots, 16383 \right\} \tag{5}$$

where M is the total number of all training images.

3.3. *"On-Line Mapping" Stage*

This stage first establishes a dynamic adaptive color prior model through the fusion and competition between category pattern, scene pattern, and test image pattern. Subsequently, indexing and pooling operations are employed to generate different levels of color prior saliency features, which are finally combined with the convolutional neural network features to jointly achieve the object detection task.

3.3.1. Dynamic Adaptive Color Prior Model

In the cognitive world, the human cerebral cortex reflexively obtains the prior information of typical objects to assist object detection. In this paper, the color space is converted into a pattern value, so that the prior saliency value of the color pattern of the corresponding category object can be retrieved in order to assist in identifying important objects. Specifically, the color prior model of the n-th category object can be expressed as:

$$W^n = \{w_i^n, i = 0, 1, \cdots, 16383\}, n \in [1, N] \tag{6}$$

where w_i^n is the prior saliency value of the i-th color pattern of the n-th object class and N is the total number of object categories.

In the object detection scene, the object usually only occupies a small part of the image, and it is "significant" relative to the test image. Therefore, only colors that are "significant" in the category pattern distribution and "rare" in both the scene pattern distribution and the test image pattern distribution should be enhanced. That is to say, the saliency value of the color prior needs to be adaptively adjusted with the test image. For the test image I, we calculate its asymmetric color pattern image B according to formula (1), and then calculate the pattern distribution Q^t of the test image, as below:

$$Q^t = Normalize\{Hist\lfloor B \rfloor\} = \{q_i^t, i = 0, 1, \cdots, 16383\} \tag{7}$$

Subsequently, the pattern weight can be obtained by fusion and competition among the object pattern, scene pattern, and test image pattern:

$$W^n = \left\{w_i^n = \frac{p_i^n}{\eta \cdot q_i^0 + (1-\eta) \cdot q_i^t}, i = 0, 1, \cdots, 16383\right\}, n = 1, 2, \cdots, N \tag{8}$$

where η is the forgetting factor of the scene pattern. The forgetting factor is introduced, because, in addition to the scene pattern, it also needs to consider the influence of the test image on the pattern weight.

3.3.2. Feature Map Generation

For the input image $I \in R^{w \times h \times 3}$, the pattern operation Φ is used to map it to the pattern image $B \in Z^{w \times h}$. For each pixel $x \in R^2$ in the pattern image, $B(x) \in Z$ is the color pattern of the pixel. Taking $B(x)$ as the index and reading the corresponding prior saliency values $\{w_{B(x)}^n, n = 1, 2, \cdots, N\}$ from the color prior model W^n as its feature, we get N prior features for each pixel of the input image. Finally, the prior saliency values of all the pixels are organized in the form of multi-channel images to obtain priori saliency features $F \in R^{w \times h \times N}$ with N channels.

Subsequently, the obtained color prior saliency images are treated as features maps and they are fused with those extracted by the convolutional neural network to complete the subsequent object detection task. Taking Faster R-CNN as an example, the convolutional neural network uses ResNet-101, and it employs the feature pyramid network to extract multiple levels of features namely P2, P3, P4, P5, and P6. For the color prior model, we utilize the image pyramid technique to pool the saliency maps into multiple corresponding levels of saliency maps, namely C2, C3, C4, C5, and C6. We regard them as feature maps and fuse them with multiple hierarchical features extracted by the convolutional neural network together for the following object detection modules. In the detection task, the fused features are firstly input to the region proposal network to obtain regions of interests (ROIs). Subsequently, ROI pooling operation is utilized to extract regional features for each ROI. Finally, the acquired regional features are inputs to fully connected networks to output category probabilities and bounding box parameters.

4. Experiments

4.1. Datasets

This paper uses the cognitive-driven color prior model to improve the performance of the object detection algorithm. We verify the effectiveness of the proposed algorithm on the PASCAL VOC dataset. We use the AP (Average Precision) of each category object to measure the detection precision of the algorithm in different categories of objects, and then use the average precision of all categories to measure the advantages and disadvantages of the overall performance of the algorithm.

In the field of computer vision, PASCAL VOC is a set of standardized high-quality data sets, which are mainly used for tasks, such as object recognition, image segmentation, and object detection. PASCAL VOC marked a total of 20 categories of objects, namely: people, birds, cats, dogs, cattle, sheep, horses, chairs, bottles, potted plants, dining tables, TVs, sofas, bicycles, airplanes, buses, boats, cars, trains, and motorcycles. Figure 2 shows a typical example of the PASCAL VOC2007 data set. It contains 9963 images and a total of 24,640 labeled objects. It is composed of three parts, namely ghd train/val/test. The PASCAL VOC2012 data set is an upgraded version of PASCAL VOC2007. Figure 3 shows a typical example. The training set has 11,540 images and a total of 27,450 labeled objects. This paper uses the train/val parts of VOC2007 and VOC2012 as the training set, and the test part of VOC2007 as the test set to verify the proposed algorithm.

Figure 2. PASCAL VOC2007 datasets.

Figure 3. PASCAL VOC2012 datasets.

4.2. Implementation Details

The programming software for algorithm verification is the VsCode integrated development environment with Ubuntu 18.04 operating system and Cuda environment. We use an Intel i9-9920X CPU, 4 NVIDIA GeForce RTX2080TL GPUs and 96G DDR3 memories. The deep learning framework is Pytorch.

We use Faster R-CNN+FPN [42] to undertake ablation experiments on the PASCAL VOC data set, where the backbone network adopts ResNet-101 with pre-training weights on the ImageNet data set. The training period is set to 12, the batch size to 1, the initial learning rate to 0.001, and other parameters are consistent with the open source project [31].

The color prior model involves an important parameter η. In order to make the color prior model achieve the best results, we conduct a qualitative analysis on the parameter η, and finally take η = 0.9 for ablation experiments and comparison experiments.

We select four classic target detection frameworks for comparison experiments, including SSD [33], RetinaNet [43], Cascade-R-CNN [44], and Libra-R-CNN [45], in order to verify the generality of the color prior model. The implementation of these algorithms mainly refers to the mmdetection library. Mmdetection is an open source library that is based on Pytorch and it currently supports many mainstream object detection models. Among them, Libra-R-CNN [45] selects ResNet-50 [7] as the backbone network; SSD [33] sets the input size to 300 and selects VGG16 [46] as the backbone network; Cascade-R-CNN [44] selects ResNet-50 [7] as the backbone network; and, RetinaNet [43] selects ResNet-50 [7] as the backbone network.

4.3. Overall Performance Verification of the Algorithm

We select Faster R-CNN+FPN as the baseline and compare the AP of the object detection network with/without color priors on the VOC datasets in order to verify the effectiveness of the color prior model. Faster R-CNN selects ResNet-101 to extract the basic features of the input image, and it uses FPN to construct high-level semantic features at various scales. FPN adopts a top-down hierarchical structure with side links. The advantage is that it can make reasonable use of the inherent multi-scale and hierarchical structure of deep convolutional networks, so that features at different scales have strong semantic information. Through CNN+FPN, the test image obtains multiple feature layers, namely $\{P2, P3, P4, P5, P6\}$, and it generates saliency maps through the color prior model. The acquired saliency maps are then pooled into the corresponding multiple hierarchical features namely $\{C2, C3, C4, C5, C6\}$ through the image pyramid technique (the feature maps are illustrated in Figure 4). The two sets of features are spliced and input into the detection head RPN for the subsequent detection modules.

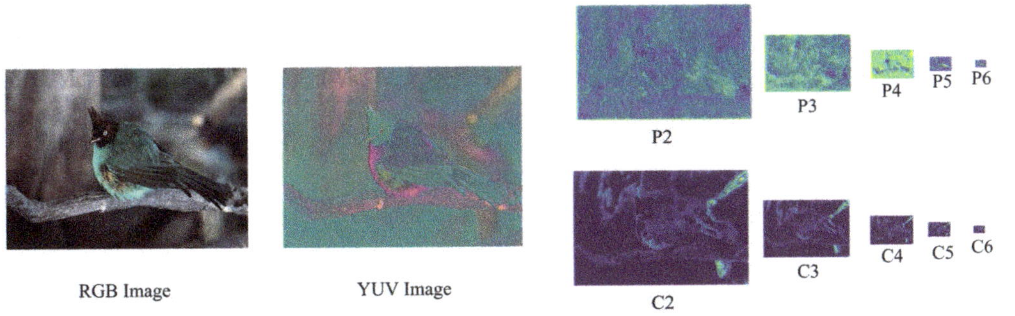

Figure 4. Test image and multi-level features.

Table 1 shows the performance comparison of the object detection algorithm with/without the color prior model. On the VOC2007 data set and VOC07+12 data set, the mAP of the object detection network combined with the color prior model surpasses the baselines by 1.1% and 1.0%, respectively. It shows that the color prior model is effective and it can improve the performance of the object detection algorithm on different data sets. Table 2 shows the speed comparison of whether to use the color prior model. We implement the color prior model in Python language in order to quickly verify the experimental effect. Because pixel-level image operations in Python are inefficient, the speed of current version is slow. We will reimplement it in C language in future work to improve the efficiency.

Table 1. Performance comparison of whether to use the color prior model.

	Backbone	Datasets	ColorPriors	mAP
Faster R-CNN+FPN	ResNet-101	VOC2007	×	0.756
Faster R-CNN+FPN	ResNet-101	VOC2007	✓	0.764
Faster R-CNN+FPN	ResNet-101	VOC07+12	×	0.803
Faster R-CNN+FPN	ResNet-101	VOC07+12	✓	0.811

Table 2. Speed comparison of whether to use the color prior model (datasets:VOC2007).

	Backbone	ColorPriors	Train Speed/Epoch	Test Speed/Epoch
Faster R-CNN+FPN	ResNet-101	×	0.5 h	0.08 h
Faster R-CNN+FPN	ResNet-101	✓	0.66 h	0.16 h

Figure 5 shows the result images of the benchmark algorithm and the proposed algorithm in a typical experiment of object detection. It can be seen from the figure that the benchmark algorithm has an error detection (left column in Figure 5), class error (middle column in Figure 5), and missed detection (right column in Figure 5). However, the object to be detected still has certain color characteristics, and the correct detection result can be obtained by combining the color prior model, which proves that the cognitive-driven color prior model can enhance the performance of the target detection algorithm.

Figure 5. The typical result images of target detection, the first row is from the benchmark algorithm, and the second row is from the proposed algorithm.

4.4. Ablation Study

We conduct ablation experiments according to whether the following design ideas are included:

- YUV asymmetric color pattern: Another option is to use the RGB color pattern.
- Whether to use forgetting factor η.
- Fusion methods of color features and network features: a variety of feature fusion methods are designed and compared, and the experiments are performed to verify that the best feature fusion method is selected.

Table 3 shows the performance comparison of object detection algorithms using YUV asymmetric color pattern and RGB color pattern. The RGB color pattern is formed by quantizing the R, G, and B components into 32 levels and splicing them, thus each pattern occupies 16 bits. Meanwhile, the YUV asymmetric color pattern that is proposed in this article only occupies 14 bits, which reduces the storage capacity of the color prior model and increases the calculation speed. It can be seen from the table that the mAP of using the YUV asymmetric color pattern has advantages in both the VOC2007 data set and the VOC07+12 data set, indicating that it is more effective than the RGB color pattern.

Table 3. Performance comparison of object detection algorithms using different color spaces.

	Backbone	Datasets	ColorPriors	Color Pattern	mAP
Faster R-CNN+FPN	ResNet-101	VOC2007	✓	RGB	0.762
Faster R-CNN+FPN	ResNet-101	VOC2007	✓	YUV	0.764
Faster R-CNN+FPN	ResNet-101	VOC07+12	✓	RGB	0.808
Faster R-CNN+FPN	ResNet-101	VOC07+12	✓	YUV	0.811

In order to obtain more robust color prior information, we introduce the forgetting factor η, using test image features to dynamically and adaptively adjust the memory prior. Table 4 shows the performance comparison of object detection algorithms with/without the forgetting factor. From the table it can be seen that the color prior model using the forgetting factor is better than the case where the forgetting factor is not used on the VOC2007 data set and the VOC07+12 data set, which proves that the forgetting factor is useful for the color prior model.

Table 4. Performance comparison of object detection algorithms with/without forgetting factor.

	Backbone	Datasets	ColorPriors	η	mAP
Faster R-CNN+FPN	ResNet-101	VOC2007	✓	×	0.760
Faster R-CNN+FPN	ResNet-101	VOC2007	✓	✓	0.764
Faster R-CNN+FPN	ResNet-101	VOC07+12	✓	×	0.807
Faster R-CNN+FPN	ResNet-101	VOC07+12	✓	✓	0.811

In the feature fusion stage, we compared various fusion strategies, as shown in Table 5, where C represents the color prior features, P represents the neural network features, C_P represents splicing C and P, PH represents multiplying P by the channel mean value of C pixel by pixel as spatial saliency weighting, PH_P represents splicing PH and P, and PH+P represents the addition of PH and P. Through comparative experiments, it can be seen that the feature fusion method of C_P has the best effect. The PH_P and PH+P fusion methods can also improve the performance of the detection network, but the effect is not as good as the C_P method. The PH method reduces the performance of the object detection network. The reason may be that the spatial saliency weighting inhibits part of the effective feature expression of the neural network. We will continue to pay attention to the research in this area in future experiments.

Table 5. Performance comparison of object detection algorithms using different feature fusion strategies.

	Backbone	Datasets	ColorPriors	Fusion Strategy	mAP
Faster R-CNN+FPN	ResNet-101	VOC2007	✓	C_P	0.764
Faster R-CNN+FPN	ResNet-101	VOC2007	✓	PH	0.737
Faster R-CNN+FPN	ResNet-101	VOC2007	✓	PH_P	0.759
Faster R-CNN+FPN	ResNet-101	VOC2007	✓	PH+P	0.759
Faster R-CNN+FPN	ResNet-101	VOC07+12	✓	C_P	0.811
Faster R-CNN+FPN	ResNet-101	VOC07+12	✓	PH	0.779
Faster R-CNN+FPN	ResNet-101	VOC07+12	✓	PH_P	0.805
Faster R-CNN+FPN	ResNet-101	VOC07+12	✓	PH+P	0.804

4.5. Comparative Study

We select four classic object detection networks for grouping experiments in order to verify the universality of the color prior model. Table 6 shows the comparative experimental results on VOC2007 and VOC07+12 data sets. It can be seen from the table that: (1) color prior model can be used to improve the accuracy for all the above-mentioned object detection networks, which proves that the proposed method is universal to different object detection networks. (2) The object detection networks utilizing the color prior model obtained improved performances when compared with those of the original frameworks on both the VOC2007 data set and VOC07+12 data set. This proves that the color prior model has strong versatility under different data sets.

Table 6. Comparative experiments of different target detection networks.

	Backbone	ColorPriors	mAP Using VOC07	mAP Using VOC07+12
Cascade R-CNN	ResNet-50	×	0.726	0.781
	ResNet-50	✓	0.732	0.788
SSD300	VGG16	×	0.707	0.775
	VGG16	✓	0.712	0.782
Libra R-CNN	ResNet-50	×	0.743	0.808
	ResNet-50	✓	0.748	0.813
RetinaNet	ResNet-50	×	0.712	0.793
	ResNet-50	✓	0.717	0.797

5. Conclusions

This paper proposes an object detection method utilizing the color prior model. Specifically, we first learn from the visual attention mechanism to calculate the scene pattern distribution and category pattern distributions from annotated datasets, and save them in the form of tables off-line. For the on-line phase, the scene pattern distribution, category pattern distribution, and test image pattern distribution are competed and fused to generate adaptive color pattern weights, based on which color prior features can be efficiently obtained through indexing. Finally, the acquired color prior features are fused with CNN features for the subsequent object detection modules. The proposed color prior model is cognitively driven and it has no training parameters, so it has strong generalization ability. The experiments show that color priors can effectively improve the performance of object detection networks with different structures. In future work, we will investigate the effects of other color patterns, explore the self adaptability method of the forgetting factor, and study the fusion strategy between the color prior model and the object detection network at different stages.

Author Contributions: Conceptualization, P.G. and S.L.; methodology, P.G., X.L. and S.L.; software, P.G.; formal analysis, P.G. and S.L.; investigation, P.G. and X.L.; resources, X.L. and S.L.; data curation, P.G.; writing—original draft preparation, P.G.; writing—review and editing, X.L. and S.L.; supervision, S.L.; project administration, X.L. and S.L.; funding acquisition, X.L. and S.L. All authors have read and agreed to the published version of the manuscript.

Funding: This work is partly supported by National Natural Science Foundation of China (Grant No. U19B2033, Grant No.62076020), and National Key R&D Program (Grant No. 2019YFF0301801).

Institutional Review Board Statement: Not applicable.

Informed Consent Statement: Not applicable.

Data Availability Statement: Not applicable.

Conflicts of Interest: The authors declare no conflict of interest.

References

1. Zhao, Z.Q.; Zheng, P.; Xu, S.T.; Wu, X. Object detection with deep learning: A review. *IEEE Trans. Neural Netw. Learn. Syst.* **2019**, *30*, 3212–3232. [CrossRef]
2. Chen, Z.; Li, J.; You, X. Learn to focus on objects for visual detection. *Neurocomputing* **2019**, *348*, 27–39. [CrossRef]
3. Gupta, D.; Anantharaman, A.; Mamgain, N.; Balasubramanian, V.N.; Jawahar, C. A multi-space approach to zero-shot object detection. In Proceedings of the IEEE Winter Conference on Applications of Computer Vision, Snowmass, CO, USA, 1–5 March 2020; pp. 1209–1217.
4. Liu, Y.; Wang, Y.; Wang, S.; Liang, T.; Zhao, Q.; Tang, Z.; Ling, H. Cbnet: A novel composite backbone network architecture for object detection. In Proceedings of the AAAI Conference on Artificial Intelligence, New York, NY, USA, 7–12 February 2020; Volume 34, pp. 11653–11660.
5. Oksuz, K.; Cam, B.C.; Akbas, E.; Kalkan, S. Generating positive bounding boxes for balanced training of object detectors. In Proceedings of the IEEE Winter Conference on Applications of Computer Vision, Snowmass, CO, USA, 1–5 March 2020; pp. 894–903.
6. Qin, Z.; Li, Z.; Zhang, Z.; Bao, Y.; Yu, G.; Peng, Y.; Sun, J. ThunderNet: Towards real-time generic object detection on mobile devices. In Proceedings of the IEEE International Conference on Computer Vision, Seoul, Korea, 27 October–2 November 2019; pp. 6718–6727.
7. He, K.; Zhang, X.; Ren, S.; Sun, J. Deep residual learning for image recognition. In Proceedings of the IEEE Conference on Computer Vision and Pattern Recognition, Las Vegas, NV, USA, 27–30 June 2016; pp. 770–778.
8. Gardner, M.; Grus, J.; Neumann, M.; Tafjord, O.; Dasigi, P.; Liu, N.; Peters, M.; Schmitz, M.; Zettlemoyer, L. Allennlp: A deep semantic natural language processing platform. *arXiv* **2018**, arXiv:1803.07640.
9. Nassif, A.B.; Shahin, I.; Attili, I.; Azzeh, M.; Shaalan, K. Speech recognition using deep neural networks: A systematic review. *IEEE Access* **2019**, *7*, 19143–19165. [CrossRef]
10. Tan, M.; Pang, R.; Le, Q.V. Efficientdet: Scalable and efficient object detection. In Proceedings of the IEEE/CVF Conference on Computer Vision and Pattern Recognition, Seattle, WA, USA, 13–19 June 2020; pp. 10781–10790.
11. Carion, N.; Massa, F.; Synnaeve, G.; Usunier, N.; Kirillov, A.; Zagoruyko, S. End-to-end object detection with transformers. In Proceedings of the European Conference on Computer Vision, Glasgow, UK, 23–28 August 2020; Springer: Berlin/Heisenberg, Germany, 2020; pp. 213–229.
12. Wu, X.; Sahoo, D.; Hoi, S.C. Recent advances in deep learning for object detection. *Neurocomputing* **2020**, *396*, 39–64. [CrossRef]
13. Jia, P.; Liu, F. Lightweight feature enhancement network for single-shot object detection. *Sensors* **2021**, *21*, 1066. [CrossRef]
14. Gong, B.; Ergu, D.; Cai, Y.; Ma, B. Real-time detection for wheat head applying deep neural network. *Sensors* **2021**, *21*, 191. [CrossRef]
15. Sepulveda, P.; Usher, M.; Davies, N.; Benson, A.A.; Ortoleva, P.; De Martino, B. Visual attention modulates the integration of goal-relevant evidence and not value. *Elife* **2020**, *9*, e60705. [CrossRef] [PubMed]
16. Khachatryan, H.; Rihn, A.; Behe, B.; Hall, C.; Campbell, B.; Dennis, J.; Yue, C. Visual attention, buying impulsiveness, and consumer behavior. *Mark. Lett.* **2018**, *29*, 23–35. [CrossRef]
17. Shen, Y.; Li, S.; Zhu, C.; Chang, H. A fast top-down visual attention method to accelerate template matching. *Comput. Model. New Technol.* **2014**, *18*, 86–93.
18. Lowe, D.G. Distinctive image features from scale-invariant keypoints. *Int. J. Comput. Vis.* **2004**, *60*, 91–110. [CrossRef]
19. Zitnick, C.L.; Dollár, P. Edge boxes: Locating object proposals from edges. In Proceedings of the European Conference on Computer Vision, Zurich, Switzerland, 6–12 September 2014; Springer: Berlin/Heisenberg, Germany, 2014; pp. 391–405.
20. Lowe, D.G. Object recognition from local scale-invariant features. In Proceedings of the IEEE International Conference on Computer Vision, Corfu, Greece, 20–27 September 1999; pp. 1150–1157.
21. Ojala, T.; Pietikainen, M.; Maenpaa, T. Multiresolution gray-scale and rotation invariant texture classification with local binary patterns. *IEEE Trans. Pattern Anal. Mach. Intell.* **2002**, *24*, 971–987. [CrossRef]

22. Dalal, N.; Triggs, B. Histograms of oriented gradients for human detection. In Proceedings of the IEEE Computer Society Conference on Computer Vision and Pattern Recognition, San Diego, CA, USA, 20–25 June 2005; pp. 886–893.
23. Cortes, C.; Vapnik, V. Support-vector networks. *Mach. Learn.* **1995**, *20*, 273–297. [CrossRef]
24. Viola, P.; Jones, M. Rapid object detection using a boosted cascade of simple features. In Proceedings of the IEEE Conference on Computer Vision and Pattern Recognition, Kauai, HI, USA, 8–14 December 2001; pp. 511–518.
25. Chahal, K.S.; Dey, K. A survey of modern object detection literature using deep learning. *arXiv* **2018**, arXiv:1808.07256.
26. Zou, Z.; Shi, Z.; Guo, Y.; Ye, J. Object detection in 20 years: A survey. *arXiv* **2019**, arXiv:1905.05055.
27. Krizhevsky, A.; Sutskever, I.; Hinton, G.E. Imagenet classification with deep convolutional neural networks. *Adv. Neural Inf. Process. Syst.* **2012**, *25*. [CrossRef]
28. Li, W.; Li, F.; Luo, Y.; Wang, P. Deep domain adaptive object detection: A survey. In Proceedings of the Symposium Series on Computational Intelligence (SSCI), Canberra, Australia, 1–4 December 2020; IEEE: New York, NY, USA, 2020; pp. 1808–1813.
29. Cazzato, D.; Cimarelli, C.; Sanchez-Lopez, J.L.; Voos, H.; Leo, M. A Survey of Computer Vision Methods for 2D Object Detection from Unmanned Aerial Vehicles. *J. Imaging* **2020**, *6*, 78. [CrossRef]
30. Girshick, R. Fast r-cnn. In Proceedings of the IEEE International Conference on Computer Vision, Santiago, Chile, 7–13 December 2015; pp. 1440–1448.
31. Ren, S.; He, K.; Girshick, R.; Sun, J. Faster r-cnn: Towards real-time object detection with region proposal networks. *IEEE Trans. Pattern Anal. Mach. Intell.* **2016**, *39*, 1137–1149. [CrossRef]
32. Redmon, J.; Divvala, S.; Girshick, R.; Farhadi, A. You only look once: Unified, real-time object detection. In Proceedings of the IEEE Conference on Computer Vision and Pattern Recognition, Las Vegas, NV, USA, 27–30 June 2016; pp. 779–788.
33. Liu, W.; Anguelov, D.; Erhan, D.; Szegedy, C.; Reed, S.; Fu, C.Y.; Berg, A.C. Ssd: Single shot multibox detector. In Proceedings of the European Conference on Computer Vision, Amsterdam, The Netherlands, 8–16 October 2016; pp. 21–37.
34. Girshick, R.; Donahue, J.; Darrell, T.; Malik, J. Rich feature hierarchies for accurate object detection and semantic segmentation. In Proceedings of the IEEE Conference on Computer Vision and Pattern Recognition, Columbus, OH, USA, 23–28 June 2014; pp. 580–587.
35. Itti, L.; Koch, C.; Niebur, E. A model of saliency-based visual attention for rapid scene analysis. *IEEE Trans. Pattern Anal. Mach. Intell.* **1998**, *20*, 1254–1259. [CrossRef]
36. Itti, L.; Koch, C. Computational modelling of visual attention. *Nat. Rev. Neurosci.* **2001**, *2*, 194–203. [CrossRef]
37. Xie, Y.; Lu, H. Visual saliency detection based on Bayesian model. In Proceedings of the IEEE International Conference on Image Processing, Brussels, Belgium, 11–14 September 2011; pp. 645–648.
38. Yang, C.; Zhang, L.; Lu, H.; Ruan, X.; Yang, M.H. Saliency detection via graph-based manifold ranking. In Proceedings of the IEEE Conference on Computer Vision and Pattern Recognition, Portland, OR, USA, 23–28 June 2013; pp. 3166–3173.
39. Jiang, Z.; Davis, L.S. Submodular salient region detection. In Proceedings of the IEEE Conference on Computer Vision and Pattern Recognition, Portland, OR, USA, 23–28 June 2013; pp. 2043–2050.
40. Liu, T.; Yuan, Z.; Sun, J.; Wang, J.; Zheng, N.; Tang, X.; Shum, H.Y. Learning to detect a salient object. *IEEE Trans. Pattern Anal. Mach. Intell.* **2010**, *33*, 353–367.
41. Zhang, L.; Tong, M.H.; Marks, T.K.; Shan, H.; Cottrell, G.W. SUN: A Bayesian framework for saliency using natural statistics. *J. Vis.* **2008**, *8*, 32. [CrossRef] [PubMed]
42. Lin, T.Y.; Dollár, P.; Girshick, R.; He, K.; Hariharan, B.; Belongie, S. Feature pyramid networks for object detection. In Proceedings of the IEEE Conference on Computer Vision and Pattern Recognition, Honolulu, HI, USA, 21–26 July 2017; pp. 2117–2125.
43. Lin, T.Y.; Goyal, P.; Girshick, R.; He, K.; Dollár, P. Focal loss for dense object detection. In Proceedings of the IEEE International Conference on Computer Vision, Venice, Italy, 22–29 October 2017; pp. 2980–2988.
44. Cai, Z.; Vasconcelos, N. Cascade r-cnn: Delving into high quality object detection. In Proceedings of the IEEE Conference on Computer Vision and Pattern Recognition, Salt Lake City, UT, USA, 18–23 June 2018; pp. 6154–6162.
45. Pang, J.; Chen, K.; Shi, J.; Feng, H.; Ouyang, W.; Lin, D. Libra r-cnn: Towards balanced learning for object detection. In Proceedings of the IEEE Conference on Computer Vision and Pattern Recognition, Long Beach, CA, USA, 15–20 June 2019; pp. 821–830.
46. Simonyan, K.; Zisserman, A. Very deep convolutional networks for large-scale image recognition. *arXiv* **2014**, arXiv:1409.1556.

sensors

MDPI

Article

An Efficient Plaintext-Related Chaotic Image Encryption Scheme Based on Compressive Sensing

Zhen Li [1,2], Changgen Peng [1,*], Weijie Tan [1] and Liangrong Li [2]

1 College of Computer Science and Technology, State Key Laboratory of Public Big Data, Guizhou University, Guiyang 550025, China; zli6@gzu.edu.cn (Z.L.); wjtan@gzu.edu.cn (W.T.)

2 College of Big Data and Information Engineering, Guizhou University, Guiyang 550025, China; lrli@gzu.edu.cn

* Correspondence: cgpeng@gzu.edu.cn

Abstract: With the development of mobile communication network, especially 5G today and 6G in the future, the security and privacy of digital images are important in network applications. Meanwhile, high resolution images will take up a lot of bandwidth and storage space in the cloud applications. Facing the demands, an efficient and secure plaintext-related chaotic image encryption scheme is proposed based on compressive sensing for achieving the compression and encryption simultaneously. In the proposed scheme, the internal keys for controlling the whole process of compression and encryption is first generated by plain image and initial key. Subsequently, discrete wavelets transform is used in order to convert the plain image to the coefficient matrix. After that, the permutation processing, which is controlled by the two-dimensional Sine improved Logistic iterative chaotic map (2D-SLIM), was done on the coefficient matrix in order to make the matrix energy dispersive. Furthermore, a plaintext related compressive sensing has been done utilizing a measurement matrix generated by 2D-SLIM. In order to make the cipher image lower correlation and distribute uniform, measurement results quantified the 0∼255 and the permutation and diffusion operation is done under the controlling by two-dimensional Logistic-Sine-coupling map (2D-LSCM). Finally, some common compression and security performance analysis methods are used to test our scheme. The test and comparison results shown in our proposed scheme have both excellent security and compression performance when compared with other recent works, thus ensuring the digital image application in the network.

Keywords: image encryption; compressive sensing; plaintext related; chaotic system

Citation: Li, Z.; Peng, C.; Tan, W.; Li, L. An Efficient Plaintext-Related Chaotic Image Encryption Scheme Based on Compressive Sensing. *Sensors* **2021**, *21*, 758. https://doi.org/10.3390/s21030758

Academic Editor: Hyoungsik Nam

Received: 30 December 2020

Accepted: 20 January 2021

Published: 23 January 2021

1. Introduction

Nowadays, digital images are becoming one of the most important data formats in our daily life. The risk of information leakage is inevitable when we share photos with others on the social network platform. Therefore, the security of digital images attracts a great of scholars' attention.

The image encryption is an important method in image security. Many text structure encryption schemes, such as advanced encryption standard (AES), data encryption standard (DES), etc., have poor performance on image encryption. These schemes cannot break the correlation among adjacent pixels that may leak some geometric distribution of plain image. Image data are different from text data, which have some specific features, so the image encryption scheme must be designed according to these characteristics. At the beginning, the researchers used some special transformation matrixes, such as magic cube transformation, Arnold cat map, etc., in order to permutate the plain image without under security keys controlling. However, they are against Kerckhoffs's principle, which requires the cryptosystem to be a white box, except for security keys.

Chaos theory as a cornerstone of nonlinear dynamic that is wildly used in many fields was first proposed by Lorenz [1]. The chaotic system has many good characteristics, such as randomness, ergodicity, sensitivity to initial values, and parameters [2], so it is suitable for the design of the cryptosystem. Matthews [3] first introduced a chaotic system into designing image cryptosystem. After that, a variety of image encryption algorithms have been put forward in the sspatial domain. These image encryption methods can be roughly classified as: (1) image encryption that is based on transformation matrixes [4–6]. This kind of algorithm mainly used the transformation matrix like magic cube transformation and Arnold cat map to permutation, and then used the chaotic system for diffusion. (2) The image encryption scheme that is based on deoxyribonucleic acid (DNA) encoding and chaotic system [7–9]. In this kind of scheme, the authors usually used DNA rules to encode the plain-image and controlled permutation and diffusion process on encoded data by the chaotic system. (3) Image encryption that is based on chaotic S-box [10–12]. Researchers usually used chaotic systems to design an S-box and encrypted image data by the nonlinear component. (4) Other spatial domain image encryption schemes [13–16].

Generally, the online transmission of image data requires a larger bandwidth. Therefore, image compression is very important for network applications, which can improve the efficiency of image transmission. In general, image compression not only utilizes the correlation between adjacent pixels, but it also encodes the non-uniform distribution of image pixels. However, image encryption will totally break the correlation among adjacent pixels and make its distribution uniform. Thus, the cipher image is not suitable for image compression. Not only that, the loss of image compression can also make the image impossible to be decrypted. Therefore, the compression must be executed early or at the same time as encryption.

Compressed sensing (CS) is a kind of effective data compression technology [17] when the data satisfy sparsity in a certain domain. When compared with Nyquist theory, CS can recover the entire signal from a smaller number of measurements [18]. In CS theory, when the signal is sparse in a transformation domain, a measurement matrix can be used to project the signal randomly, and then the original signal can be reconstructed by convex optimization algorithm. Fortunately, the images are sparse in many transform domains and are well suited to apply to CS theory. Therefore, based on compressed sensing, how to perform the image encryption also is a topical issue. Chai et al. [19] proposed an image encryption scheme that is based on magnetic-controlled memristive chaotic system and compressive sensing. This scheme first transform image to discrete wavelet transform (DWT) domain. Subsequently, some permutations have been done with this coefficient matrix. Finally, the compressive sensing used a measurement matrix that was generated by a chaotic system. In this scheme, although the generation of measurement matrix is related to plain image information, the plain image sensitivity of the scheme is still not good enough. In addition, the uneven energy distribution of cipher images generated by the scheme may also cause the leakage of some plain image information. Zhu et al. [20] proposed an image encryption scheme, which uses the random Gaussian matrix generated by Chebyshev mapping to execute compressive sensing. Chai et al. [21] proposed a chaotic image encryption scheme, which uses elementary cellular automata and block compressive sensing. In this scheme, a plain image is transformed by DWT at first, then compressive processing under the measurement matrix generated by a parameter-varying chaotic system is done. The plain image sensitivity of this scheme is good, because all of the initial values of chaotic system are related to the plain image. However, the randomness of cipher image seems not good enough. Zhu et al. [22] proposed an image encryption scheme, which is based on nonuniform sampling by block CS. In this scheme, the discrete cosine transform (DCT) is used to generate the coefficients matrix, and then perform compressive sensing processing by two measurement matrices that are generated by the logistic map. Finally, undertake the diffusion and permutation under logistic map controlling. However, the whole diffusion process is related to the result that is calculated in previous pixels; thus, the robustness must not be as good as mentioned. Gong et al. [23] proposed an image compression and encryption algorithm. In this scheme, the plain image is first permuted by the Arnold transform to reduce the block effect in the compression process, and then the coefficient matrix is compressed and encrypted by CS, simultaneously. The keys in this scheme are generated by a plain image without any external keys; it means that each cipher image corresponds to a unique key, which is not conducive to key distribution management and batch image encryption. Kayalvizhi et al. [24] proposed an image encryption scheme, which is based on compressive sensing, fractional order hyper chaotic Chen system, and DNA operations. In this scheme, block compressive sensing is executed to the plain image, and then execute DNA encoding to the measurement matrix. After that, complete some diffusion operation in DNA sequences. The whole process is not related to the plain image and the sensitivity of plain image has weak resistance to differential attack. Moreover, DNA encoding and decoding may consume a large amount of computing time, which results in the low efficiency of this algorithm.

To conquer the drawback what mentioned above, a plaintext related image encryption scheme is given using compressive sensing and two hyper chaotic systems. The detailed contributions are as follows:

- In order to make the image cryptosystem more sensitive to the plain image, a plain image information-related method is proposed, which makes the plaintext information involved in the whole control process of compressive sensing and encryption, and make the image cryptosystem have excellent performance in resisting differential attack.
- The generation method of the measurement matrix for compression encryption is presented, which is based on a chaotic system and the information of plain images, and make the CS process fully related to the plain image. In other words, different plain images correspond to different measurement matrices. Additionally, a permute and diffuse operation is used for the measurement matrix, which makes the pixels of the cipher image present lower correlation and uniform distribution.
- The peak signal to noise ratio and structural similarity index measurement is used to evaluate compression performance, and many common security analyses methods are carried out, such as key space analysis, differential attack, statistical analysis, key sensitivity analysis, etc., in order to evaluate security performance.

This paper is organized, as follows: in Section 2, the preliminary for this paper is given, such as compressive sensing and chaotic system. In Section 3, an efficient image encryption scheme that is based on chaos and compressive sensing is introduced. In Section 4, some common compression analyses and security analyses of the proposed image cryptosystem are given. In Section 5, we conclude this paper.

2. Preliminary

2.1. Compressive Sensing

The aim of CS model [25] is to recover a sparse image signal $\boldsymbol{X} \in \mathbb{R}^{n\times n}$ from fewer measurements $\boldsymbol{Y} \in \mathbb{R}^{m\times n}$ is given by:

$$\boldsymbol{Y} = \boldsymbol{\Phi X} = \boldsymbol{\Phi \Psi P} \tag{1}$$

where $\boldsymbol{\Phi} \in \mathbb{R}^{m\times n}$ is a measurement sensing matrix whose distribution satisfies Gaussian distribution. Let $\boldsymbol{A} = \boldsymbol{\Phi\Psi}$, where the columns satisfy the linearly independent condition. When $\boldsymbol{A}$ satisfies a certain condition, i.e., Restricted Isometry Constant (RIC), the restricted isometry property(RIP), the CS theory shows that only a sufficiently sparse signal $\boldsymbol{P}$ can be recovered with a high probability exactly from $\boldsymbol{Y}$. The linear measurement process is expressed as a regularized form, as

$$min\|vec(\boldsymbol{P})\|_0 \quad s.t. \quad \|\boldsymbol{AP} - \boldsymbol{Y}\|_2 \le \eta \tag{2}$$

where $\|\cdot\|_0$ denotes the l_0 norm as a sparsity constraint and η is a constant. This form aims to find the most sparse solution that fits the observation model well. However, it is Nondeterministic Polynomial (NP)-hard problem to solve Equation (2). A convex relaxation method is to apply the l_1 norm of the l_0 norm, as follows:

$$min\|vec(\boldsymbol{P})\|_1 \quad s.t. \quad \|\boldsymbol{AP}-\boldsymbol{Y}\|_2 \leq \eta \tag{3}$$

Theoretical analysis has shown that the l_1 norm can also approach the most sparse solution under some conditions [25,26]. Equation (3) can be solved by some optimization algorithms, such as the gradient descent method (GDM) [27] abd orthogonal matching pursuit (OMP) [28].

2.2. Chaotic System

The chaotic systems in our proposed scheme are used to control the permutation and diffusion process, and to generate a measure matrix of compressive sensing, which are the key points of encryption and compression performance. Therefore, our cryptosystem is required to choose hyper chaotic systems that have better chaotic characteristics.

The two-dimensional Sine improved Logistic iterative chaotic map (2D-SLIM) [29] is given by

$$\begin{cases} x_{i+1} = \sin(by_i)\sin(50/x_i) \\ y_{i+1} = a(1-2x_{i+1}^2)\sin(50/y_i) \end{cases} \tag{4}$$

where a and b are the system parameters. When $a \in (0,3]$ and $b = 2\pi$ or when $a = 1$ and $b \in [4,7]$, the system becomes hyper chaotic.

The two-dimensional Logistic-Sine-coupling map (2D-LSCM) [30] is given by

$$\begin{cases} x_{i+1} = \sin(\pi(4\theta x_i(1-x_i) + (1-\theta)\sin(\pi y_i))) \\ y_{i+1} = \sin(\pi(4\theta y_i(1-y_i) + (1-\theta)\sin(\pi x_{i+1}))) \end{cases} \tag{5}$$

where θ is the control parameter. When $\theta \in (0,1)$, the system has hyper chaotic behavior.

Remark 1. *In our proposed scheme, we used 2D-SLIM and 2D-LSCM, two discrete hyper chaotic systems, to control the encryption process. In fact, other discrete hyper chaotic systems can also be extended in our scheme, and the only difference in those selection is the size of key space. Furthermore, the reason why we select the two different hyper chaotic systems to control the encryption and compression process is that it can avoid, as much as possible, some unexpected situations occurring, such as dynamical degradation of chaotic systems [31,32], weak real keys, etc.*

3. Our Proposed Scheme

The image cryptosystem is proposed in this section. First, a plaintext related internal keys generation method is introduced in Section 3.1. Afterwards, we present the encryption scheme in Section 3.2. Finally, we propose the decryption scheme in Section 3.3.

3.1. Plaintext-Related Internal Keys Generation

In this subsection, we proposed a method for generating the internal keys that are related to plaintext. The internal keys are used to generate the initial values and parameters of hyper chaotic systems that are used to control all processes of encryption and decryption. Therefore, the plaintext-related internal key generation method can make our proposed image cryptosystem more plaintext sensitive to resisting differential attack. There are two parts in plaintext-related internal key generation: plaintext information extraction and internal keys generation.

Algorithm 1 shows the plain image information extraction algorithm.

The detailed description are as follows:

Step 1: input plain image matrix ***P*** and initial key ***K*** into algorithm, and begin.

Step 2: expand the plain image matrix ***P*** into a vector ***P(:)*** in rows, and then change this vector to a string ***SP***.

Step 3: input string ***SP*** into hash function SHA256, and denote the hash value as ***HP***.

Step 4: input initial key ***K*** into hash function SHA256, and denote the hash value as ***HK***.

Step 5: put hash values ***HP*** and ***HK*** together and input them into hash function SHA256. The hash value is extracted plain image information ***EPI***.

Step 6: output the extracted plain image information ***EPI***, and finished.

Algorithm 1 Plain image information extraction

Input: Plain image matrix ***P*** and initial key ***K***

Output: Extracted plain image information ***EPI***.

String SP ← ***P(:)***

HP ← SHA256(***SP***)

HK ← SHA256(***K***)

EPI ← SHA256(***[HP,HK]***)

Algorithm 2 shows the internal keys generation algorithm.

Algorithm 2 Internal keys generation

Input: Extracted plain image information ***EPI***, initial key ***K***

Output: Internal keys ***[K1,K2,K3,K4]***.

HK ← SHA256(***K***)

INKEY ← SHA256(***[EPI,HK]***)

I1 ← ***INKEY***(1:64)

I2 ← ***INKEY***(65:128)

I3 ← ***INKEY***(129:192)

I4 ← ***INKEY***(193:256)

CN1 ← $mod(\boldsymbol{I1}/10^8, 256)$ ***CN2*** ← $mod(\boldsymbol{I2}/10^8, 256)$

CN3 ← $mod(\boldsymbol{I3}/10^8, 256)$ ***CN4*** ← $mod(\boldsymbol{I4}/10^8, 256)$

K1 ← ***BitCyclicShift***(***INKEY***, ***CN1***)

K2 ← ***BitCyclicShift***(***INKEY***, ***CN2***)

K3 ← ***BitCyclicShift***(***INKEY***, ***CN3***)

K4 ← ***BitCyclicShift***(***INKEY***, ***CN4***)

The detailed description are as follows:

Step 1: input extracted plain image information ***EPI*** and initial key ***K***, and begin.

Step 2: input initial key ***K*** into hash function SHA256, and denote the hash value as ***HK***.

Step 3: put extracted plain image information ***EPI*** and hash value ***HK*** together and input them into hash function SHA256. Denote the hash value as ***INKEY***.

Step 4: split ***INKEY*** into four parts, and everypart with 64 bits, denoted as ***I1***, ***I2***, ***I3*** and ***I4***.

Step 5: calculate the control values as ***CN1*** $= mod(\boldsymbol{I1}/10^8, 256)$, ***CN2*** $= mod(\boldsymbol{I2}/10^8, 256)$, ***CN3*** $= mod(\boldsymbol{I3}/10^8, 256)$, and ***CN4*** $= mod(\boldsymbol{I4}/10^8, 256)$.

Step 6: bit cyclic shift ***INKEY CN1*** bits to right direction, and generate 256 bits internal key ***K1***, after that, at same operation to bit cyclic shift ***INKEY*** under ***CN2***, ***CN3***, ***CN4*** control, and generate internal keys ***K2,K3,K4***.

Step 7: output internal keys ***[K1,K2,K3,K4]***, and finished.

3.2. Encryption Scheme

In this subsection, we will introduce our proposed encryption scheme. The encryption scheme takes, as inputs, plain image ***P***, initial key ***K***, and compression ratio **CR**, and put outputs, such as cipher image and some additional ciphertext information. The compression ratio (CR) means the ratio of the number of pixels in the compressed image to that in the original image. Figure 1 shows the block diagram of the encryption scheme, and the detailed description are as follows:

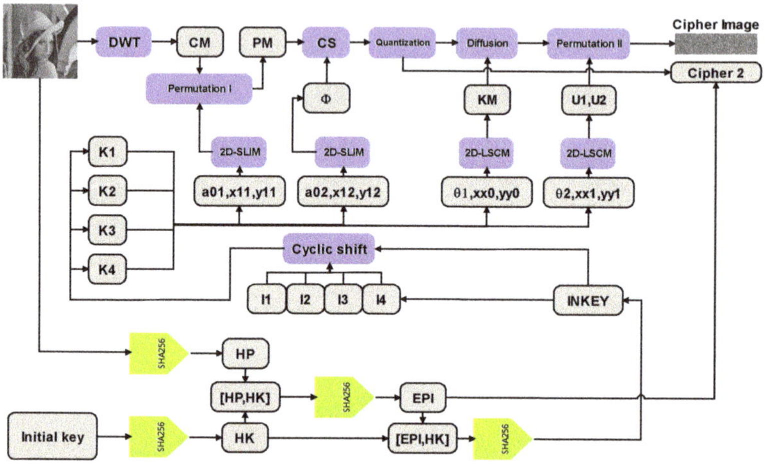

Figure 1. The block diagram of the encryption scheme.

Step 1: input plain image ***P*** $(N \times N)$, initial key ***K*** and compression ratio **CR**, and the encryption process begins.

Step 2: input plain image matrix ***P*** and initial key ***K*** into Algorithm 1 to get extracted plain image information ***EPI***. After that, input extracted plain image information ***EPI*** and initial key ***K*** into Algorithm 2 to generate internal keys ***[K1,K2,K3,K4]***.

Step 3: input plain image ***P*** into discrete wavelet transform (DWT) to sparse representation, and we denote sparse coefficient matrix as ***CM***.

Step 4: input coefficient matrix ***CM*** and internal key ***K1*** into Algorithm 3 to make plaintext energy evenly distributed.

Algorithm 3 Permutation I algorithm

Input: Coefficient matrix ***CM*** and internal key ***K1***

Output: Permutated coefficient matrix ***PM***.

$\mathbf{I1} \leftarrow \mathbf{K1}(1{:}64)$; $\mathbf{I2} \leftarrow \mathbf{K1}(65{:}128)$; $\mathbf{I3} \leftarrow \mathbf{K1}(129{:}192)$

$a01 \leftarrow fix(mod(\mathbf{I1}/10^6, 3)) + mod(\mathbf{I1}/10^{14}, 1); b \leftarrow 2\pi$

$x11 \leftarrow mod(\mathbf{I2}/10^{14}, 1); y11 \leftarrow mod(\mathbf{I3}/10^{14}, 1)$

Put $a01, b, x11, y11$ into Equation (4) to generate a sequence S by iterating.

$\mathbf{X} \leftarrow mod(fix((S + 100) \times 10^{10}), N \times N) + 1$

Remove the repeated elements from $\mathbf{X}$, put the absent numbers at the end.

Change ***CM*** to a vector ***CMA*** in rows.

$len \leftarrow length(\mathbf{CMA})$

for $i = 1$ **to** $fix(len/2)$ **do**

  $\mathbf{CMA}(\mathbf{X}(i)) \leftrightarrow \mathbf{CMA}(\mathbf{X}(len - i + 1))$

end for

$\mathbf{PM} \leftarrow reshape(\mathbf{CMA}, N, N)$

Step 5: Calculate the threshold value ***TS*** by Algorithm 4.

Algorithm 4 Calculate threshold algorithm

Input: Coefficient matrix ***CM*** and compression ratio **CR**

Output: Threshold value ***TS***.

Change ***CM*** to a vector ***CMA*** in rows.

Arrange the vector ***CMA*** from the smallest to the largest

$len \leftarrow length(\mathbf{CMA})$

$\mathbf{TS} \leftarrow \mathbf{CMA}(floor(len - N \times N \times \mathrm{CR}/7))$

Step 6: if the element in the permutated coefficient matrix ***PM*** absolute value less than threshold ***TS***, then set this element to 0. The new generated matrix is denoted as ***PM2***.

Step 7: input matrix ***PM2***, compression ratio **CR**, and internal key ***K2*** into Algorithm 5 to obtain the measurements ***CSM***.

Algorithm 5 Compressive sensing algorithm

Input: The matrix ***PM2***, compression ratio ***CR***, and internal key ***K2***

Output: The compressive sensing measurements ***CSM***.

$\boldsymbol{I1} \leftarrow \boldsymbol{K2}(1{:}64)$

$\boldsymbol{I2} \leftarrow \boldsymbol{K2}(65{:}128)$

$\boldsymbol{I3} \leftarrow \boldsymbol{K2}(129{:}192)$

$a02 \leftarrow fix(mod(\boldsymbol{I1}/10^6, 3)) + mod(\boldsymbol{I1}/10^{14}, 1);$

$b \leftarrow 2\pi$

$x12 \leftarrow mod(\boldsymbol{I2}/10^{14}, 1)$

$y12 \leftarrow mod(\boldsymbol{I3}/10^{14}, 1)$

$M \leftarrow fix(N \times \boldsymbol{CR})$

Put $a02, b, x12, y12$ into Equation (4) to generate a sequence ***S*** by iterating $M \times N$ times.

$\Phi 0 \leftarrow reshape(\boldsymbol{S}, M, N)$

$\Phi \leftarrow \sqrt{2/M} \cdot \Phi 0$, where $\sqrt{2/M}$ is used for normalization [33].

$\boldsymbol{CSM} \leftarrow PM2 \times \Phi$

Step 8: quantize the compressive sensing measurements ***CSM*** to the range of $[0, 255]$ and generate quantized matrix ***QM*** by

$$\boldsymbol{QM} = \text{round}\ (255 \times (\boldsymbol{CSM} - \boldsymbol{MAX})/(\boldsymbol{MAX} - \boldsymbol{MIN})) \tag{6}$$

where $round(x)$ represents the nearest integer with x, and ***MIN*** and ***MAX*** are the minimum and maximum numbers of ***CSM***.

Step 9: input quantized matrix ***QM*** and internal keys ***K3,K4*** into Algorithm 6 to do diffusion and permutation II.

Step 10: output cipher image ***C1*** and additional cipher information $C2 = [\boldsymbol{EPI}, \boldsymbol{MAX}, \boldsymbol{MIN}]$. The encryption process is finished.

Algorithm 6 Diffusion and permutation II algorithm

Input: Quantized matrix QM and internal keys **K3,K4**

Output: Cipher image **C1**.

$\mathbf{I1} \leftarrow \mathbf{K3}(1{:}64)$; $\mathbf{I2} \leftarrow \mathbf{K3}(65{:}128)$; $\mathbf{I3} \leftarrow \mathbf{K3}(129{:}192)$;
$\mathbf{I4} \leftarrow \mathbf{K4}(1{:}64)$; $\mathbf{I5} \leftarrow \mathbf{K4}(65{:}128)$; $\mathbf{I6} \leftarrow \mathbf{K4}(129{:}192)$.
$\theta 1 \leftarrow mod(\mathbf{I1}/10^{14}, 1)$; $xx0 \leftarrow mod(\mathbf{I2}/10^{14}, 1)$; $yy0 \leftarrow mod(\mathbf{I3}/10^{14}, 1)$.
$\theta 2 \leftarrow mod(\mathbf{I4}/10^{14}, 1)$; $xx1 \leftarrow mod(\mathbf{I5}/10^{14}, 1)$; $yy1 \leftarrow mod(\mathbf{I6}/10^{14}, 1)$.
if $0.33 < \theta 1 < 0.66$ **then**
    $\theta 1 \leftarrow mod(\theta 1 + 0.33, 1)$
end if
if $0.33 < \theta 2 < 0.66$ **then**
    $\theta 2 \leftarrow mod(\theta 2 + 0.33, 1)$
end if
$[M, N] \leftarrow size(\mathbf{QM})$
Put $\theta 1, xx0, yy0$ into Equation (5) to generate a sequence **S0** by iterating $M \times N$ times.
$\mathbf{KM} \leftarrow floor(mod(\mathbf{S0} \times 10^{13}, 256))$
$\mathbf{DM} \leftarrow \mathbf{QM} \oplus \mathbf{KM}$
Put $\theta 2, xx1, yy1$ into Equation (5) to generate a sequence **X1** and **Y1** by iterating $max(M, N)$ times.
$\mathbf{U1} \leftarrow floor(mod(\mathbf{X1}(1:M) \times 10^5, (M-1))) + 1$
$\mathbf{U2} \leftarrow floor(mod(\mathbf{Y1}(1:N) \times 10^5, (N-1))) + 1$
for $i = 1$ **to** M **do**
    $\mathbf{DM}(i,:) \leftarrow CircleShift(\mathbf{DM}(i,:), \mathbf{U1}(i))$
end for
for $i = 1$ **to** N **do**
    $\mathbf{DM}(:,i) \leftarrow CircleShift(\mathbf{DM}(:,i), \mathbf{U2}(i))$
end for
$\mathbf{C1} \leftarrow \mathbf{DM}$

3.3. Decryption Scheme

The decryption process is the inverse process of encryption, and it takes input as cipher image **C1**, additional cipher information **C2**, and initial key **K**, and put the output as recovering plain image. Figure 2 shows the block diagram of the decryption scheme, and the detailed description is as follows:

Step 1: input cipher image $\mathbf{C1}(M \times N)$, additional cipher information **C2**, and initial key **K** and the decryption process begins.

Step 2: input **EPI** and initial key **K** into Algorithm 2 to generate internal keys ***[K1,K2,K3,K4]***.

Step 3: input cipher image **C1** and internal keys **K3,K4** into Algorithm 7 to do reverse permutation II and reverse diffusion.

Algorithm 7 Reverse permutation II and reverse diffusion algorithm

Input: Cipher image ***C1*** and internal keys ***K3,K4***

Output: Reverse permutation and diffusion matrix ***RPDM***.

I1 ← ***K3***(1:64); ***I2*** ← ***K3***(65:128); ***I3*** ← ***K3***(129:192);
I4 ← ***K4***(1:64); ***I5*** ← ***K4***(65:128); ***I6*** ← ***K4***(129:192).

$\theta 1 \leftarrow mod(\boldsymbol{I1}/10^{14}, 1); xx0 \leftarrow mod(\boldsymbol{I2}/10^{14}, 1); yy0 \leftarrow mod(\boldsymbol{I3}/10^{14}, 1)$.

$\theta 2 \leftarrow mod(\boldsymbol{I4}/10^{14}, 1); xx1 \leftarrow mod(\boldsymbol{I5}/10^{14}, 1); yy1 \leftarrow mod(\boldsymbol{I6}/10^{14}, 1)$.

if $0.33 < \theta 1 < 0.66$ **then**
  $\theta 1 \leftarrow mod(\theta 1 + 0.33, 1)$
end if
if $0.33 < \theta 2 < 0.66$ **then**
  $\theta 2 \leftarrow mod(\theta 2 + 0.33, 1)$
end if

$[M, N] \leftarrow size(\boldsymbol{C1})$

Put $\theta 2, xx1, yy1$ into Equation (5) to generate a sequence ***X1*** and ***Y1*** by iterating $max(M, N)$ times.

$\boldsymbol{U1} \leftarrow floor(mod(\boldsymbol{X1}(1:M) \times 10^5, (M-1))) + 1$

$\boldsymbol{U2} \leftarrow floor(mod(\boldsymbol{Y1}(1:N) \times 10^5, (N-1))) + 1$

for $i = 1$ **to** N **do**
  $\boldsymbol{C1}(:, i) \leftarrow CircleShift(\boldsymbol{C1}(:, i), \text{-}\boldsymbol{U2}(i))$
end for
for $i = 1$ **to** M **do**
  $\boldsymbol{C1}(i, :) \leftarrow CircleShift(\boldsymbol{C1}(i, :), \text{-}\boldsymbol{U1}(i))$
end for

Put $\theta 1, xx0, yy0$ into Equation (5) to generate a sequence ***S0*** by iterating $M \times N$ times.

$\boldsymbol{KM} \leftarrow floor(mod(\boldsymbol{S0} \times 10^{13}, 256))$

$\boldsymbol{RPDM} \leftarrow \boldsymbol{C1} \bigoplus \boldsymbol{KM}$

Step 4: Do reverse quantization to the matrix ***RPDM***, and generate reverse quantized matrix ***RQM*** by

$$\boldsymbol{RQM} = \frac{\boldsymbol{RPDM} \times (\boldsymbol{MAX} - \boldsymbol{MIN})}{255} + \boldsymbol{MIN} \tag{7}$$

Step 5: Input matrix ***RQM*** and internal keys ***K2*** into Algorithm 8 to reconstruct matrix ***RCM***.

Step 6: input reconstruct matrix ***RCM*** and internal key ***K1*** into Algorithm 9 to undertake reverse permutation I.

Step 7: input matrix ***RPM*** into inverse discrete wavelet transform (IDWT) in order to recover plain image ***P***.

Step 8: output recover plain image ***P*** and the decryption process is finished.

Algorithm 8 Matrix reconstruction algorithm

Input: Matrix $\boldsymbol{RQM}$ and internal keys $\boldsymbol{K2}$

Output: Reconstruct matrix $\boldsymbol{RCM}$.

$\boldsymbol{I1} \leftarrow \boldsymbol{K2}(1{:}64)$
$\boldsymbol{I2} \leftarrow \boldsymbol{K2}(65{:}128)$
$\boldsymbol{I3} \leftarrow \boldsymbol{K2}(129{:}192)$
$a02 \leftarrow fix(mod(\boldsymbol{I1}/10^6, 3)) + mod(\boldsymbol{I1}/10^{14}, 1);\ b \leftarrow 2\pi$
$x12 \leftarrow mod(\boldsymbol{I2}/10^{14}, 1)$
$y12 \leftarrow mod(\boldsymbol{I3}/10^{14}, 1)$
$[M, N] \leftarrow size(\boldsymbol{RQM})$
Put $a02, b, x12, y12$ into Equation (4) to generate a sequence $\boldsymbol{S}$ by iterating $M \times N$ times.
$\Phi 0 \leftarrow reshape(\boldsymbol{S}, M, N)$
$\Phi \leftarrow \sqrt{2/M} \cdot \Phi 0$, where $\sqrt{2/M}$ is used for normalization.
for $i = 1$ to N **do**
  $\boldsymbol{RCM}(:, i) \leftarrow OMP(\boldsymbol{RQM}, \Phi, N)$
end for

Algorithm 9 Reverse permutation I algorithm

Input: Reconstruct matrix $\boldsymbol{RCM}$ and internal key $\boldsymbol{K1}$

Output: Reverse permutated matrix $\boldsymbol{RPM}$.

$\boldsymbol{I1} \leftarrow \boldsymbol{K1}(1{:}64)$
$\boldsymbol{I2} \leftarrow \boldsymbol{K1}(65{:}128)$
$\boldsymbol{I3} \leftarrow \boldsymbol{K1}(129{:}192)$
$a01 \leftarrow fix(mod(\boldsymbol{I1}/10^6, 3)) + mod(\boldsymbol{I1}/10^{14}, 1);$
$b \leftarrow 2\pi$
$x11 \leftarrow mod(\boldsymbol{I2}/10^{14}, 1)$
$y11 \leftarrow mod(\boldsymbol{I3}/10^{14}, 1)$
Put $a01, b, x11, y11$ into Equation (4) to generate a sequence $\boldsymbol{S}$ by iterating.
$\boldsymbol{X} \leftarrow mod(fix((\boldsymbol{S} + 100) \times 10^{10}), N \times N) + 1$
Remove the repeated elements from $\boldsymbol{X}$, put the absent numbers at the end.
Change $\boldsymbol{RCM}$ to a vector $\boldsymbol{RCMA}$ in rows.
$len \leftarrow length(\boldsymbol{RCMA})$
for $i = 1$ **to** $fix(len/2)$ **do**
  $\boldsymbol{RCMA}(\boldsymbol{X}(i)) \leftrightarrow \boldsymbol{RCMA}(\boldsymbol{X}(len - i + 1))$
end for
$\boldsymbol{RPM} \leftarrow reshape(\boldsymbol{RCMA}, N, N)$

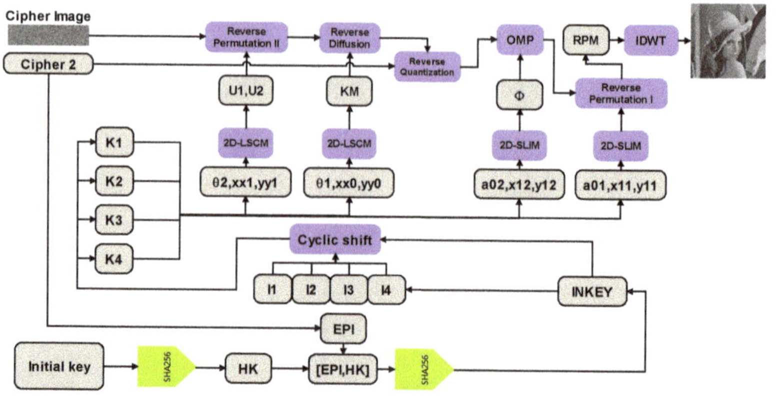

Figure 2. The block diagram of the decryption scheme.

4. Simulation and Analysis

In this section, we will evaluate our proposed image cryptosystem. The simulations and performance evaluations are implemented in MATLAB R2016a. Our hardware environment for tests was a personal computer with Inter(R) Core i7-6700k CPU 4.00 GHz, 32 GB memory, and the operation system is Windows 7 home edition. For simulation and tests, the initial key is selected as 'a2b235c5dd4345d2445e33e25ef255f524235ec' in hexadecimal, and one of the parameters of 2D-SLIM, which is given in Equation (4), is set as $b = 2\pi$. We first select 512×512 8-bit level gray images 'Lena', 'Pepper', and 'Cameraman' for simulation, and encrypt them with CR = $0.1, 0.2, \cdots, 0.9$, respectively. Figure 3 shows the simulation result. In the following subsections, we first discussed the performance of compression, and then provided the common security analysis result of our proposed image cryptosystem. Finally, we compared our work with other recent works in order to make our proposed image cryptosystem more convincing.

4.1. Compression Analyses

4.1.1. Peak Signal to Noise Ratio (PSNR)

For measuring the difference between the decrypted image and the original image to evaluate the recovery quality, we use the peak signal to noise ratio (PSNR) as a measurement for evaluation. PSNR is given by

$$\mathrm{PSNR} = 10\log_{10}\frac{255\times 255}{\frac{1}{N^2}\sum_{i=1}^{N}\sum_{j=1}^{N}(I(i,j)-I'(i,j))^2} \tag{8}$$

where I and I' are the decrypted image and original image, respectively. In this test, we first encrypt 256×256 and 512×512 plain images at different CRs, and then decrypt these cipher images to obtain recovery images. Finally, we calculate PSNR between plain images and recovery images. Table 1 and Figure 4 show the test result. According to the results, the PSNR values between plain images and recovery images are increasing with the growth of CRs, and the minimum of the PSNR is 31.7675 dB when the image is 256×256 and be encrypted in CR = 0.2. Therefore, our proposed scheme has a very good compression recovery performance.

4.1.2. Structural Similarity Index Measurement (SSIM)

The structural similarity index measurement (SSIM) is another important indictor for evaluating the compression performance. The SSIM value can be calculated by

$$\mathrm{SSIM} = \frac{(2\mu_I\mu_{I'}+C_1)(2\sigma_{II'}+C_2)}{(\mu_I^2+\mu_{I'}^2+C_1)(\sigma_I^2+\sigma_{I'}^2+C_2)} \tag{9}$$

where $C_1 = (k_1 \times L)^2$, $C_2 = (k_2 \times L)^2$, $k_1 = 0.01$, $k_2 = 0.03$, $L = 255$. The μ_I and $\mu_{I'}$ are the average values of the decrypted image I' and the original image I. The σ_I and $\sigma_{I'}$ are the variance values, and $\sigma_{II'}$ is the covariance value between I and I'. In this test, we also first encrypt 256×256 and 512×512 plain images at different CRs, and then decrypt these cipher images to obtain recovery images. Finally, we calculate the SSIM value between recovery images and plain images. Table 2 and Figure 5 show the SSIM result. The SSIM values are also increasing with CRs and the minimum value is also over 0.7, according to the results. It means that, in this indicator, our scheme also has very good compression recovery performance.

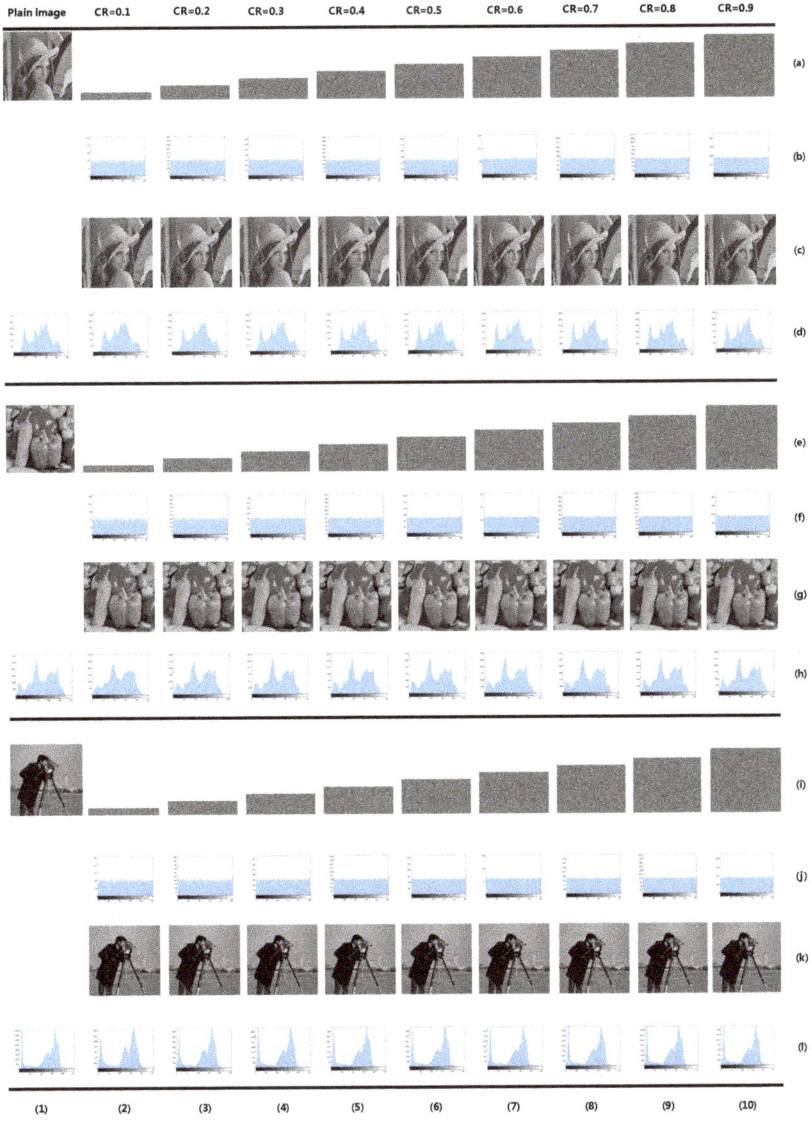

Figure 3. Encryption simulation. Rows (**a–d**) are the simulation results of image 'Lena', Rows (**e–h**) are the simulation results of image 'Pepper' and Rows (**i–l**) are the simulation results of image 'Cameraman'. Column (**1**) are plain images. Columns (**2–10**) are the simulation results of CR = 0.1, $\cdots$, CR = 0.9, respectively.

Table 1. The peak signal to noise ratio (PSNR) test results.

Images	CR = 0.2	CR = 0.3	CR = 0.4	CR = 0.5	CR = 0.6	CR = 0.7	CR = 0.8	CR = 0.9
Lena (256 × 256)	32.8060	34.3257	35.5494	36.2231	37.0150	37.7089	38.4807	39.1588
Lena (512 × 512)	35.4609	36.6669	37.5536	38.4038	38.8188	39.3051	39.6735	40.0186
Pepper (256 × 256)	32.3384	33.9538	35.2645	36.3140	37.2489	37.9024	38.5522	39.4550
Pepper (512 × 512)	35.1376	36.1261	36.8545	37.1843	37.5528	37.9703	38.1828	38.2197
Airplane (256 × 256)	32.1119	33.1697	34.0963	34.7593	35.2559	35.6004	36.3997	36.9712
Airplane (512 × 512)	34.1957	35.0839	35.6081	36.2459	37.0352	37.2760	37.4211	37.7294
Boat (256 × 256)	31.7675	32.5097	33.3848	33.9783	34.5981	35.2112	35.7384	36.2905
Boat (512 × 512)	32.8568	33.9273	34.4103	35.1400	35.7692	36.4291	36.7441	37.3474
Cameraman (256 × 256)	31.9799	33.1808	33.8290	34.4296	35.0475	35.5772	36.1571	36.7097
Cameraman (512 × 512)	36.0659	37.7784	39.2547	39.8190	40.6606	40.7227	40.7004	41.4673
Barbara (256 × 256)	32.3834	33.3900	34.2665	35.1264	35.7329	36.4359	36.9932	37.6480
Barbara (512 × 512)	32.3296	33.3472	34.1002	34.9419	35.8160	36.5540	37.3483	37.7838

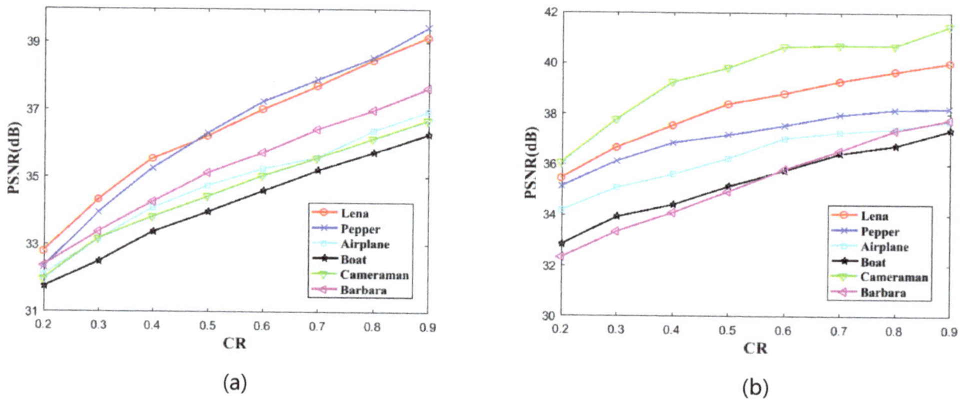

Figure 4. The Peak Signal to Noise Ratio (PSNR) test result. (**a**) shows the results of images with the size of 256 × 256. (**b**) shows the results of images with the size of 512 × 512.

4.2. Key Space Analysis

The image cryptosystem requires enough key space to resist the brute-force attack. In our proposed image cryptosystem, the two 2D-SLIM and two 2D-LSCM hyper chaotic systems are used for controlling the permutation and diffusion process and for generating a measure matrix of compressive sensing. Hence, the real keys are two system parameters $a01 \in (0,3], a02 \in (0,3]$ and four initial values $x11 \in (0,1), y11 \in (0,1), x12 \in (0,1), y12 \in (0,1)$ of two 2D-SLIM systems, and two system parameters $\theta 1 \in (0,0.33) \bigcup (0.66,1), \theta 2 \in (0,0.33) \bigcup (0.66,1)$ and four initial values $xx0 \in (0,1), yy0 \in (0,1), xx1 \in (0,1), yy1 \in (0,1)$ of two 2D-LSCM systems. The change step of each initial value and parameters are 10^{-15}, the key space can be calculated as $S = (3 \times 10^{15})^2 \times 0.66^2 \times (10^{15})^{10} = 3.9204 \times 10^{180} \approx 2^{600}$. Usually, if the key space is more than 2^{100}, then we can consider that the image cryptosystem is good at resisting the brute-force attack [34].

4.3. Differential Attack

Differential attack is a method for analyzing keys from two cipher images that are encrypted by two tiny different plain images. The plain image sensitivity is an important feature for an image cryptosystem to resist differential attack. There are two measurements for evaluating the plain image sensitivity: the number of pixels change rate (NPCR) and unified average changing intensity (UACI) [12,13]. The NPCR and UACI are given by Equations (11) and (12), respectively.

$$\mathrm{NPCR} = \frac{\sum_{i=1}^{M}\sum_{j=1}^{N} D(i,j)}{M \times N} \times 100\%, \tag{10}$$

where

$$D(i,j) = \begin{cases} 0, & C_1(i,j) = C_2(i,j) \\ 1, & C_1(i,j) \neq C_2(i,j) \end{cases}, \tag{11}$$

$$\mathrm{UACI} = \frac{1}{M \times N}\left(\sum_{i=1}^{M}\sum_{j=1}^{N} \frac{|C_1(i,j) - C_2(i,j)|}{255}\right) \times 100\%, \tag{12}$$

where $C_1(i,j)$ and $C_2(i,j)$ are denoted as two cipher images that are generated by encrypting one-pixel different two plain images. M and N are the height and width of images, respectively. In order to evaluate the NPCR and UACI results, the critical values are given by Wu et al. [35]. The critical value of NPCR is given by:

$$\mathcal{N}_{\alpha}^{*} = \frac{Q - \Phi^{-1}(\alpha)\sqrt{Q/H}}{Q+1}, \tag{13}$$

where H represents the total pixel numbers of image, Q represents the largest value that the pixels allowed in the image, and α is the significance level. When the test NPCR value is larger than critical value $\mathcal{N}_{\alpha}^{*}$, we can consider that the proposed system has good plain image sensitivity.

Table 2. The Structural Similarity Index Measurement (SSIM) test results.

Images	CR = 0.2	CR = 0.3	CR = 0.4	CR = 0.5	CR = 0.6	CR = 0.7	CR = 0.8	CR = 0.9
Lena (256 × 256)	0.7740	0.8476	0.8859	0.9018	0.9150	0.9262	0.9357	0.9424
Lena (512 × 512)	0.8278	0.8549	0.8713	0.8862	0.8909	0.8983	0.9034	0.9081
Pepper (256 × 256)	0.7603	0.8338	0.8716	0.8947	0.9097	0.9165	0.9227	0.9337
Pepper (512 × 512)	0.7911	0.8141	0.8309	0.8368	0.8458	0.8568	0.8616	0.8631
Airplane (256 × 256)	0.7303	0.8018	0.8429	0.8643	0.8713	0.8705	0.8901	0.8988
Airplane (512 × 512)	0.7988	0.8190	0.8254	0.8405	0.8586	0.8587	0.8581	0.8615
Boat (256 × 256)	0.7015	0.7670	0.8184	0.8425	0.8640	0.8816	0.8946	0.9050
Boat (512 × 512)	0.7330	0.7860	0.8012	0.8247	0.8424	0.8587	0.8656	0.8793
Cameraman (256 × 256)	0.6885	0.7740	0.8086	0.8332	0.8525	0.8664	0.8780	0.8873
Cameraman (512 × 512)	0.8567	0.8850	0.9076	0.9104	0.9202	0.9184	0.9149	0.9261
Barbara (256 × 256)	0.7509	0.8061	0.8424	0.8706	0.8864	0.9034	0.9136	0.9250
Barbara (512 × 512)	0.7644	0.8214	0.8471	0.8693	0.8886	0.8992	0.9122	0.9142

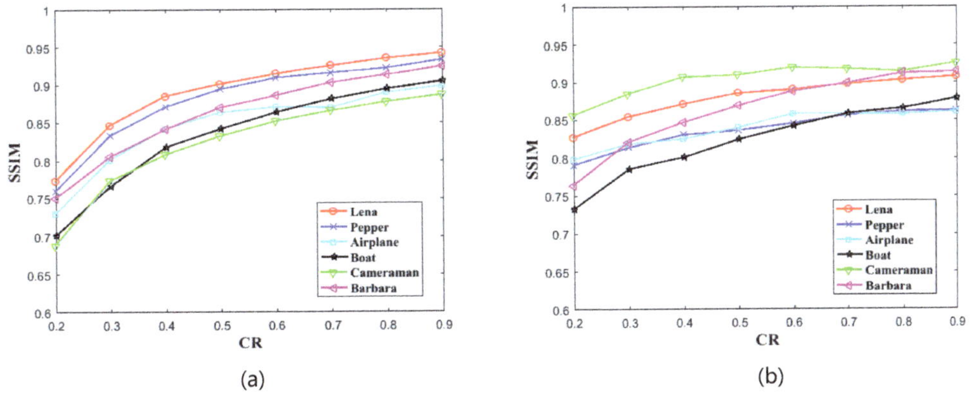

Figure 5. The Structural Similarity Index Measurement (SSIM) test result. (**a**) is shown the results of images with the size of 256 × 256. (**b**) is shown the results of images with the size of 512 × 512.

The UACI critical interval $(\mathcal{U}_{\alpha}^{*-}, \mathcal{U}_{\alpha}^{*+})$ is given by:

$$\begin{cases} u_{\alpha}^{*-} = \mu_u - \Phi^{-1}(\frac{\alpha}{2})\sigma_u, \\ u_{\alpha}^{*+} = \mu_u + \Phi^{-1}(\frac{\alpha}{2})\sigma_u, \end{cases} \tag{14}$$

where

$$\mu_u = \frac{Q+2}{3Q+3}, \tag{15}$$

and

$$\sigma_u = \sqrt{\frac{(Q+2)(Q^2+2Q+3)}{18(Q+1)^2 QH}}. \tag{16}$$

If the UACI value falls into interval $(\mathcal{U}_{\alpha}^{*-}, \mathcal{U}_{\alpha}^{*+})$, then we can consider the two test images have enough difference. We assume significance level $\alpha = 0.05$. When the test image is 512×512, the NPCR critical value is $\mathcal{N}_{0.05}^{*} = 99.5893\%$ and the UACI critical interval is $(\mathcal{U}_{0.05}^{*-}, \mathcal{U}_{0.05}^{*+}) = (33.3730\%, 33.5541\%)$. When the test image is 256×256, the NPCR critical value is $\mathcal{N}_{0.05}^{*} = 99.5693\%$ and the UACI critical interval is $(\mathcal{U}_{0.05}^{*-}, \mathcal{U}_{0.05}^{*+}) = (33.2824\%, 33.6447\%)$.

In this test, we complete the test 100 times for each CRs and calculate the average value, respectively. Table 3 and Figure 6 show the tests results. According to the data and figures, the NPCR and UACI test values are floating with CR changes. Nonetheless, all of the test values are basically within the critical values. The test result has shown that our proposed scheme is plain image enough in resisting the differential attack.

4.4. Statistical Analysis

4.4.1. Histogram Analysis

A histogram can reflect the statistical feature of cipher image; the histogram is closer to uniform the better security performance. The histogram is shown in Figure 3. In Figure 3 rows (a, e, i), there are some cipher images that are encrypted in deferent CRs, and the corresponding histograms are shown in Figure 3 rows (b, f, j). The corresponding decrypted images are shown in Figure 3 rows (c, g, k). Figure 3 rows (d, h, l) shows the histograms of recovery image. The chi-squared test is used to evaluate the uniformity of cipher image's histogram. Table 4 provides the chi-squared test results of cipher images, when the significance level is $\alpha = 0.05$. According to the results, our scheme has enough good diffused property to resist the statistical attack.

Table 3. Number of pixels change rate (NPCR) and unified average changing intensity (UACI) results.

Image	CR = 0.2		CR = 0.5		CR = 0.8	
	NPCR(%)	UACI(%)	NPCR(%)	UACI(%)	NPCR(%)	UACI(%)
Lena (256 × 256)	99.6127	33.4770	99.6112	33.3843	99.6076	33.5224
Lena (512 × 512)	99.6086	33.3894	99.6078	33.4552	99.6086	33.5457
Pepper (256 × 256)	99.6179	33.3606	99.6132	33.4766	99.6097	33.5329
Pepper (512 × 512)	99.6050	33.4691	99.6103	33.4910	99.6087	33.4590
Airplane (256 × 256)	99.6142	33.5532	99.6110	33.6366	99.6099	33.5275
Airplane (512 × 512)	99.6135	33.5396	99.6112	33.5445	99.6097	33.5375
Boat (256 × 256)	99.6058	33.2956	99.6171	33.5142	99.6073	33.5132
Boat (512 × 512)	99.6119	33.3949	99.6073	33.4537	99.6081	33.4332
Cameraman (256 × 256)	99.6164	33.6333	99.6054	33.5684	99.6058	33.3942
Cameraman (512 × 512)	99.6078	33.4550	99.6100	33.4309	99.6095	33.5411
Barbara (256 × 256)	99.6125	33.5789	99.6036	33.5636	99.6125	33.5594
Barbara (512 × 512)	99.6088	33.4506	99.6085	33.4766	99.6102	33.4737

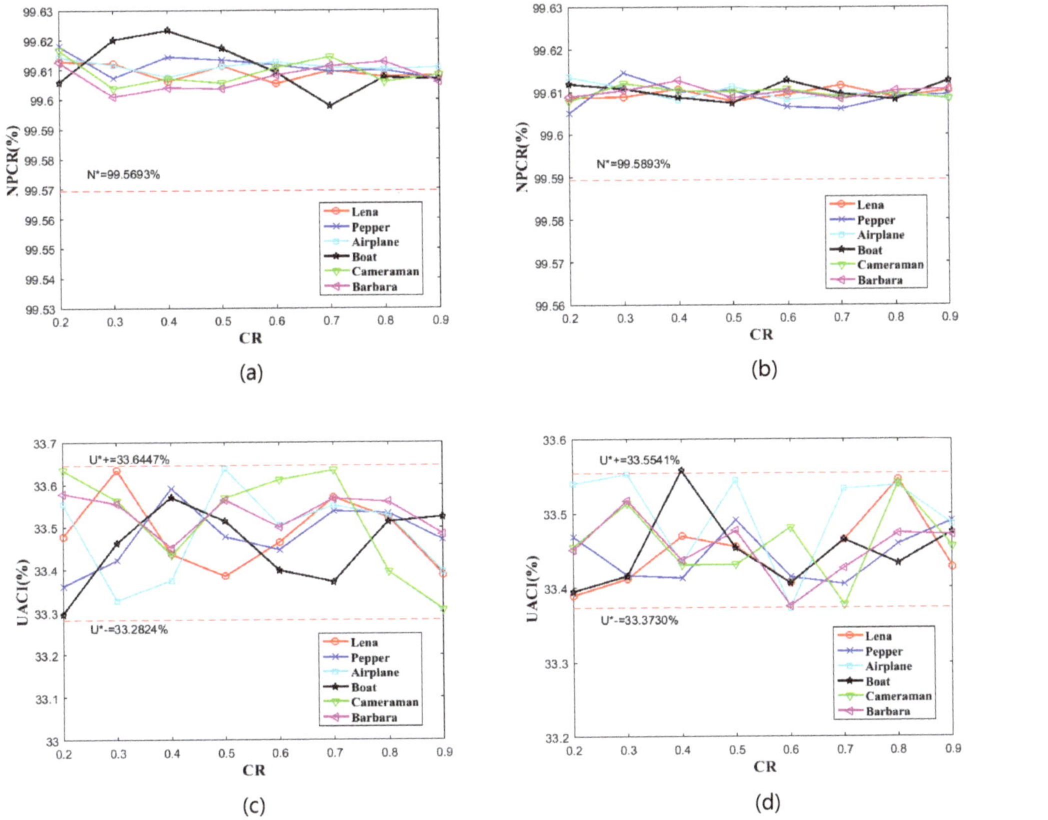

Figure 6. The number of pixels change rate (NPCR) and unified average changing intensity (UACI) test result. (**a**,**c**) show the results of images with the size of 256 × 256. (**b**,**d**) show the results of images with the size of 512 × 512. (**a**,**b**) show the results of NPCR tests. (**c**,**d**) show the results of UACI tests.

Table 4. Histogram uniformity evaluation by chi-squared test (*p*-value).

Images	CR = 0.2	CR = 0.3	CR = 0.4	CR = 0.5	CR = 0.6	CR = 0.7	CR = 0.8	CR = 0.9
Lena (256 × 256)	0.7183	0.9713	0.7252	0.3415	0.4128	0.8003	0.8271	0.7778
Lena (512 × 512)	0.2174	0.1773	0.1874	0.7268	0.1592	0.8726	0.4672	0.1685
Pepper (256 × 256)	0.3111	0.2117	0.2766	0.2717	0.3582	0.9891	0.6976	0.7083
Pepper (512 × 512)	0.6795	0.8444	0.9260	0.9765	0.1287	0.4755	0.9823	0.5199
Airplane (256 × 256)	0.7867	0.1359	0.5739	0.3754	0.1576	0.9868	0.3256	0.9142
Airplane (512 × 512)	0.9632	0.5859	0.9817	0.9213	0.8393	0.8304	0.2412	0.8186
Boat (256 × 256)	0.9429	0.4776	0.4672	0.1594	0.4086	0.7931	0.9114	0.7712
Boat (512 × 512)	0.5941	0.8562	0.3941	0.8378	0.8965	0.2870	0.7980	0.9894
Cameraman (256 × 256)	0.5087	0.7340	0.7240	0.2415	0.2752	0.2131	0.3656	0.3492
Cameraman (512 × 512)	0.4639	0.5619	0.3150	0.7148	0.4347	0.3540	0.8971	0.1729
Barbara (256 × 256)	0.4548	0.5059	0.2825	0.2531	0.3462	0.2301	0.1449	0.3510
Barbara (512 × 512)	0.5269	0.2898	0.9238	0.3938	0.3166	0.2771	0.5193	0.4375

4.4.2. Correlation Coefficient

As we all know, encryption is a process breaking the correlation of adjacent pixels. Therefore, correlation coefficient analysis is an important measurement for evaluating the permutation performance of image cryptosystem. The less correlation in cipher image, the better permutation performance.

The correlation coefficient can be calculated by Equation (17).

$$r_{ab} = \frac{\text{cov}(a,b)}{\sqrt{D(a)D(b)}}, \tag{17}$$

where a and b are two adjacent pixels' gray values, and

$$\text{E}(a) = \frac{1}{N}\sum_{i=1}^{N} a_i, \tag{18}$$

$$\text{D}(a) = \frac{1}{N}\sum_{i=1}^{N}(a_i - \text{E}(a))^2, \tag{19}$$

$$\text{cov}(a,b) = \frac{1}{N}\sum_{i=1}^{N}(a_i - \text{E}(a))(b_i - \text{E}(b)). \tag{20}$$

In this test, we first randomly select 10,000 pairs of adjacent pixels in the test image, and then calculate the correlation coefficient among these pixels. The test plain images are 512 × 512 and the cipher images are encrypted on $CR = 0.5$. Table 5 shows the test results. The correlation distributions are shown in Figure 7 and the rows **(1–6)** correspond to images of 'Lena', 'Pepper', 'Airplane', 'Boat', 'Cameraman', and 'Barbara', respectively; Column **(a)** shows the corresponding plain images; Columns **(b–d)** correspond to the plain images' distributions of 'horizontal direction', 'vertical direction', and 'diagonal direction', respectively; Column **(e)** is the corresponding cipher images by encryption; Columns **(f–h)** correspond to the cipher images' distributions of 'horizontal direction', 'vertical direction', and 'diagonal direction', respectively.

Table 5. Correlation coefficients.

Image	Horizontal		Vertical		Diagonal	
	Plain	**Cipher**	**Plain**	**Cipher**	**Plain**	**Cipher**
Lena (512 × 512)	0.9852	0.00057	0.9714	0.0028	0.9587	−0.0014
Pepper (512 × 512)	0.9801	0.0017	0.9778	0.0013	0.9643	0.00077
Airplane (512 × 512)	0.9653	0.0046	0.9678	0.0065	0.9369	0.0026
Boat (512 × 512)	0.9723	0.0012	0.9403	−0.0091	0.9238	0.0071
Cameraman (512 × 512)	0.9898	0.0019	0.9833	0.0043	0.9711	0.0056
Barbara (512 × 512)	0.9596	−0.0039	0.8610	0.0022	0.8406	0.00067

4.5. Key Sensitivity Analysis

In this subsection, we will test the key sensitivity of our proposed image cryptosystem. In our scheme, there are 12 real keys, which are parameters and initial values of 2D-SLIM and 2D-LSCM systems. For this test, we select 'Barbara' 512 × 512 as the test image. There are two tests for key sensitivity analysis. The first test we encrypted plain image on $CR = 0.5$, and then decrypted with tiny modified keys. Figure 8 shows the test results. The second test is that we encrypt plain images with tiny modified keys and then compare the corresponding cipher images with without modified cipher images. Figure 9 shows the test results.

We quantitatively measure the difference between cipher images using NPCR and UACI. Table 6 shows the result. As the results of tests, our proposed scheme is very sensitive to the real keys.

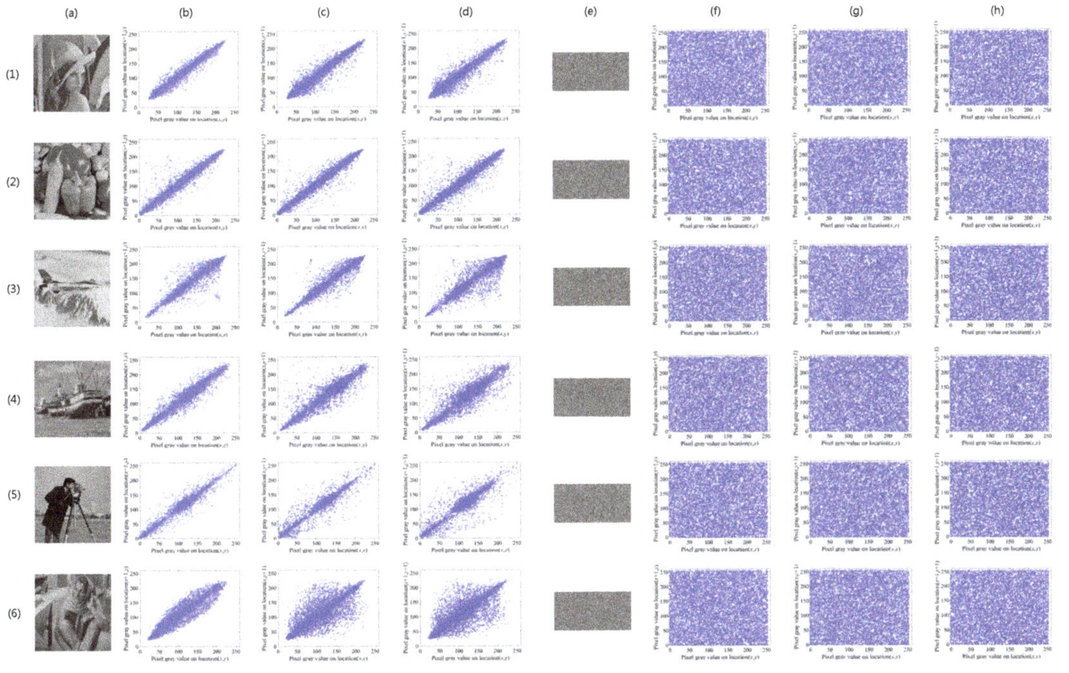

Figure 7. Correlation distributions. Rows (**1–6**) correspond to images of 'Lena', 'Pepper', 'Airplane', 'Boat', 'Cameraman', and 'Barbara', respectively; Column (**a**) shows the corresponding plain images; Columns (**b–d**) correspond to the plain images' distributions of 'horizontal direction', 'vertical direction', and 'diagonal direction', respectively; Column (**e**) is the corresponding cipher images by encryption; Columns (**f–h**) correspond to the cipher images' distributions of 'horizontal direction', 'vertical direction', and 'diagonal direction', respectively.

Figure 8. Decryption with tiny modified keys.

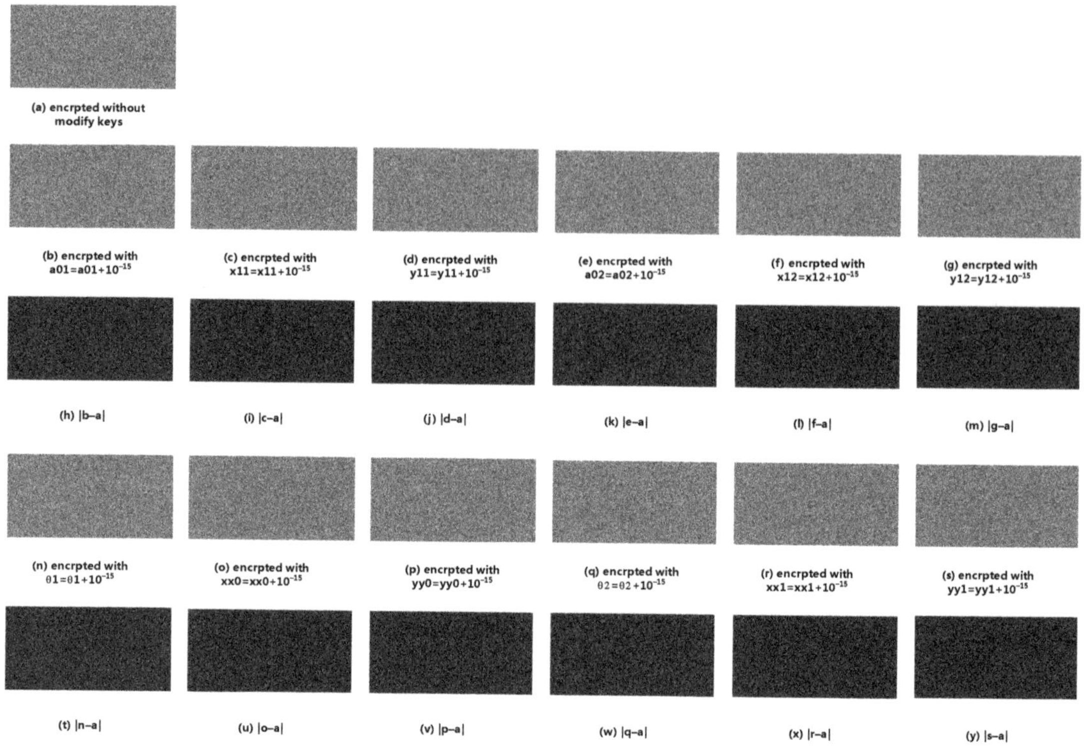

Figure 9. Comparison from encrypted images through tiny modified keys.

Table 6. Quantitative analysis of key sensitivity.

Cipher Image with Modify Key	NPCR (%)	UACI (%)
$a01 = a01 + 10^{-15}$	97.4136	30.0742
$x11 = x11 + 10^{-15}$	98.8632	33.8622
$y11 = y11 + 10^{-15}$	97.9836	32.8013
$a02 = a02 + 10^{-15}$	96.9574	31.1781
$x12 = x12 + 10^{-15}$	97.0551	31.6505
$y12 = y12 + 10^{-15}$	98.6200	32.6244
$\theta 1 = \theta 1 + 10^{-15}$	99.6124	33.5163
$xx0 = xx0 + 10^{-15}$	99.5987	33.4155
$yy0 = yy0 + 10^{-15}$	99.5819	33.5568
$\theta 2 = \theta 2 + 10^{-15}$	99.5796	33.5179
$xx1 = xx1 + 10^{-15}$	99.6002	33.4799
$yy1 = yy1 + 10^{-15}$	99.6185	33.5206

4.6. Information Entropy

Global Shannon entropy (GSE) is used to evaluate the randomness of the whole image. The GSE is given by

$$\mathrm{H(s)} = \sum_{i=0}^{2^K-1} P(s_i) \log_2 \frac{1}{P(s_i)}, \tag{21}$$

where K is the gray level of the test image and $P(s_i)$ means the probability of s_i. The GSE of 8-bit gray image is 8 bits in the ideal case. Table 7 shows the GSE results in different CRs.

Table 7. Global Information entropy.

Images	CR = 0.2	CR = 0.3	CR = 0.4	CR = 0.5	CR = 0.6	CR = 0.7	CR = 0.8	CR = 0.9
Lena (512 × 512)	7.9962	7.9974	7.9980	7.9986	7.9987	7.9990	7.9991	7.9990
Pepper (512 × 512)	7.9969	7.9977	7.9985	7.9987	7.9989	7.9990	7.9990	7.9992
Airplane (512 × 512)	7.9964	7.9978	7.9982	7.9981	7.9988	7.9988	7.9990	7.9992
Boat (512 × 512)	7.9967	7.9978	7.9983	7.9987	7.9990	7.9989	7.9991	7.9992
Cameraman (512 × 512)	7.9964	7.9976	7.9981	7.9986	7.9988	7.9989	7.9992	7.9991
Barbara (512 × 512)	7.9965	7.9975	7.9984	7.9985	7.9985	7.9988	7.9991	7.9992

In order to further measure the randomness of cipher image, Wu et al. [36] introduced a method of local Shannon entropy (LSE). To calculate LSE, k non-overlapping image blocks $B_1, B_2, \cdots, B_k$ with T_B pixels are randomly selected from image I, and then the LSE is defined by:

$$\overline{\mathrm{H}_{k,T_B}}(I) = \sum_{i=1}^{k} \frac{H(B_i)}{k}, \tag{22}$$

where $H(B_i)$ is the GSE of image block B_i. For this test, the parameters $(k, T_B) = (30, 1936)$ are selected. In this situation, the ideal value of LSE is 7.902469317. When the significance $\alpha = 0.05$, the tests passed when the test results fell into the interval $(7.901901305, 7.903037329)$. Table 8 shows the LSE test results. Figure 10 shows the global and local entropy analysis. According to this figure, the global entropies are increasing with CRs and the local entropies are floating with CRs. Nevertheless, the minimum of global entropy of 7.9962 entails sufficient security and the local entropies basically fall in security interval. Therefore, the information entropy results show that the cipher images that are generated by our proposed image cryptosystem have excellent randomness.

Table 8. The local Shannon entropy test.

Images	CR = 0.2	CR = 0.3	CR = 0.4	CR = 0.5	CR = 0.6	CR = 0.7	CR = 0.8	CR = 0.9
Lena (512 × 512)	7.9022	7.9025	7.9020	7.9028	7.9025	7.9030	7.9024	7.9027
Pepper (512 × 512)	7.9023	7.9027	7.9025	7.9029	7.9025	7.9021	7.9029	7.9023
Airplane (512 × 512)	7.9027	7.9021	7.9024	7.9028	7.9032	7.9019	7.9029	7.9025
Boat (512 × 512)	7.9022	7.9029	7.9030	7.9020	7.9021	7.9028	7.9027	7.9020
Cameraman (512 × 512)	7.9018	7.9024	7.9031	7.9023	7.9031	7.9020	7.9021	7.9027
Barbara (512 × 512)	7.9028	7.9026	7.9030	7.9026	7.9025	7.9028	7.9020	7.9025

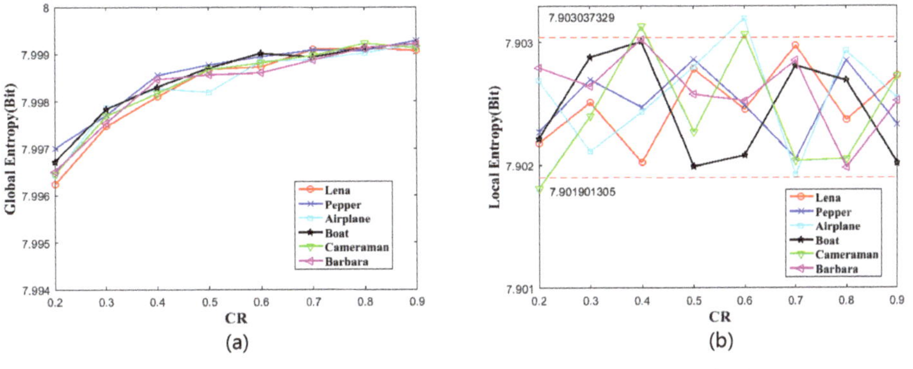

Figure 10. The information entropy analysis. (**a**) is global entropy. (**b**) is local entropy.

4.7. Robust Analysis

The robustness of the image cryptosystem means that some useful information can still be recovered when the cipher image is disturbed by noise or part of the data is lost during transmission. The robustness of the image cryptosystem in real communication applications is very important. In the test, some noise and different data loss amounts are added to Lena cipher images that are encrypted on CR = 0.5 to evaluate the robustness of our proposed image cryptosystem. Figure 11 shows the test results. Most of the information in the plain image can still be identified from the decrypted image, as shown in Figure 11. Figure 11a–f are cipher images with 1×10^{-4}, 2×10^{-4}, 3×10^{-4}, 4×10^{-4}, 5×10^{-4}, 1×10^{-3} salt & pepper noise, respectively. Figure 11g–l are corresponding decrypted images. Figure 11m–o are the cipher images with 16×16 data lost. Figure 11s–u are corresponding decrypted images. Figure 11p–r are the cipher images with 32×32 data lost. Figure 11v–x are corresponding decrypted images. It is shown that the algorithm has good robustness and it can be applied to practical scenarios.

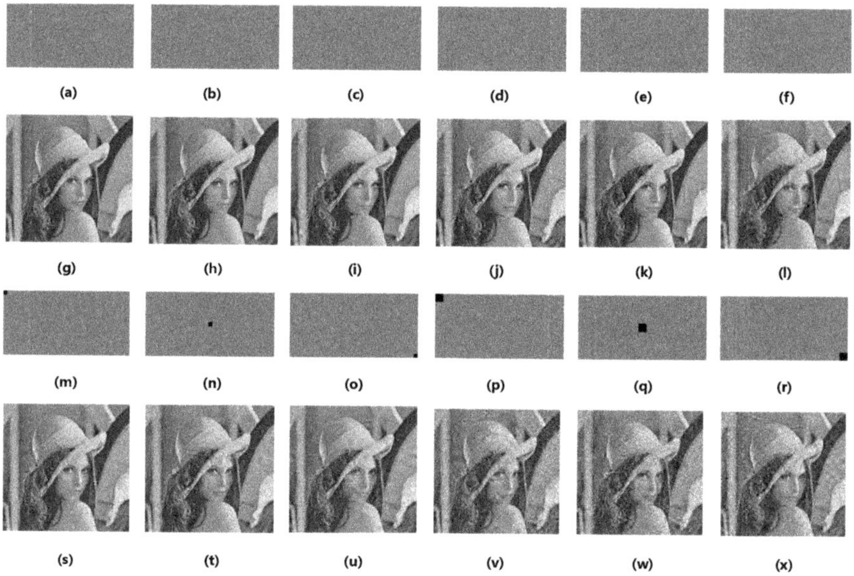

Figure 11. The robustness analysis result. **(a–f)** are cipher images with 1×10^{-4}, 2×10^{-4}, 3×10^{-4}, 4×10^{-4}, 5×10^{-4}, 1×10^{-3} salt & pepper noise, respectively. **(g–l)** are corresponding decrypted images. **(m–o)** are the cipher images with 16×16 data lost. **(s–u)** are corresponding decrypted images. **(p–r)** are the cipher images with 32×32 data lost. **(v–x)** are corresponding decrypted images.

4.8. Time Complexity Analysis

In order to evaluate the efficiency of our proposed image cryptosystem, we give the time complexity analysis and the time consuming of the simulation in this subsection. In this paper, we use two 2D-SLIM and two 2D-LSCM hyper chaotic systems to control the processes of compressive sensing and encryption, and it needs a total of $\Theta(3MN + N)$ iterations of computing floating point number. As we all know, there are many factors that affect the results of the actual test, such as hardware and software environments, programming languages, code optimization, parallel processing, programming skills, etc. Therefore, we give our simulation results of the time consumption under the environments that are mentioned at beginning of Section 4 and while using parallel computing technology. In our test, we encrypt and decrypt the same image 100 times, taking the average time. The encryption time of 256×256 Lena is 0.082 s and the decryption with CR = 0.25, 0.5, 0.75

are 0.82 s, 1.23 s and 2.32 s, respectively. The encryption time of 512 × 512 Pepper is 0.336 s and the decryptions with CR = 0.25, 0.5, and 0.75 are 3.23 s, 7.42 s, and 13.36 s, respectively.

4.9. Comparison with Other Works

In this section, we compare our proposed image cryptosystem with other recent works. For this comparison, we encrypt 512 × 512 8-bit gray level plain images 'Lena' with CR = 0.5. Table 9 shows the comparison result. The cipher image that is generated in Ref. [21] has poor randomness, because the global entropy is too low. Moreover, it also missing local entropy and plaintext sensitivity assessment. Ref. [22] presents the small key space and it is missing the information entropy assessment. Ref.[23] also has a small key space. Ref.[24] is missing the key space and a local entropy assessment. Our proposed image cryptosystem has the advantage of more comprehensive security performance, according to the comparison results.

Table 9. The comparison result.

Algorithms	Cipher Correlation Coefficients			Global	Local	Key Space	Plaintext Sensitivity		PSNR (dB)
	Horizontal	Vertical	Diagonal	Entropy	Entropy		NPCR (%)	UACI (%)	
Our work	0.00057	0.0028	−0.0014	7.9986	7.9028	2^{600}	99.6078	33.4552	38.3438
Ref. [21]	0.0061	0.0018	−0.0024	5.0508	-	1.15×10^{105}	-	-	-
Ref. [22]	−0.0016	−0.0010	−0.0015	-	-	2^{149}	99.6061	33.4150	35.51
Ref. [23]	0.0016	0.0081	−0.0016	7.9974	7.9027	2^{176}	99.6201	33.5247	30.8184
Ref. [24]	−0.0028	−0.0096	−0.0030	7.9960	-	-	99.60	34.17	39.04

5. Conclusions

In this paper, an efficient and secure plaintext-related chaotic image encryption scheme that is based on compressive sensing was proposed, which can simultaneously achieve the compression and encryption. In the proposed scheme, we generate the plaintext-sensitive internal keys to control the whole process of compression and encryption, which can make all processes have enough sensitivity to the plain image. The permutation that was controlled by the two-dimensional Sine improved Logistic iterative chaotic map (2D-SLIM) has been applied to the coefficient matrix in order to make the energy of matrix dispersive. A plaintext related compressive sensing was used to reduce the data storage capacity while the privacy of image is guaranteed. Additionally, we make sure the cipher image lower correlation and distribute uniform by quantifying the measurement results to 0∼255 and doing permutation and diffusion under the controlling by two-dimensional Logistic-Sine-coupling map (2D-LSCM). Finally, some common compression and security performance analysis methods are used for testing our scheme. The tests and comparison results have shown that our proposed scheme has both excellent security and compression performance in order to ensure the digital image application in the network. The image encryption combining compressive sensing is still under constant research, and there are still many problems that need to be further studied and solved. In the next stage, we will focus on the multi-image aggregation encryption and parallel block compressed sensing.

Author Contributions: Formal analysis, Z.L.; funding acquisition, C.P.; investigation, Z.L.; methodology, Z.L.; project administration, C.P.; software, W.T. and L.L.; writing—original draft preparation, Z.L.; writing—review and editing, C.P., W.T., and L.L. All authors have read and agreed to the published version of the manuscript.

Funding: This research received no external funding.

Institutional Review Board Statement: Not applicable.

Informed Consent Statement: Not applicable.

Data Availability Statement: The data used to support the findings of this study are included within the article.

Acknowledgments: Our research is supported by the National Natural Science Foundation of China (U1836205, 61662009, and 61772008), the open Foundation of Guizhou Provincial Key Laboratory of Public Big Data (2017BDKFJJ023, 2017BDKFJJ026), the Science and Technology Foundation of Guizhou (Guizhou Science-Contract-Major-Program [2018]3001, Guizhou-Science-Contract-Major-Program [2018]3007, Guizhou-Science-Contract-Major-Program [2017]3002, Guizhou-Science-Contract-Support [2019]2004, Guizhou-Science-Contract-Support [2018]2162, Guizhou-Science-Contract-Support [2018]2159, Guizhou-Science-Contract [2017]1045, Guizhou-Science-Contract 1049, Guizhou-Science-Contract [2019]1249), and Scientific Research Foundation of Guizhou province, China (QKHPTRC[2017]5788). The cultivation project of Guizhou University ([2019] 56). The Project of Innovative Group in Guizhou Education Department ([2013]09). The Youth Science and Technology Talents Growth Project of the Guizhou Provincial Department of Education (Guizhou-Education-Contract-KY-Word [2018]260).

Conflicts of Interest: The authors declare no conflict of interest.

Abbreviations

The following abbreviations are used in this manuscript:

2D-SLIM	two-dimensional Sine improved Logistic iterative chaotic map
2D-LSCM	two-dimensional Logistic-Sine-coupling map
DES	Data Encryption Standard
AES	Advanced Encryption Standard
DNA	deoxyribonucleic acid
CS	Compressed sensing
CR(s)	compression ratio(s)
RIP	restricted isometry property
RIC	Restricted Isometry Constant
DWT	Discrete Wavelet Transform
PSNR	Peak signal to noise ratio
SSIM	Structural similarity index measurement
NPCR	number of pixels change rate
UACI	unified average changing intensity

References

1. Lorenz, E. Deterministic Non-period Flows. *J. Atmos. Sci.* **1963**, *20*, 130–141. [CrossRef]
2. Wang, C.; Xia, H.; Zhou, L. A Memristive Hyperchaotic Multiscroll Jerk System with Controllable Scroll Numbers. *Int. J. Bifurc. Chaos* **2017**, *27*, 1750091. [CrossRef]
3. Matthews, R. On the derivation of a "chaotic" encryption algorithm. *Cryptologia* **1989**, *13*, 29–42. [CrossRef]
4. Batool, S.I.; Waseem, H.M. A novel image encryption scheme based on Arnold scrambling and Lucas series. *Multimed. Tools Appl.* **2019**, *78*, 27611–27637. [CrossRef]
5. Putra, R.K. Implementasi Arnold's Cat Map dan Logistic Map dalam Pemanfaatan Enkripsi dan Dekripsi Citra Digital. Ph.D. Thesis, Institut Teknologi Sepuluh Nopember, Surabaya, Indonesia, 2020.
6. Joshi, A.B.; Kumar, D.; Gaffar, A.; Mishra, D. Triple color image encryption based on 2D multiple parameter fractional discrete Fourier transform and 3D Arnold transform. *Opt. Lasers Eng.* **2020**, *133*, 106139. [CrossRef]
7. Zheng, J.; Luo, Z.; Zeng, Q. An efficient image encryption algorithm based on multi chaotic system and random DAN coding. *Multimed. Tools Appl.* **2020**, *79*, 29901–29921. [CrossRef]
8. Zefreh, E.Z. An image encryption scheme based on a hybrid model of DNA computing, chaotic systems and hash functions. *Multimed. Tools Appl.* **2020**, *79*, 24993–25022. [CrossRef]
9. Li, Z.; Peng, C.; Tan, W.; Li, L. A Novel Chaos-Based Image Encryption Scheme by Using Randomly DNA Encode and Plaintext Related Permutation. *Appl. Sci.* **2020**, *10*, 7469. [CrossRef]
10. Zhang, Y. The unified image encryption algorithm based on chaos and cubic S-Box. *Inf. Sci.* **2018**, *450*, 361–377. [CrossRef]
11. Wang, X.; Çavuşoğlu, Ü.; Kacar, S.; Akgul, A.; Pham, V.T.; Jafari, S.; Alsaadi, F.E.; Nguyen, X.Q. S-box based image encryption application using a chaotic system without equilibrium. *Appl. Sci.* **2019**, *9*, 781. [CrossRef]
12. Liu, H.; Kadir, A.; Niu, Y. Chaos-based color image block encryption scheme using S-box. *AEU Int. J. Electron. Commun.* **2014**, *68*, 676–686. [CrossRef]
13. Li, Z.; Peng, C.; Li, L.; Zhu, X. A novel plaintext-related image encryption scheme using hyper-chaotic system. *Nonlinear Dyn.* **2018**, *94*, 1319–1333. [CrossRef]
14. Ouyang, D.; Shao, J.; Jiang, H.; Nguang, S.K.; Shen, H.T. Impulsive synchronization of coupled delayed neural networks with actuator saturation and its application to image encryption. *Neural Netw.* **2020**, *128*, 158–171. [CrossRef] [PubMed]

15. Zhu, H.; Zhang, X.; Yu, H.; Zhao, C.; Zhu, Z. An image encryption algorithm based on compound homogeneous hyper-chaotic system. *Nonlinear Dyn.* **2017**, *89*, 61–79. [CrossRef]
16. Li, Z.; Peng, C.; Tan, W.; Li, L. A Novel Chaos-Based Color Image Encryption Scheme Using Bit-Level Permutation. *Symmetry* **2020**, *12*, 1497. [CrossRef]
17. Donoho, D.L. Compressed sensing. *IEEE Trans. Inf. Theory* **2006**, *52*, 1289–1306. [CrossRef]
18. Baraniuk, R.G. Compressive sensing [lecture notes]. *IEEE Signal Process. Mag.* **2007**, *24*, 118–121. [CrossRef]
19. Chai, X.; Zheng, X.; Gan, Z.; Han, D.; Chen, Y. An image encryption algorithm based on chaotic system and compressive sensing. *Signal Process.* **2018**, *148*, 124–144. [CrossRef]
20. Zhu, S.; Zhu, C. A new image compression-encryption scheme based on compressive sensing and cyclic shift. *Multimed. Tools Appl.* **2019**, *78*, 20855–20875. [CrossRef]
21. Chai, X.; Fu, X.; Gan, Z.; Zhang, Y.; Lu, Y.; Chen, Y. An efficient chaos-based image compression and encryption scheme using block compressive sensing and elementary cellular automata. *Neural Comput. Appl.* **2020**, *32*, 4961–4988. [CrossRef]
22. Zhu, L.; Song, H.; Zhang, X.; Yan, M.; Zhang, L.; Yan, T. A novel image encryption scheme based on nonuniform sampling in block compressive sensing. *IEEE Access* **2019**, *7*, 22161–22174. [CrossRef]
23. Gong, L.; Qiu, K.; Deng, C.; Zhou, N. An image compression and encryption algorithm based on chaotic system and compressive sensing. *Opt. Laser Technol.* **2019**, *115*, 257–267. [CrossRef]
24. Kayalvizhi, S.; Malarvizhi, S. A novel encrypted compressive sensing of images based on fractional order hyper chaotic Chen system and DNA operations. *Multimed. Tools Appl.* **2020**, *79*, 3957–3974. [CrossRef]
25. Elad, M. Optimized Projections for Compressed Sensing. *IEEE Trans. Signal Process.* **2007**, *55*, 5695–5702. [CrossRef]
26. Shen, H.; Li, X.; Zhang, L.; Tao, D.; Zeng, C. Compressed Sensing-Based Inpainting of Aqua Moderate Resolution Imaging Spectroradiometer Band 6 Using Adaptive Spectrum-Weighted Sparse Bayesian Dictionary Learning. *IEEE Trans. Geosci. Remote Sens.* **2014**, *52*, 894–906. [CrossRef]
27. Jin, J.; Gu, Y.; Mei, S. A Stochastic Gradient Approach on Compressive Sensing Signal Reconstruction Based on Adaptive Filtering Framework. *IEEE J. Sel. Top. Signal Process.* **2010**, *4*, 409–420. [CrossRef]
28. Tan, W.; Feng, X.; Tan, W.; Liu, G.; Ye, X.; Li, C. An Iterative Adaptive Dictionary Learning Approach for Multiple Snapshot DOA Estimation. In Proceedings of the 2018 14th IEEE International Conference on Signal Processing (ICSP), Beijing, China, 12–16 August 2018; pp. 214–219.
29. Xu, Q.; Sun, K.; Cao, C.; Zhu, C. A fast image encryption algorithm based on compressive sensing and hyperchaotic map. *Opt. Lasers Eng.* **2019**, *121*, 203–214. [CrossRef]
30. Hua, Z.; Jin, F.; Xu, B.; Huang, H. 2D Logistic-Sine-coupling map for image encryption. *Signal Process.* **2018**, *149*, 148–161. [CrossRef]
31. Li, S.; Chen, G.; Mou, X. On the dynamical degradation of digital piecewise linear chaotic maps. *Int. J. Bifurc. Chaos* **2005**, *15*, 3119–3151. [CrossRef]
32. Curiac, D.I.; Iercan, D.; Dranga, O.; Dragan, F.; Banias, O. Chaos-based cryptography: End of the road? In Proceedings of the International Conference on Emerging Security Information, Systems, and Technologies (SECUREWARE 2007), Valencia, Spain, 14–20 October 2007; pp. 71–76.
33. Chai, X.; Gan, Z.; Chen, Y.; Zhang, Y. A visually secure image encryption scheme based on compressive sensing. *Signal Process.* **2017**, *134*, 35–51. [CrossRef]
34. Wang, X.; Wang, Y.; Zhu, X.; Luo, C. A novel chaotic algorithm for image encryption utilizing one-time pad based on pixel level and DNA level. *Opt. Lasers Eng.* **2020**, *125*, 105851. [CrossRef]
35. Wu, Y.; Noonan, J.P.; Agaian, S.; others. NPCR and UACI randomness tests for image encryption. *Cyber J. Multidiscip. J. Sci. Technol. J. Sel. Areas Telecommun.* **2011**, *1*, 31–38.
36. Wu, Y.; Zhou, Y.; Saveriades, G.; Agaian, S.; Noonan, J.P.; Natarajan, P. Local Shannon entropy measure with statistical tests for image randomness. *Inf. Sci.* **2013**, *222*, 323–342. [CrossRef]

sensors

MDPI

Article

Wildfire Risk Assessment of Transmission-Line Corridors Based on Naïve Bayes Network and Remote Sensing Data

Weijie Chen [1], You Zhou [1,*], Enze Zhou [2], Zhun Xiang [2], Wentao Zhou [1] and Junhan Lu [1]

1 School of Electrical & Information Engineering, Changsha University of Science & Technology, Changsha 410114, China; chenweijie8558@163.com (W.C.); zhouwentao199734@163.com (W.Z.); L18924167973@163.com (J.L.)

2 Electric Power Research Institute, Guangdong Power Grid, Guangzhou 510080, China; zhou_dky@163.com (E.Z.); xiang_dky@163.com (Z.X.)

* Correspondence: zhouyou243@csust.edu.cn

Abstract: Considering the complexity of the physical model of wildfire occurrence, this paper develops a method to evaluate the wildfire risk of transmission-line corridors based on Naïve Bayes Network (NBN). First, the data of 14 wildfire-related factors including anthropogenic, physiographic, and meteorologic factors, were collected and analyzed. Then, the relief algorithm is used to rank the importance of factors according to their impacts on wildfire occurrence. After eliminating the least important factors in turn, an optimal wildfire risk assessment model for transmission-line corridors was constructed based on the NBN. Finally, this model was carried out and visualized in Guangxi province in southern China. Then a cost function was proposed to further verify the applicability of the wildfire risk distribution map. The fire events monitored by satellites during the first season in 2020 shows that 81.8% of fires fall in high- and very-high-risk regions.

Keywords: wildfire; risk assessment; Naïve bayes; transmission-line corridors

Citation: Chen, W.; Zhou, Y.; Zhou, E.; Xiang, Z.; Zhou, W.; Lu, J. Wildfire Risk Assessment of Transmission-Line Corridors Based on Naïve Bayes Network and Remote Sensing Data. *Sensors* **2021**, *21*, 634. https://doi.org/10.3390/s21020634

Received: 16 December 2020
Accepted: 14 January 2021
Published: 18 January 2021

1. Introduction

Due to the inhomogeneous distribution of energy resources and power loads in China, a large number of overhead transmission lines pass through forests and mountains to achieve an optimal allocation of power resources [1,2]. Wildfires are prone to occur in transmission-line corridors during the high-incidence periods. If a wildfire occurs under an overhead transmission line, the high temperature and smoke in wildfire would reduce the insulating strength of air gap drastically and induce a trip fault of transmission line. In addition, the reclosing usually fails under wildfire conditions with continuous high temperature, which seriously endangers the reliable operation of power grid [3–5].

To improve the wildfire prevention capability of transmission lines, scholars carried out much research, including distribution analysis of widfire occurrence [6,7], transmission-line trip mechanisms caused by wildfire [8,9], fire-spot monitoring algorithms [10–13] and wildfire risk assessment methods [14–16]. Remote sensing satellite is an economical and effective way to monitor wildfires in transmission-line corridors continuously. After identifying the flame strength of wildfire, the tripping risk would be evaluated by comparing the height of fire and transmission line [17,18]. However, this method needs numerous land-surface and environmental parameters to estimate the possible height of wildfire. Once a wildfire occurs near the transmission lines, it may cause a trip within tens of minutes. Furthermore, multiple wildfires usually occur simultaneously during the period of the Spring Festival, Qingming Festival, and autumn harvest in China. It is difficult for operation and maintenance personnel to rush the field and put out the fire in time. Therefore, it is necessary to assess occurrence probability of wildfire to propose differentiated wildfire prevention strategies.

The wildfire risk assessment was initially proposed by forest and fire departments to supporting decisions or policies for general fire, forest management, and fire suppression. Forestry departments in the US and Canada evaluated the large-scale wildfire risk by combining meteorological factors [19,20]. Nevertheless, these meteorology-based assessment methods cannot meet the requirement of spatial accuracy for transmission-line corridors due to the geographical differences. In 2016, State Grid Corporation of China issued a standard of drawing guidelines (DG) for region distribution map of wildfires near overhead transmission lines. In this standard, fire-spot densities and vegetation burning hazard grades are introduced to construct a risk assessment matrix [21]. However, the wildfire occurrence risk is affected by multi-dimension factors [22]. Besides meteorology and vegetation, physiographic and human-related factors are believed to have an important role in affecting wildfire occurrences [23]. Unfortunately, due to the complexity of wildfire occurrence there is still no physical model that could assess wildfire risk with specific variables.

Bayesian network is an effective approach to estimate uncertainty in risk evaluation in terms of the likelihood of risks and hazards [24]. In this paper, we aim to propose a wildfire risk assessment model based on Naïve Bayes Network (NBN) and remote sensing data. The region of Guangxi province which locates in southern China, is selected as the study area. A total of 14 sub-categories of wildfire-related factors including anthropogenic, physiographic, and meteorologic factors are collected. Then the spatial data are divided into grids of 1 km $\times$ 1 km to meet the spatial accuracy requirement of power grid. Considering the historical wildfire occurrences, the grids are divided based on whether there have been wildfires. After the importance evaluation by the relief algorithm, an NBN-based model is built with the optimal factor subset to map the wildfire risk distribution of Guangxi province. The wildfire risks are then divided into four levels based on the wildfire occurrence probability. In addition, a cost function is proposed to evaluate the applicability of wildfire risk map.

2. Study Area and Data Collection

2.1. Study Area

Guangxi province is in South China with the latitude of 20°54′–26°24′ and the longitude of 104°28′–112°04′. The total area is 237,600 square kilometers. It is located on the southeastern edge of the Yunnan–Guizhou Plateau. In addition, the terrain is high in the northwest and low in the southeast. The region of Guangxi province is dominated by subtropical and tropical monsoon climate, in which the precipitation is synchronous with high temperature. With a large forest area, Guangxi is vulnerable to wildfire disasters due to its changeable climate, complex topography, and various vegetation.

2.2. Wildfire-Related Factors

2.2.1. Anthropogenic Factors

Based on survey, more than 90% of wildfires are linked directly or indirectly to intentional and unintentional human activities [25]. It mainly includes productive fires such as wasteland and slash burning, and non-productive fires such as smoking in the wild and incensing on the grave [26]. To represent the influence of human activities, five kinds of anthropogenic sub-factors, which are Distance to Roads (DR), Distance to Settlements (DS), Population Density, Gross Domestic Product (GDP) and Historical Fire-Spot Density, are selected.

Human settlements and roads are the main areas of human activities, in where there are more human-caused fires. Population density reflects the regional population aggregation. A high population density generally corresponds to more human activities and consequently high probability of wildfire occurrence [27]. However, this law may be only suitable to explain the fire events in forests and rural areas. In large urbans, the higher population density may cause the less fire occurrence due to lack of fuels [28]. GDP represents the economic status of regions, which affects the human's fire habits. The data of roads, settlements, population density were downloaded from the website of the Resource and Environment Science and Data Center (RESDE) with a resolution of 1 km × 1 km. In addition, the DR and DS of grids were calculated by using ArcGIS 10.4.

Historical Fire-Spot Density represents the spatial distribution of wildfires in the past few years. It is related to not only human activities but also other meteorological and physiographic factors. The database of historical fire-spots was monitored by polar orbiting meteorological satellites from 2010 to 2019, which is provided by the National Meteorological Center. The fire-spot data during 2010–2014 is used to calculate the Historical Fire-Spot Density of grids, whereas the remains (from 2015 to 2019) are used to train and test the NBN model. To calculate the Historical Fire-Spot Density, the study area is divided into grids of 2.5 km × 2.5 km first. Then the fire-spots from 2010–2014 were allocate into grids based on their longitude and latitude. In addition, the final Historical Fire-Spot Density of grids was obtained by using Kriging interpolation method from the resolution of 2.5 km × 2.5 km into 1 km × 1 km.

2.2.2. Physiographic Factors

The physiographic factors include the land cover and landscape parameters of grids. The land cover factors are consisted of the Land-Usage Type, Vegetation Type, Fuel Load and Normalized Difference Vegetation Index (NDVI), whereas the landscape parameters are average Elevation, Slope, and Aspect of underlying surface.

The vegetations provide the fuel basis for wildfires' development. Different types of vegetations differ from their burning capacity in fire ignition and spread. Wildfires prone to ignite and spread rapidly at woodlands, shrubberies and meadows [29]. Therefore, the Land-Usage Type and Vegetation Type are taken to describe the flammability of the underlying surface. To construct the NBN model, the Land-Usage Type and Vegetation Type of study area are classified into four levels according to their flammability, which are shown in Tables 1 and 2, respectively. The Fuel Loading is represented by the drying weight of fuel per unit area, which affects the spread velocity and flame intensity of fires. The NDVI represents the coverage of surface vegetation. It is another representative parameter about fuel contents on the underlying surface.

Table 1. Classification of Vegetation Types.

Level	Description
1	Desert, Swamp, Cultivated plants
2	Meadow, Grassland, Alpine vegetation
3	Broad-leaved forest, Shrub
4	Coniferous forest, theropencedrymion

Table 2. Classification of Land-Usage Types.

Level	Description
1	Paddy field, Dry land, Water area, Unused land, Urban-rural fringe, Industrial and mining land, Residential land
2	Meadow, Grassland, Alpine vegetation
3	Broad-leaved forest, Shrub
4	Coniferous forest, Theropencedrymion

Complex topography influences not only the distribution of vegetations, but also the spreading behavior of fires directly. The elevation brings differences of temperature and humidity to affect the composition of vegetations. Moreover, human population tends to cluster at the region with low elevation and gentle slopes, which increases the fire activities. In addition, the slope also has a direct impact on the spreading speed of wildfires [30]. The increase of slope leads to the faster surface runoff, which is beneficial to the drying of vegetations. Slope aspect determines the amount of solar radiation, therefore the humidity of atmosphere and vegetations.

The data of Fuel Loading was obtained from the National Meteorological Center. In addition, other land cover factors and landscape parameters were downloaded from RESDE. All the data resolution are 1 km × 1 km.

2.2.3. Meteorologic Factors

Annual Precipitation and Temperature which have great influences on vegetation growth are selected as the meteorologic factors. In the high precipitation regions, the growth of vegetations is flourishing, which provides fuel conditions for wildfires to burn. However, the reduced transpiration in these areas increases the humidity of vegetation and reduces the flammability. The water-holding capacity of soils and air humidity are also increased. The fires' ignition and spread are therefore restrained. The higher temperature in forests generally benefits plant growth. In addition, the increase of temperature also accelerates the transpiration of vegetations, which promoting the rapid drying of vegetations. The data of Annual Precipitation and Annual Temperature were obtained from the RESDE, with a resolution of 1 km × 1 km.

2.3. Sample Preparation and Pre-Processing

The influencing degree of wildfires in transmission-line corridors varies with the distance. Generally, the fire that is 1 km away from the transmission-line is regarded as the highest risk, whereas the fire that is 3 km away is assumed to be no impact on the operation of transmission-line [31]. To meet the spatial accuracy requirement, the study area is divided into 1 km × 1 km grids. Then all the wildfire-related factors are allocated into grids. Specifically, the grids which have monitored fire-spots from 2015–2019 are taken as the fire samples. Considering the ignition, spread, and extinction of wildfire last for several hours, those monitoring fire-spots within 4 h and 3 km are regarded as a same fire-spot. Meanwhile, a same number of the non-fire samples are sampled randomly from the rest grids of 3 km away from the grids of the fire samples.

Compared to processing continuous factors, the Bayesian model has a higher efficiency and better robustness for discrete factors [24]. Therefore, the equal frequency method integrating empirical knowledge was used to discretize the factors. The standards of factor discretization are shown in Table 3.

Table 3. Factor discretization standards.

Factors	Discrete Intervals
GDP (10,000 yuan/km^2)	(0, 194), (194, 400), (400, 638), (638, ∞)
Fuel load(t/km^2)	(0, 1), (1, 1.3), (13, 23.3), (23.3, ∞)
NDVI	(0, 0.8), (0.8, 0.86), (0.86, 0.90), (0.90, 1]
Population density (people/km^2)	(0, 91.9), (91.9, 144.7), (144.7, 303.1), (303.1, ∞)
Elevation (m)	(0, 145), (145, 295), (295, 575), (575, ∞)
Fire-spot density (unit/(100 km^2·year))	(0, 1), (1, 2.4), (2.4, 4.5), (4.5, ∞)
Annual precipitation (mm)	(0, 173.1), (173.1, 195.1), (195.1, 209.6), (209.6, ∞)
Annual temperature (°C)	(0, 19.9), (19.9, 21.5), (21.5, 22.8), (22.8, ∞)
Slope (°)	(0, 3.3), (3.3, 10), (10, 18.3), (18.3, 90]
Aspect (°)	North (0°, 45°) ∪ (315°, 360°), East (4 5°, 135°), South (135°, 225°), West (225°, 315°)
DS (m)	(0, 356.5), (356.5, 635.7), (635.7, 1041.8), (1041.8, ∞)
DR (m)	(0, 832.6), (832.6, 2043.7), (2043.7, 3860.3), (3860.3, ∞)

2.4. Spatial Distribution of Factors

The factor distributions of the study area are visualized by using the ArcGIS software, as shown in Figure 1. It can be seen from the spatial distribution map that the historical wildfire high-incidence areas mainly distribute in the northwest, east, and central parts of Guangxi province. The distribution of GDP is similar to that of population density. The population concentrates in the vicinity of large municipal districts such as Nanning City and Liuzhou City in the central part, and Wuzhou City in the east. The elevation, slope, and NDVI in Guangxi province are relatively similarly distributed. Guangxi province has a wide distribution of karst landforms with mountains and hills. The west and north of Guangxi province are adjacent to the Yunnan–Guizhou Plateau, in where the elevation and slope values are relatively large. As the elevation increases, the climate, hydrothermal conditions change, leading to the plant no longer flourish in these regions. In addition, the higher elevation and steeper slope at the northern of Guangxi province make the less human settlements and higher vegetation coverages and higher NDVI values in this region. For the meteorological distribution, the annual temperature in the northern plateau is slightly lower. That is caused by the increase of latitude and elevation in plateau. The precipitation distribution shows that a circular decreasing trend from the east to the west. In addition, the annual precipitation in the northeast of Guangxi province is the largest. In Guangxi province, sparse forests and shrub forests are the main vegetation and land-Usage Type. In addition, coniferous forests and theropencedrymions mainly locates in the eastern region of Guangxi province, whereas the western region is dominated by broad-leaved forests and shrubs.

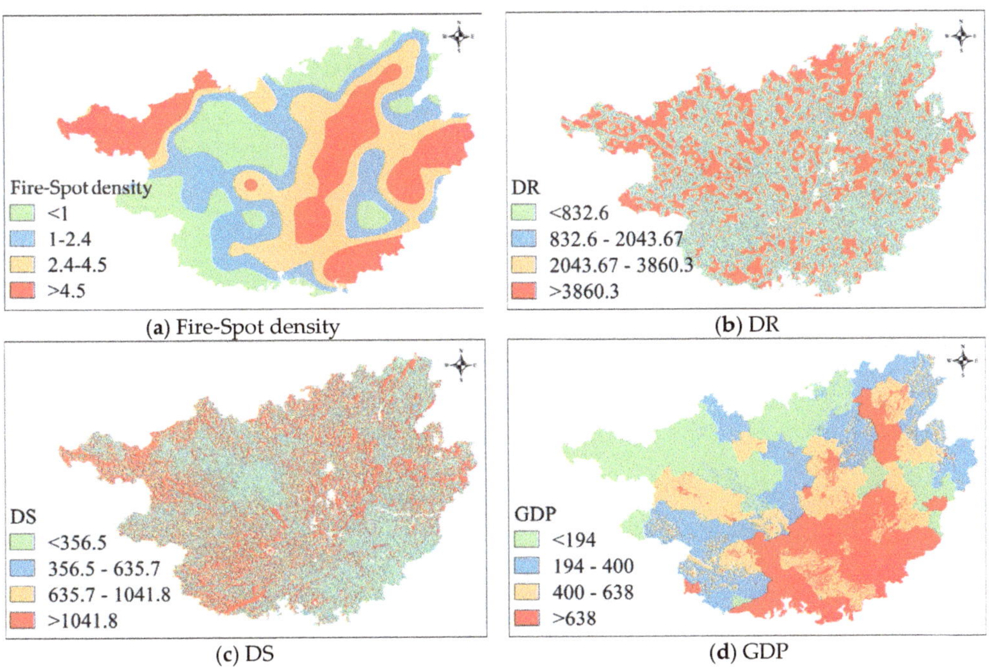

(**a**) Fire-Spot density (**b**) DR (**c**) DS (**d**) GDP

Figure 1. *Cont.*

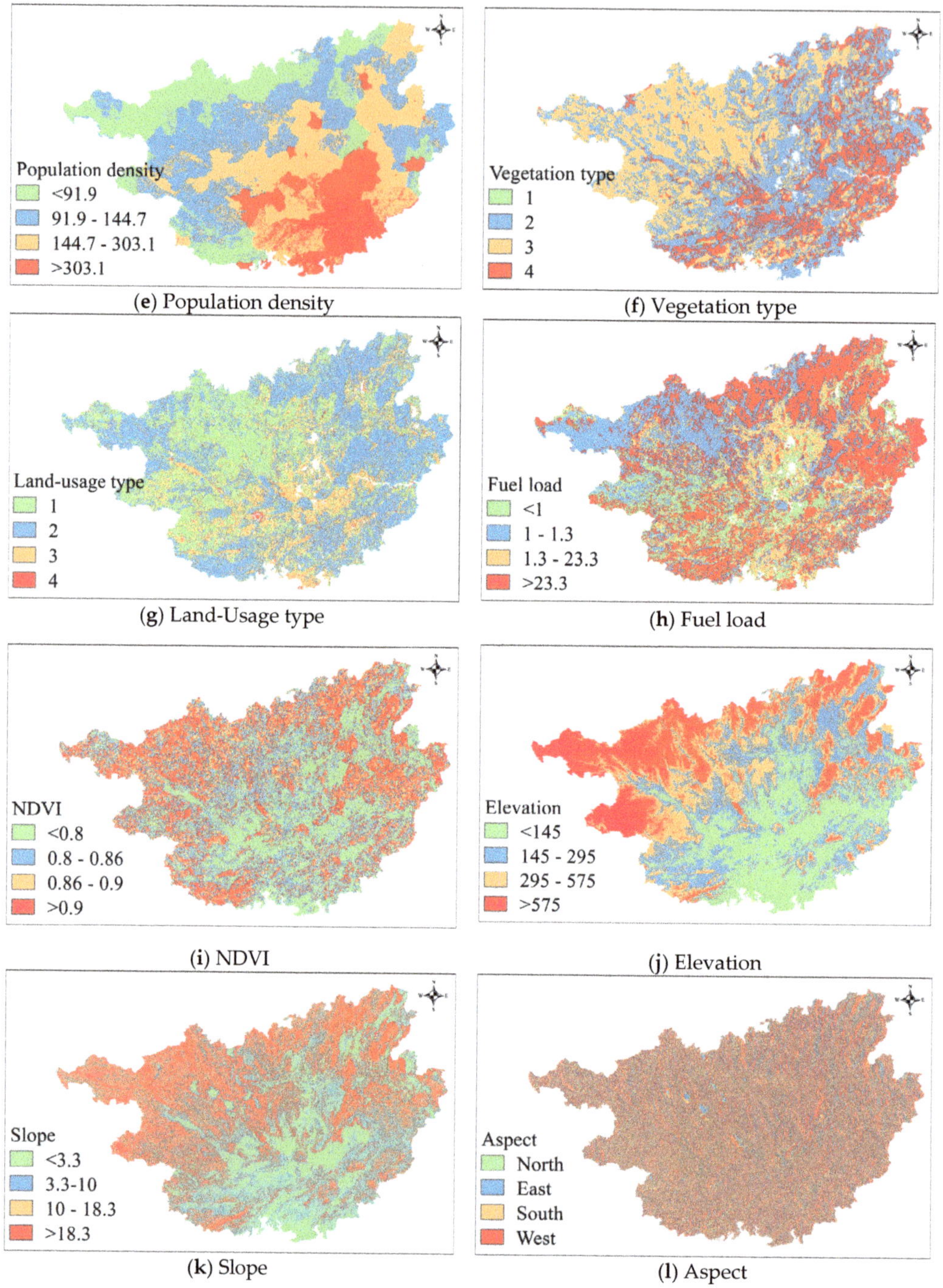

Figure 1. *Cont.*

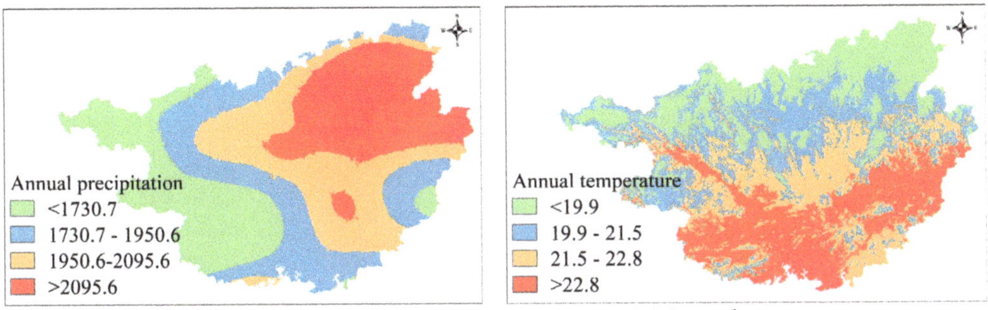

(**m**) Annual precipitation (**n**) Annual temperature

Figure 1. Spatial factor distribution of Guangxi province.

3. Importance of Wildfire-Related Factors

Wildfire occurrences are influenced by many feature factors. However, the relationship among these factors and wildfire is complex and differs from the location of regions [32]. Some of wildfire-related factors may play an important role in some regions but contribute little on the occurrence probability of wildfires in other regions. In addition, they even introduce noisy information for risk analysis in this region. Moreover, the redundancy of data increases the complexity of model and reduces the evaluation performance. A feature selection method, the relief algorithm, was used to appraise the contribution importance of wildfire-related factors on the wildfire occurrences before model building.

The relief algorithm was first proposed by Kira and Rendell [33], which is a factor weighting algorithm for binary classification based on the correlation between factors and the sample classification. Figure 2 gives the basic idea about how the relief algorithm appraises the importance of two factors. For a random sample S_i and its factor x_k,$D(x_k, S_i, S_i^{NM})$ represent the distance between and S_i and its closest different-class sample S_i^{NM}. In addition, $D(x_k, S_i, S_i^{NH})$ represent the distance between S_i and its closest same-class sample S_i^{NH}. If the same-class distance is greater than the different-class distance, the factor is more useful to classify the sample and should be given a higher weight. In Figure 2, if the random S_i is a fire sample, then the factor x_1 has a smaller distance from its same-class samples than that of the factor x_2, indicating the factor x_1 is more important than the factor x_2.

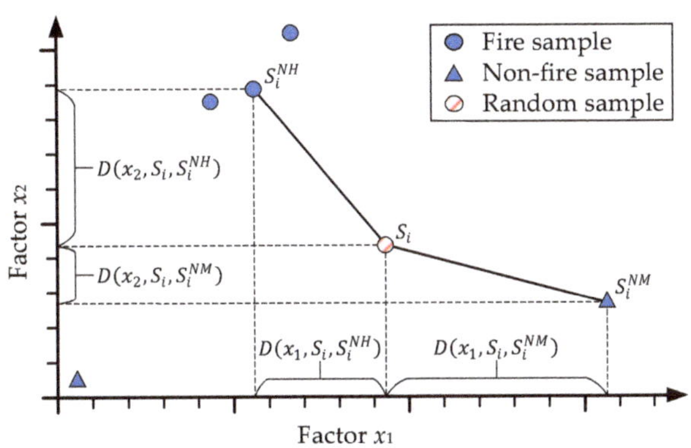

Figure 2. Basic idea of Relief algorithm.

The basic steps of the relief algorithm are as follows:

(1) Take a sample S_i randomly from the dataset $D = (S_1, S_2, \ldots, S_n)$.
(2) Go through the dataset and find the closest same-class sample S_i^{NH} and different-class sample S_i^{NM} of the S_i.
(3) Calculate the distance $D(x_j, S_i, S_i^{NM})$ between the closest samples about a certain factor x_j. If the factor x_j is a discrete variable,

$$D(x_j, S_i, S_i^{NM}) = \begin{cases} 0 & x_j \neq x_j^{NM} \\ 1 & x_j = x_j^{NM} \end{cases} \quad (1)$$

else,

$$D(x_j, S_i, S_i^{NM}) = \left| \frac{x_j - x_j^{NM}}{\max(x_j) - \min(x_j)} \right| \quad (2)$$

In this study, the most of factors are continuous variables, except the Vegetation Type and Land-Usage Type.

(4) Update the weight of factor x_j after multiple sampling

$$\omega_j^* = \omega_j + \sum\left(\frac{D(x_j, S_i, S_i^{NM})}{m} - \frac{D(x_j, S_i, S_i^{NH})}{m}\right) \quad (3)$$

where ω_j and ω_j^* are the initial and updated weight, respectively. m is the number of random sampling. The calculated weights of the 14 wildfire-related factors are listed in Table 4.

Table 4. Weights of wildfire-related factors based on the Relief.

Wildfire-Related Factor	Weight
DS	0.1265
Vegetation Type	0.1227
DR	0.1182
Annual precipitation	0.1043
Fire-spot density	0.0997
Land-Usage Type	0.0922
Elevation	0.0873
NDVI	0.0789
Aspect	0.0554
Fuel load	0.0376
Population density	0.0297
Annual temperature	0.0245
Slope	0.0134
GDP	0.0096

Based on relief algorithm, the DS, Vegetation Type, and DR shows the most three important impacts on distinguishing the fire and non-fire samples. This is mainly because of the plentiful of fire activities near the living and transportation regions of populations. In addition, the type of vegetation affects the ignition and spread of wildfires. Population Density, Annual Temperature, Slope, and GDP are the least four important factors. The population in Guangxi province is concentrated in the regions of cities, whereas the rest large parts of region are subtropical forests. In urban area, it is hard to inflame due to the lack of fuels and timely fire-fighting behavior. On the other hand, in sparsely populated forests, the happen probability of wildfire is also low due to the low human activities. The wildfire is likely to be happened at the interfacial region of forests and human settlements, which leads the population and GDP to fade into insignificance in this study. The study area is the provincial power grid in southern of China with a small temperature difference. Thus, the average temperature has little influence on the wildfire occurrence.

4. Naïve Bayes Network-Based Wildfire Risk Assessment

4.1. Bayes Theorem and Independence Assumption

Bayes' theorem was initially proposed by Thomas Bayes in the 18th century. It expresses the relationship between the conditional probabilities of two events statistically [34]. It has been widely used in uncertain fields such as disaster prediction, medical diagnosis, speech recognition, and so on. The Bayes' theorem is as follows:

$$P(X_i|Y) = \frac{P(Y|X_i) \cdot P(X_i)}{\sum\limits_j P(Y|X_j) \cdot P(X_j)} \quad (4)$$

where $P(X_i)$ is the priori probability which are obtained from the past experience or data distribution. $P(Y|X_i)$ is the probability of event Y occurring under the condition of known event X_i. $P(X_i|Y)$ is the probability of X_i when the result Y is known. Based on this theorem, the probabilities of wildfire-related factors and then the wildfire occurrence probability under certain conditions could be estimated statistically. However, the estimation of joint probability $P(Y|X_i)$ is difficult due to the limited number of samples. Therefore, the Naïve Bayes Network (NBN), in which the factors are assumed to be independent with each other, is used to model the risk of wildfire occurrences. Even through it sacrifices the interaction of factors but still gets an acceptable performance of model in many applications.

4.2. Model Construction

The construction of a Bayes network includes the structure learning and the parameter learning. Due to the independence assumption, the structure of NBN is simplified into a directed acyclic graph with factor nodes connecting to a class node. In addition, the parameter learning process is as follows:

(1) Sample preparation from the grids. The fire samples are the grids where fire-spots have been monitored by satellites in the years from 2015 to 2019. A total of 20,348 fires were recorded. In addition, a same number of non-fire samples were extracted randomly excluding buffer zones of 3 km around the fire samples. Then the samples were graded according to the discretization standards in Table 3. The training subset was randomly chosen from 70% of the samples, the remaining samples were used as the testing subset to evaluate the model's performance. In addition, the spatial distribution of samples in the testing subset are shown in Figure 3.

(2) By using the training subset, the probabilities of factors are obtained under fire and non-fire condition based on the maximum likelihood estimation.

$$P(x_{ij}|Y) = \frac{n_{ij}}{\sum\limits_{k=1}^{4} n_{ik}} \quad (5)$$

where x_{ij} is the ith wildfire-related factor fall in jth level; and n_{ij} is the number of x_{ij}.k represented the discretization level of factors. $P(x_{ij}|Y=0)$ and $P(x_{ij}|Y=1)$ are the probabilities of x_{ij} for the girds with non-fires and fires, separately.

(3) The conditional probability $P(Y=1|x_1, x_2, \ldots, x_n)$ and $P(Y=0|x_1, x_2, \ldots, x_n)$ of samples in the testing subset were calculated based on the Bayes' theorem. In addition, the final wildfire occurrence probability $P(Y)$ was obtained after normalization.

$$P(Y) = \frac{P(Y=1|x_1, x_2, \ldots, x_n)}{P(Y=1|x_1, x_2, \ldots, x_n) + P(Y=0|x_1, x_2, \ldots, x_n)} \quad (6)$$

(4) To test and optimize model's performance, the value of 0.5 was selected as the threshold to divide the samples into two classes: "Prone to fire" and "Prone to non-fire". The least important factors were then eliminated in turn to study the influence of factors composition on the model's performance.

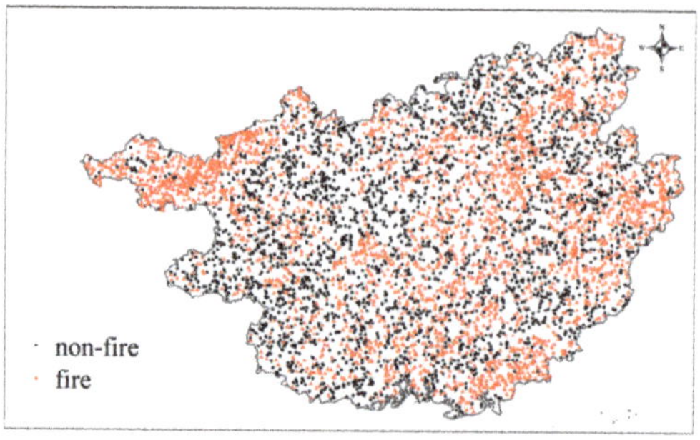

Figure 3. Samples of testing subset in the Guangxi province.

4.3. Model Assessment

The confusion matrix is used to evaluate the assessment performance of NBN model. It compares the predicted fire tendency of the testing subset with the actual wildfire events to measure the following four indexes, as shown in Table 5.

Table 5. Definition of confusion matrix.

Samples in Testing Subset		Predicted Results	
		Prone to Fire	**Prone to Non-fire**
Actual events	Fire	TP	FN
	Non-fire	FP	TN

TP: True Positive, represents the number of fire events correctly predicted as "Prone to fire".

TN: True Negative, indicates the grids where are non-fire and that were identified as "Prone to non-fire", correctly.

FN: False Negative, reflects the fire events that were identified as "Prone to non-fire", mistakenly.

FP: False Positive, recounts the grids where are non-fire but that are predicted to be "Prone to fire".

It is obviously that larger shares of TP and TN indicate a better predicted performance of model. Therefore, the indexes of *Accuracy* P_a, *Recall* P_r, *Precision* P_p and a more balanced index F-score are introduced on the basis of the confusion matrix.

$$P_a = \frac{TP + TN}{TP + TN + FP + FN} \tag{7}$$

$$P_r = \frac{TP}{TP + FN} \tag{8}$$

$$P_p = \frac{TP}{TP + FP} \tag{9}$$

$$F = \frac{(1 + \beta^2) P_p P_r}{\beta^2 P_p + P_r} \tag{10}$$

where β reflect the attention degree of power grids on fault tolerance. Considering the wildfire would induce outage of transmission lines, bring huge economic losses, and even casualties, this study takes β as 3.

As mentioned in Table 4, the weights of wildfire-related factors have been evaluated by using relief algorithm. Then the NBN models were re-trained and re-tested by eliminating the least important wildfire-related factor one by one. The assessment performances of models are shown in Figure 4.

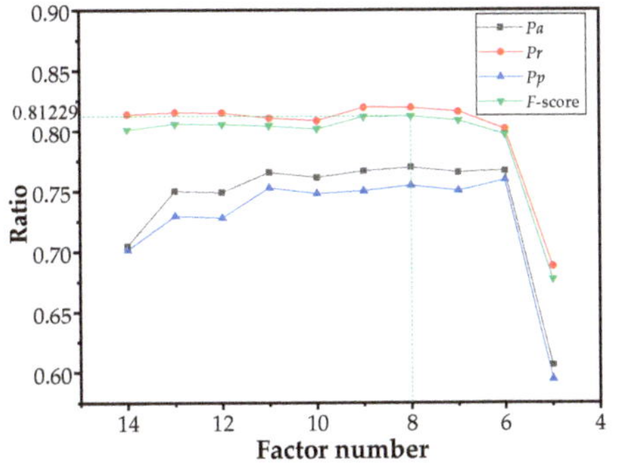

Figure 4. Variation of assessment indexes with wildfire-related factors.

With the reduced number of wildfire-related factors, the impact of noise, which is bring by unimportant factors, on the model is gradually reduced, which leading to the improvement of *Accuracy* P_a and *Precision* P_p. The *Accuracy* P_a is only 70.14% when all 14 wildfire-related factors are used. It reaches the maximum of 75.93% when the number of factors is reduced to six. The *Recall* P_r and *F*-score remain about 81% until the used factor number less than six. However, the performance of NBN model deteriorates significantly when the number of wildfire-related factors reduces from six to five, indicating that the remaining factors have greatest impacts on the wildfire occurrence.

The *F*-score reaches the highest value of 81.23% when eight wildfire-related factors are used. The results of confusion matrix are listed in Table 6. 81.92% of actual fire events in the testing subset are predicted correctly. The eight wildfire-related factors include three anthropogenic factors (DS, DR, Elevation and Fire-spot density), three physiographic factors (Vegetation Type, land-Usage Type, and NDVI) and only one meteorological factor (Annual Precipitation). The composition of important wildfire-related factors indicates an important role of human activities in increasing the risk of wildfires. Based on survey, more than 90% of wildfires in Guangxi province are human-caused, deliberately and unintentionally.

Table 6. Results of confusion matrix when 8 wildfire-related factors are used

		Predicted Results		
		Prone to Fire	**Prone to Non-fire**	**Total**
Actual events	Fire	4839	1161	6000
	Non-fire	1641	4359	6000
	Total	6480	5520	—
Accuracy P_a			76.65%	
Recall P_r			80.65%	
Precision P_p			74.68%	
F-score			81.23%	

5. Visualization of Wildfire Risk and Discussion

To guide the wildfire prevention of transmission corridors, the wildfire occurrence probability is calculated in all 1 km × 1 km grids of the study area. The wildfire risks of grids are then divided into four levels based on the probability: Low-($0\% \leq p < 25\%$), Medium-($25\% \leq p < 50\%$), High-($50\% \leq p < 75\%$) and Very-high-($p \geq 75\%$) risk. The conditional probability distribution of wildfire-related factors and an example of probability inference for grids are shown in Figure 5. The remarkable difference of conditional probability distribution gives a visualized explanation of why and how these wildfire-related factors affect wildfire occurrence.

Figure 5. Conditional probability distribution of wildfire-related factors and an example of probability inference.

The NBN-based wildfire risk distribution of Guangxi province is mapped by using the ArcGIS software (Figure 6a). For comparison, another wildfire risk distribution map is drawn according to the DG of State Grid Corporation of China (Figure 6b) [21]. In addition, a total of 527 fire events, which were monitored by remoting satellite during the first season in 2020, allocate on the maps.

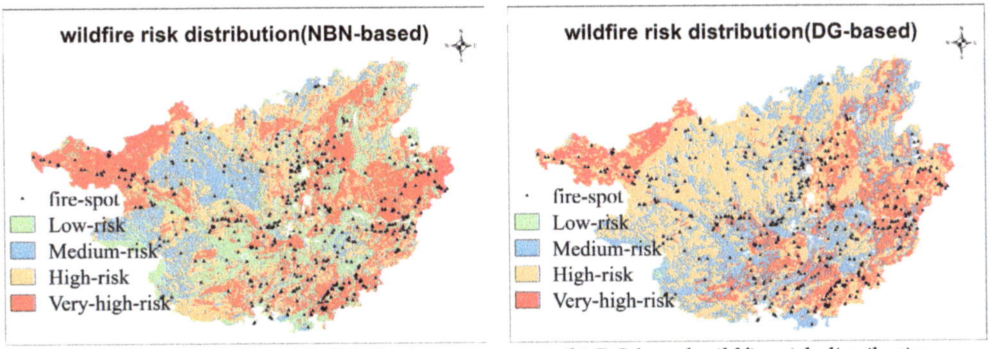

(**a**) NBN-based wildfire risk distribution (**b**) DG-based wildfire risk distribution

Figure 6. Wildfire risk distribution in Guangxi province.

It can be observed in the NBN-based map that the very-high-risk regions mainly locate at the northwest, south, and east of Guangxi province, which has a high spatial consistency with the historical fire-spot density distribution. In addition, the northwest of Guangxi province is in the border of Yunnan–Guizhou Plateau. The higher elevation and less annual precipitation result in low moisture in air and vegetation, which is the flammable condition for wildfires. In the eastern and southern regions, the population density is higher, and the settlements are widely distributed, therefore more fire-using activities. In addition, low-risk regions distribute at the middle of Guangxi province, which are generally municipal districts with few fuels and the deep forests without human activities.

The distribution of very-high-risk regions in DG-based map are almost the same as that in NBN-based map, but few low-risk areas are observed in the DG-based map. The area proportions of risk levels, as well as the location proportions of fire events, are summarized in Table 7. Due to the huge ratio of forest in Guangxi province, more areas are assessed as high-and very-high-risk level by both NBN-based map and DG-based map. Compared to NBN-based map, the area proportions of risk levels in DG-based map are more inhomogeneous. More than half of regions in Guangxi province are assessed to be high risk.

Table 7. Comparison of NBN-based map and DG-based map

	NBN-Based Map		DG-Based Map	
	Fire Proportion	**Area Proportion**	**Fire Proportion**	**Area Proportion**
Low-risk	8.92%	19.94%	0.00%	0.28%
Medium-risk	9.30%	16.93%	22.20%	26.40%
High-risk	36.43%	31.46%	45.35%	50.32%
Very-high-risk	45.35%	31.67%	32.45%	23.00%
R-score	6.15		6.62	

Table 7 also gives the proportions of fire events in risk level regions. 45.35% of fire events distribute in the very-high-risk regions of NBN-based map, whereas 32.45% in that of DG-based map, indicating a higher prediction precision of NBN-based map. However, it should be noticed that the higher predicted precision of NBN-based map may be caused by the bigger proportion of very-high-risk region, which are 31.67% and 23.00% for the NBN-based map and DG-based map, respectively. A larger area of high-risk region indicates a greater probability of fire event happening in the region, which leads to the better prediction of model. On the other hand, the larger area of high-risk region needs more costs for wildfire rescue and management. To balance the contradiction of prediction precision and management cost, a cost index -score is proposed.

$$R = \sum_{i=1}^{4} (k_i \times S_i + f_i \times N_i) \tag{11}$$

where k_i are the maintenance cost for the region with the ith risk level; The maintenance costs of different risk levels need further studies in power grid cases, and are simplified to 1, 2, 4, 8 for the low-risk, medium-risk, high-risk, and very-high-risk regions, respectively. S_i is the area proportion of the ith risk region. In addition, f_i represents the misjudgment cost when the wildfires are happened in the ith risk level. If a fire happened in the low-risk and medium-risk region, it is likely to be enlarged and bring disaster cost due to negligence of management. Thus, f_i are set as 8, 8, 2, and 0 for simplicity. The R-score of NBN-based map is 6.15, which is lower than that of DG-based map (6.62). The reduced cost also indicates the applicability of NBN-based map.

6. Conclusions

This study develops a spatial framework to assess and map wildfire risk of transmission-line corridors by integrating remote sensing data. The proposed NBN-based wildfire risk assessment combines empirical knowledge and machine learning into the discretization of input factors, construction of conditional probabilities of wildfire-related factors, and mapping of wildfire risk distribution. NBN as a core algorithm is used to infer the probabilities of wildfire occurrence. The remote sensing data including a total of 14 sub-categories of wildfire-related factors is assembled to construct NBN model. Based on the relief algorithm, the number of key wildfire-related factors is reduced into 8, indicating human activities and fuels at underlying surface play more important roles in wildfires occurrence. This spatial framework was implemented in a case study of Guangxi province in the south of China. In addition, a cost index, R-score, is proposed to reflect the maintenance costs and misjudgment costs. The results show that the NBN-based wildfire risk map has a higher prediction precision and lower costs for power grids than the traditional method. 45.35% of new monitored fire events distribute at the very-high-risk regions. The visual wildfire risk distribution can assist decision maker of power grids to optimize both supplies and staff resources and make strategies for responding damage control in the future.

Author Contributions: Funding acquisition, Z.X.; Investigation, E.Z.; Methodology, W.Z.; Project administration, Y.Z.; Software, W.Z.; Supervision, Z.X.; Writing—original draft, W.C.; Writing—review & editing, Y.Z. and J.L. All authors have read and agreed to the published version of the manuscript.

Funding: The research was funded by the Science and technology projects funded by China Southern Power Grid Corporation for supporting this research under Contract No. GDKJXM20198386, by the Changsha University of Science and Technology Professional Master's Innovation Project SJCX202045, by the Changsha University of Science and Technology Academic Master's Research and Innovation Project CX2020SS56, by the Hunan Province Natural Science Foundation of China under Grant No. 2019JJ50658, and by the Scientific Research Foundation of Hunan Education Department under Grant No.20B007 for scientific research.

Institutional Review Board Statement: No statement.

Informed Consent Statement: Not applicable.

Data Availability Statement: No statement.

Acknowledgments: The authors would like to thank Sanyi Sui from the Changsha University of Science and Technology for his helpful suggestions. The authors also would like to thank the anonymous reviewers for their constructive suggestions and comments.

Conflicts of Interest: The authors declare no conflict of interest.

References

1. Shi, S.; Yao, C.; Wang, S.; Han, W. A Model Design for Risk Assessment of Line Tripping Caused by Wildfires. *Sensors* **2018**, *18*, 1941. [CrossRef] [PubMed]
2. Guo, Y.; Chen, R.; Shi, J.; Wan, J.; Yi, H.; Zhong, J.J.I.G. Determination of the power transmission line ageing failure probability due to the impact of forest fire. *IET Gener. Transm. Distrib.* **2018**, *12*, 3812–3819. [CrossRef]
3. Wu, T.; Ruan, J.; Zhang, Y.; Chen, C.; Pu, Z.H.; Wang, G.L. Study on the statistic characteristics and identification of AC transmission line trips induced by forest fires. *Power Syst. Prot. Control* **2012**, *40*, 138–143.
4. Li, P.; Huang, D.; Pu, Z.; Ruan, J. Study on DC Voltage Breakdown Characteristics of Gap under Fire Conditions. In Proceedings of the Annual Report Conference on Electrical Insulation and Dielectric Phenomena, Shenzhen, China, 20–23 October 2013; pp. 338–341.
5. Fonseca, J.; Tan, A.; Silva, R.; Monassi, V.; Assuncao, L.; Junqueira, W.; Melo, M. Effects of agricultural fires on the performance of overhead transmission lines. *IEEE Trans. Power Deliv.* **1990**, *5*, 687–694. [CrossRef]
6. Chou, Y.H.; Minnich, R.A.; Salazar, L.; Power, J.; Dezzani, R.J. Spatial autocorrelation of wildfire distribution in the Idyllwild quadrangle, San Jacinto Mountain, California. *Remote Sens.* **1990**, *56*, 1507–1513.
7. Sakellariou, S.; Cabral, P.; Caetano, M.; Pla, F.; Painho, M.; Christopoulou, O.; Sfougaris, A.; Dalezios, N.; Vasilakos, C. Remotely Sensed Data Fusion for Spatiotemporal Geostatistical Analysis of Forest Fire Hazard. *Sensors* **2020**, *20*, 5014. [CrossRef] [PubMed]
8. El-Zohri, E.H.; Abdel-Salam, M.; Shafey, H.M.; Ahmed, A. Mathematical modeling of flashover mechanism due to deposition of fire-produced soot particles on suspension insulators of a HVTL. *Electr. Power Syst. Res.* **2013**, *95*, 232–246. [CrossRef]
9. You, F.; Zhang, Y.; Chen, H.X.; Zhang, L.H.; Zhu, J.P.; Zhou, J.J. Preliminary studies on flashovers of high-voltage transmission lines induced by wildfires by field survey and experimental tests. *Procedia Eng.* **2013**, *52*, 557–565. [CrossRef]
10. Giglio, L.; Descloitres, J.; Justice, C.O.; Kaufman, Y. An enhanced contextual fire detection algorithm for MODIS. *Remote Sens. Environ.* **2003**, *87*, 273–282. [CrossRef]
11. Giglio, L.; Csiszar, I.; Restás, Á.; Morisette, J.T.; Schroeder, W.; Morton, D.; Justice, C.O. Active fire detection and characterization with the advanced spaceborne thermal emission and reflection radiometer (ASTER). *Remote Sens. Environ.* **2008**, *112*, 3055–3063. [CrossRef]
12. Chiang, S.H.; Ulloa, N.I. Mapping and Tracking Forest Burnt Areas in the Indio Maiz Biological Reserve Using Sentinel-3 SLSTR and VIIRS-DNB Imagery. *Sensors* **2019**, *19*, 5423. [CrossRef]
13. Barmpoutis, P.; Papaioannou, P.; Dimitropoulos, K.; Grammalidis, N. A Review on Early Forest Fire Detection Systems Using Optical Remote Sensing. *Sensors* **2020**, *20*, 6442. [CrossRef]
14. Choobineh, M.; Ansari, B.; Mohagheghi, S. Vulnerability assessment of the power grid against progressing wildfires. *Fire Saf. J.* **2015**, *73*, 20–28. [CrossRef]
15. Sánchez, Y.; Martínez-Graña, A.; Santos Francés, F.; Mateos Picado, M. Mapping Wildfire Ignition Probability Using Sentinel 2 and LiDAR (Jerte Valley, Cáceres, Spain). *Sensors* **2018**, *18*, 826. [CrossRef]
16. Sadasivuni, R.; Cooke, W.; Bhushan, S. Wildfire risk prediction in southeastern Mississippi using population interaction. *Ecol. Model.* **2013**, *251*, 297–306. [CrossRef]
17. Lu, J.; Liu, Y.; Xu, X.; Yang, L.; Zhang, G.; He, L. Prediction and early warning technology of wildfire nearby overhead transmission lines. *High Volt. Eng.* **2017**, *43*, 314–320.
18. Dian, S.; Cheng, P.; Ye, Q.; Wu, J.; Luo, R.; Wang, C.; Hui, D.; Zhou, N.; Zou, D.; Gong, X. Integrating wildfires propagation prediction into early warning of electrical transmission line outages. *IEEE Access* **2019**, *7*, 27586–27603. [CrossRef]
19. Deeming, J.E. *National Fire-Danger Rating System;* Rocky Mountain Forest and Range Experiment Station, Forest Service: Fort Collins, CO, USA, 1972; Volume 84.
20. Wang, X.; Wotton, B.M.; Cantin, A.S.; Parisien, M.A.; Anderson, K.; Moore, B.; Flannigan, M.D. cffdrs: An R package for the Canadian forest fire danger rating system. *Ecol. Process.* **2017**, *6*, 5. [CrossRef]
21. State Grid Corporation of China. Drawing Guidelines for Region Distribution Map of Wildfires Near Overhead Transmission Lines. 2016. Available online: http://www.doc88.com/p-3837473228466.html (accessed on 21 October 2017).
22. Costafreda-Aumedes, S.; Comas, C.; Vega-Garcia, C. Human-caused fire occurrence modelling in perspective: A review. *Int. J. Wildland Fire* **2018**, *26*, 983–998. [CrossRef]
23. Rodrigues, M.; Jiménez, A.; de la Riva, J. Analysis of recent spatial–temporal evolution of human driving factors of wildfires in Spain. *Nat. Hazards* **2016**, *84*, 2049–2070. [CrossRef]
24. Albuquerque, M.T.D.; Gerassis, S.; Sierra, C.; Taboada, J.; Martín, J.; Antunes, I.M.H.R.; Gallego, J.R. Developing a new Bayesian Risk Index for risk evaluation of soil contamination. *Sci. Total Environ.* **2017**, *603*, 167–177. [CrossRef] [PubMed]
25. Depicker, A.; De Baets, B.; Baetens, J. Wildfire ignition probability in Belgium. *Nat. Hazards Earth Syst. Sci.* **2020**, *20*, 363–376. [CrossRef]
26. Lu, J.; Zhou, T.; Wu, C.; Li, B.; Tan, Y.; Zhu, Y. Fault statistics and analysis of 220 kV and above power transmission line in province-level power grid. *High Volt. Eng.* **2016**, *42*, 200–207.
27. Dlamini, W.M. A Bayesian belief network analysis of factors influencing wildfire occurrence in Swaziland. *Environ. Model. Softw.* **2010**, *25*, 199–208. [CrossRef]
28. Archibald, S.; Roy, D.P.; van Wilgen, B.W.; Scholes, R.J. What limits fire? An examination of drivers of burnt area in Southern Africa. *Glob. Chang. Biol.* **2009**, *15*, 613–630. [CrossRef]

29. O'Donnell, A.J.; Boer, M.M.; McCaw, W.L.; Grierson, P.F. Vegetation and landscape connectivity control wildfire intervals in unmanaged semi-arid shrublands and woodlands in Australia. *J. Biogeogr.* **2011**, *38*, 112–124. [CrossRef]
30. Saglam, B.; Bilgili, E.; Dincdurmaz, B.; Kadiogulari, A.I.; Kücük, Ö. Spatio-Temporal Analysis of Forest Fire Risk and Danger Using LANDSAT Imagery. *Sensors* **2008**, *8*, 3970–3987. [CrossRef]
31. Lu, J.; Liu, Y.; Wu, C.; Zhang, H.; Zhou, T. Study on satellite monitoring and alarm calculation algorithm of wild fire near transmission lines. *Proc. Csee* **2015**, *35*, 5511–5519.
32. Parisien, M.A.; Moritz, M.A. Environmental controls on the distribution of wildfire at multiple spatial scales. *Ecol. Monogr.* **2009**, *79*, 127–154. [CrossRef]
33. Kira, K.; Rendell, L.A. The Feature Selection Problem: Traditional Methods and a New Algorithm. In Proceedings of the Tenth National Conference on Artificial Intelligence, San Jose, CA, USA, 12–16 July 1992; pp. 129–134.
34. Liu, R.; Chen, Y.; Wu, J.; Gao, L.; Barrett, D.; Xu, T.; Li, L.; Huang, C.; Yu, J. Assessing spatial likelihood of flooding hazard using naïve Bayes and GIS: A case study in Bowen Basin, Australia. *Stoch. Environ. Res. Risk Assess.* **2016**, *30*, 1575–1590. [CrossRef]

sensors MDPI

Article

Shelf Auditing Based on Image Classification Using Semi-Supervised Deep Learning to Increase On-Shelf Availability in Grocery Stores

Ramiz Yilmazer [1] and Derya Birant [2,*]

[1] Graduate School of Natural and Applied Sciences, Dokuz Eylul University, Izmir 35390, Turkey; ramiz.yilmazer@ceng.deu.edu.tr
[2] Department of Computer Engineering, Dokuz Eylul University, Izmir 35390, Turkey
* Correspondence: derya@cs.deu.edu.tr

Abstract: Providing high on-shelf availability (OSA) is a key factor to increase profits in grocery stores. Recently, there has been growing interest in computer vision approaches to monitor OSA. However, the largest and well-known computer vision datasets do not provide annotation for store products, and therefore, a huge effort is needed to manually label products on images. To tackle the annotation problem, this paper proposes a new method that combines two concepts "semi-supervised learning" and "on-shelf availability" (SOSA) for the first time. Moreover, it is the first time that "You Only Look Once" (YOLOv4) deep learning architecture is used to monitor OSA. Furthermore, this paper provides the first demonstration of explainable artificial intelligence (XAI) on OSA. It presents a new software application, called SOSA XAI, with its capabilities and advantages. In the experimental studies, the effectiveness of the proposed SOSA method was verified on image datasets, with different ratios of labeled samples varying from 20% to 80%. The experimental results show that the proposed approach outperforms the existing approaches (RetinaNet and YOLOv3) in terms of accuracy.

Keywords: on-shelf availability; semi-supervised learning; deep learning; image classification; machine learning; explainable artificial intelligence

Citation: Yilmazer, R.; Birant, D. Shelf Auditing Based on Image Classification Using Semi-Supervised Deep Learning to Increase On-Shelf Availability in Grocery Stores. *Sensors* **2021**, *21*, 327. https://doi.org/s21020327

Received: 19 November 2020
Accepted: 1 January 2021
Published: 6 January 2021

1. Introduction

Machine learning techniques have been applied to different areas in the retail sector. One of them is the monitoring on-shelf availability (OSA) in grocery stores. Providing high OSA is a key factor to increase profits. When a product that a shopper looks for is not available on its designed shelf, also known as "out-of-stock" (OOS), this causes a negative impact on customer behaviors in the future. According to the research reported by Corsten and Gruen [1], when a product is not available on the designed shelf space, 31% of consumers buy the product from a different store, 26% of them buy a different brand, 19% of them buy a different size of the same brand, 15% of them buy the same product at a later time, and 9% of them buy nothing. Besides, another study [2] showed that the "out of the stocks" rate is about 8% in the United States and Europe. For this reason, OSA has a significant effect on business profit. The remaining products can be checked using an inventory management system, but it only shows the number of products in the stock. These products, available in stock, might not be on the shelves. Current inventory systems cannot understand the number of products on the shelves. OSA is checked by employees manually at most of the grocery stores. This approach is not effective and sustainable since it continuously requires human effort.

There are several studies to automate monitoring OSA. These studies consider the subject from different perspectives. One of them proposes radio frequency identification (RFID) tagging to monitor product quantity on the shelves [3], but this approach is not cost-effective to implement the technology and integrate it into existing systems [4]. Some of them applied traditional image processing techniques to detect the presence and absence of the product, such as [5], and some of them used deep learning approaches for object detection on the shelves, such as [6]. When all these previous works are examined, there are pros and cons of these approaches. When traditional image processing methods have been used, such as histogram of oriented gradient (HoG) for feature extraction and support vector machine (SVM) for a classifier, they have limited performance even if on the large datasets and their performances are hard to be increased. In addition, the visual similarity among the different products of the same brand can lead to misclassification. On the other hand, when deep learning (DL) approaches were used such as recurrent convolutional neural network (RCNN) or you only look once (YOLO), high accuracies could be achieved. The DL algorithms require annotated images to train, but the largest and well-known computer vision datasets do not provide annotation for store products, and therefore a huge effort is needed to manually label products on images.

A semi-supervised learning approach can produce results with satisfactory accuracy using an amount of labeled data and a much larger amount of unlabeled data in the training phase [7]. This paper proposes a new method that combines two concepts "semi-supervised learning" and "on-shelf availability" (SOSA) for the first time. An important advantage of the proposed SOSA method is that it solves the OSA problems where labeled image data is scarce. Labeling OSA image data is an expensive, tedious, difficult, or time-consuming process since it requires human labor. The proposed SOSA method deals with the design of on-shelf availability monitoring models in the presence of both labeled and unlabeled image data.

From another perspective, users have to understand and trust the constructed OSA model as a kind of artificial intelligence (AI) application. Besides, if their requirements change, these applications have to be managed according to their needs [8]. Our paper proposes the first demonstration of explainable artificial intelligence (XAI) on OSA.

The main contributions and novelty of this paper can be listed as follows. (1) This is the first study that combines two concepts "semi-supervised learning" and "on-shelf availability" (SOSA) for the first time. (2) It is the first time that YOLOv4 deep learning architecture is used for exploring OSA. (3) This study is also original in that it compares three deep learning approaches for monitoring OSA in the retail sector. (4) To the best of our knowledge, this is the first demonstration of explainable AI on OSA. This paper also presents a new software application, called SOSA XAI, with its capabilities and advantages.

In the experimental studies, the effectiveness of the proposed SOSA method was verified on image datasets, with different ratios of labeled samples varying from 20% to 80%. We focused on category-based detection of empty and almost empty shelves. For this purpose, we used both labeled and unlabeled images that include 49,573 products grouped under three categories. The experimental results show that the proposed approach outperforms the existing approaches (RetinaNet and YOLOv3) in terms of accuracy.

The remainder of this article is structured as follows: Section 2 summarizes the related works on OSA. In Section 3, we first briefly introduce background information of the compared algorithms and then explain semi-supervised learning of the OSA (SOSA) concept with its definitions. This section also describes our demonstration of XAI on OSA. Section 4 presents the experimental results and Section 5 presents discussions about the obtained results. Finally, concluding remarks are presented in Section 6.

2. Related Works

In the literature, researchers have approached the subject from different perspectives. They proposed both technical and managerial solutions to minimize OOS and monitor OSA. Different technical solutions have been proposed, such as analyzing textual data on the relational database management systems (RDMS) and applying computer vision-based techniques. In this paper, we focus on computer vision-based techniques, and previous works were examined based on this scope.

Moorthy et al. [5] applied image processing techniques to detect the presence and absence of the product in front of the shelf using MATLAB. In addition, they worked on the positioning of products in retail stores. The feature extraction technique was used for object detection and a speeded up robust features (SURF) algorithm was used for this purpose. Reference images were given to the system and target images were taken from video or camera devices as shelf images. Some pre-process operations were applied to images before comparison of input and target images and extracting features using the SURF algorithm. At the end of these steps, the product was evaluated for missing or misplaced. Besides, they [9] proposed the image processing approach to provide high OSA in retail. They collected images from shelves to use reference images. Reference images and target images were compared using image processing techniques. When comparison showed missing products on the shelf, the system sent messages to the manager or responsible person in the store. A similar template matching approach was used to develop OSA monitoring software in another study [10]. Moreover, some researchers [11] worked on high-resolution panoramic images and proposed a supervised learning approach. Panoramic images were created using a wide-angle fisheye camera and an accelerated-KAZE (AKAZE) feature detector was applied. Labels were detected from the panoramic images of shelves. A cascade classifier approach was used for this purpose.

Kejriwal et al. [12] worked on counting products from shelf images using robot-based equipment. Several cameras were placed on either side of the robot. The robot was moved between shelves and video data were collected from shelves. Several methods were applied to these collected datasets. The first method was used to recognize the product and the second one was used to count the product. A k-d tree was created and the SURF approach was used to recognize the product by using the nearest neighbor search. Two different techniques were used for product count. The first one is the product counting method in which repeating features were counted to understand the number of products on the image by using the SURF method. Secondly, rectangular bounding boxes were drawn around the products and these boxes were counted to understand the number of products by using the random sample consensus (RANSAC) method. Moreover, the grid search approach was applied in the neighborhood of each found product on the image. Some undetected products were tried to find in this way.

Higa et al. [6] studied product changes on the shelf. They focused on taken or returned products on the shelf. Videos, captured from a surveillance camera, were used for the study. Low-quality videos, 480 × 270 pixels, and 1 fps were used to eliminate storage problems but another problem occurred at this time. Moving objects was difficult to track from low-resolution videos. They used background subtraction and the convolutional neural network (CNN) approach to detect changes of products on the shelf. CNN was used based on CaffeNet and four classes were determined for CNN. The extended version of the study [13] proposed analyzing consecutive images when the customers stood in front of the shelf. The Hungarian method was used to analyze consecutive images. Moreover, the Canadian Institute for Advanced Research (CIFAR)-10-based network and CaffeNet-based network were used to detect change regions. In addition, a heatmap was generated to show the customer's behavior using captured images. Frequently accessed shelves were analyzed with this approach.

Some studies exist about planogram compliance checking. Planograms are created to standardize placed products on shelves for chain supermarkets. Another aim of creating a planogram is that, providing the best customer experience for promotions. Liu et al. [14] studied automatic planogram checking compliance using recurring patterns. These planograms are created by headquarters and sent to store managers. Store managers are responsible for applying planograms to store shelves. In a conventional way, they try to check planogram compliance manually. The proposed automatic checking compliance was done without using template images. Planogram was taken as XML format and parsed. Some matrix operations were applied to detect recurring patterns. The extended version of the study [15] focused on spectral graph matching and speed improvements using a divide-

and-conquer approach. Besides, Saran et al. [16] studied visual analysis for planogram compliance. They compared the reference template image to the target image. Their study focused on the presence of products, placements of products, and product count. The Hausdorff map approach was used for the presence of products. This presence of products was a group under two classes which were complete and partial cases. The Euclidean distance approach was used to identify completely missing cases. A binary distance map approach was used with the self Hausdorff map. Shelf rows were identified using the Sobel derivatives extractor and Hough line extractor. The Sobel derivatives extractor was used for vertical changes and the Hough lines extractor was used for horizontal changes. Finally, texture features and color features were used to count products from images. The color feature was used to eliminate false positives.

Briefly, the summarized related works above focus on OSA monitoring and planogram checking based on traditional image processing and deep learning techniques using different approaches. Image processing techniques have performance issues for huge datasets. Moreover, for template matching, reference images have to be stored in a database and this is not suitable for real-world applications. On the other hand, labeled images are needed using deep learning techniques for object detection on the shelves. The largest and well-known computer vision datasets do not provide labeled images for store products. This process is time-consuming and quite expensive since it requires human labor costs. This paper proposes a new method that combines two concepts "semi-supervised learning" and "on-shelf availability" (SOSA) for the first time. The main advantage of the proposed SOSA method is that satisfactory results can be achieved using a small amount of labeled data and a large amount of unlabeled data for OSA monitoring.

Monitoring on-shelf availability has very little coverage in the literature, only a few numbers of detailed analyses [5,6,9–13,16] have been performed. Table 1 shows the comparison of our study with the previous studies aforementioned. Our approach differs from the existing approaches in many respects and has numerous advantages over the rest as follows:

- First, some of the previous studies [5,9–12,16] used traditional techniques such as image processing (IP) to monitor OSA, whereas we used deep learning techniques. In IP-based approaches, a huge amount of reference images has to be stored to match the target image and for every product updating, reference images have to be updated manually. In this context, an important advantage of our method is that it does not require any reference image and therefore it does not need manual updating when products are updated. Moreover, it automatically extracts features from an input image thanks to deep learning.
- Second, the previous deep learning-based studies used different network structures such as the CaffeNet-based network [6,13] and CIFAR-10-based network [6,13], whereas we designed a novel network architecture that consists of RetinaNet, YOLOv3, and YOLOv4 detectors. Here, the advantage of our approach is that it builds three different models and selects the best one, and hence, satisfactory results can be achieved by the selection of the best model.
- Third, the previous deep learning-based studies used two-stage detectors. On the other hand, in this study, we benefit from one-stage detectors because of their speed and achieving satisfactory accuracy results for OSA monitoring.
- Four, our study differs from the rest in that we adapted the semi-supervised learning concept, and therefore we benefited from both labeled and unlabeled data. Here, the main advantage is that our method reduces the need for labeling images which is an expensive, tedious, difficult, and time-consuming process since it requires human labor. Satisfactory results can be achieved using a small number of labeled images. Moreover, the proposed method will expand the application field of machine learning in grocery stores since a large amount of OSA data generated in real-life are unlabeled.

- Finally, differently from the previous studies, we introduced an explainable AI concept into OSA. The developed new SOSA XAI software application allows users to manage, understand, and trust the model when monitoring OSA.

Table 1. Comparison of the proposed Semi-Supervised Learning on on-shelf availability (SOSA) approach and the previous approaches.

Reference No	Year	Object Detection			Learning		XAI	Methods
		Traditional Detection	Deep Learning		Supervised Learning	Semi-Supervised Learning		
			Two-Stage	One-Stage				
[5]	2015	✔			✔		×	SURF
[9]	2015	✔			✔		×	Image Processing
[10]	2015	✔			✔		×	Image Processing
[12]	2015	✔			✔		×	k-d Tree, RANSAC
[16]	2015	✔			✔		×	Hausdorff Map, Euclidean Distance, Binary Distance Map
[11]	2016	✔			✔		×	AKAZE Feature Detector, Cascade Classifier
[6]	2018		✔		✔		×	CaffeNet-based Network, CIFAR-10-based Network
[13]	2019		✔		✔		×	CaffeNet-based Network, CIFAR-10-based Network, Hungarian
Proposed Method				✔		✔	✔	RetinaNet, YOLOv3, YOLOv4

3. Materials and Methods

3.1. Deep Learning for Object Detection

One of the main tasks of computer vision is object detection. It deals with detecting objects along with their locations and classes (such as cars, fruits, food products) in images. Object detection methods can be grouped into two categories: traditional image processing methods and deep learning-based detection methods. Recently, deep learning-based object detection methods have become popular because of outstanding results. In addition, deep learning-based object detection methods can be divided into two categories as two-stage detectors such as regions with convolutional neural networks (R-CNN) [17], faster R-CNN [18], feature pyramid network (FPN) [19], and one-stage detectors such as RetinaNet [20], YOLOv3 [21], and YOLOv4 [22]. In the general structure of two-stage detectors, feature extraction is applied to the input image and generated proposed regions using different methods in the first stage. From these proposed regions, the locations (bounding boxes) and classes of the objects are determined in the second stage. In contrast to two-stage detectors, the region proposal stage is skipped and directly learns class probabilities and bounding box locations from the input image like a simple regression problem. In this study, we decided to work on one-stage detectors because of their speed and achieving satisfactory accuracy results for OSA monitoring.

The history of one-stage detectors is shown in Figure 1. Since 2012, CNN [23] has been used to be able to learn from features of images robustly. Before this year, traditional detection methods such as image processing were used for object detection tasks. After this year, computer vision techniques have improved very rapidly and novel deep learning-based detection methods have been proposed. The existing detector methods can be grouped under two categories: methods with one-stage detector and methods with two-stage detectors. The first one-stage detector method was proposed in 2013, and since then, at least one novel approach has been proposed each year. Each approach solved the object detection problem from a different perspective to increase accuracy, and each one achieved higher accuracy results from previous ones. For this reason, three of the latest published one-stage detectors (RetinaNet, YOLOv3, YOLOv4) have been selected for this study.

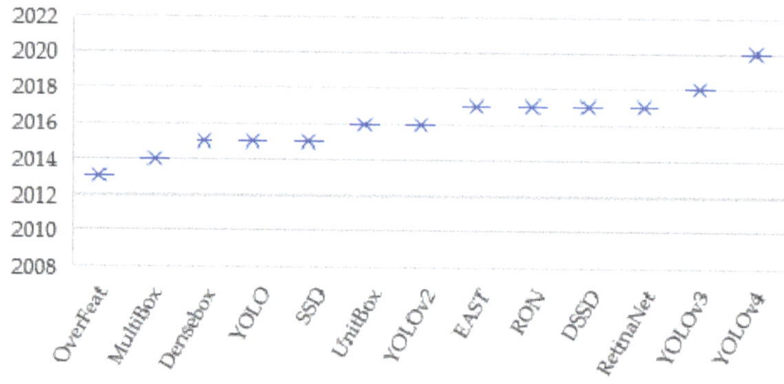

Figure 1. History of one-stage detectors.

Lin et al. [20] proposed the RetinaNet one-stage object detection model. It uses a new focal loss function to handle class imbalance issues more effectively according to alternative previous approaches. Their model has one backbone network and two subnetworks. Feature map is computed in the backbone network and the output of this network is used as input for two subnetworks. Bounding boxes and classifications are found separately in each subnetwork. FPN is used as a backbone network for RetinaNet. Their proposed architecture improved the performance of the model compared to standard top-down CNN architecture.

YOLOv3 is a one-stage detector that combines FPN, CNN, and the non-maximum suppression algorithm [21]. CNN is used for the feature extraction process. In addition, it has been integrated with the Darknet-19 and residual neural network (ResNet) network structures for feature extraction. On the other hand, a shortcut connection is added to the model. Fifty-three convolution layers are existing with dropout and batch normalization operations in the feature extraction network. The network is used as a backbone network and named Darknet-53. YOLO works on the entire image during the train and test processes.

YOLOv4 is an improved version of previous YOLO models. It combines the YOLOv3 head and path-aggregation network (PANet) for detection steps [22]. PANet is used instead of FPN in the model. The novel backbone, called cross stage partial Darknet-53 (CSPDarknet-53), is used. Some blocks are added such as spatial pyramid pooling (SPP) to increase the receptive fields on the backbone. Besides, YOLOv4 provides data augmentation to expand the training dataset to improve the accuracy of the network without extra inference time. The Mosaic data augmentation method that combines four images in the training phase was proposed.

3.2. Proposed Approach: Semi-Supervised Learning on OSA (SOSA)

This paper proposes a new approach: semi-supervised learning on OSA (SOSA). The proposed method combines two concepts "semi-supervised learning" and "on-shelf availability" for the first time to decrease OOS by automatically detecting and classifying products using shelf images. For this purpose, SOSA builds a classification model that allows the detection of "Product", "Empty Shelf", and "Almost Empty Shelf" classes by using both labeled and unlabeled image data. The aim of the SOSA approach is section-based detection of empty and almost empty shelves based on the semi-supervised learning principle.

3.2.1. The General Structure of the Proposed SOSA Method

Figure 2 shows the general structure of the proposed SOSA approach. SOSA consists of the following main steps. The first step is to train three one-stage detectors on the existing labeled image data. After that, the best-trained model is selected using the evaluation metrics such as mean average precision (mAP), F1-score, and recall measures. In the next step, the unlabeled image data is labeled by the constructed classifier via their predictions, which is commonly referred to as pseudo-labeled data. Lastly, the final classifier is built by using both originally labeled and pseudo-labeled image data. Hence, SOSA allows unlabeled image data to be introduced to the training process in an efficient manner.

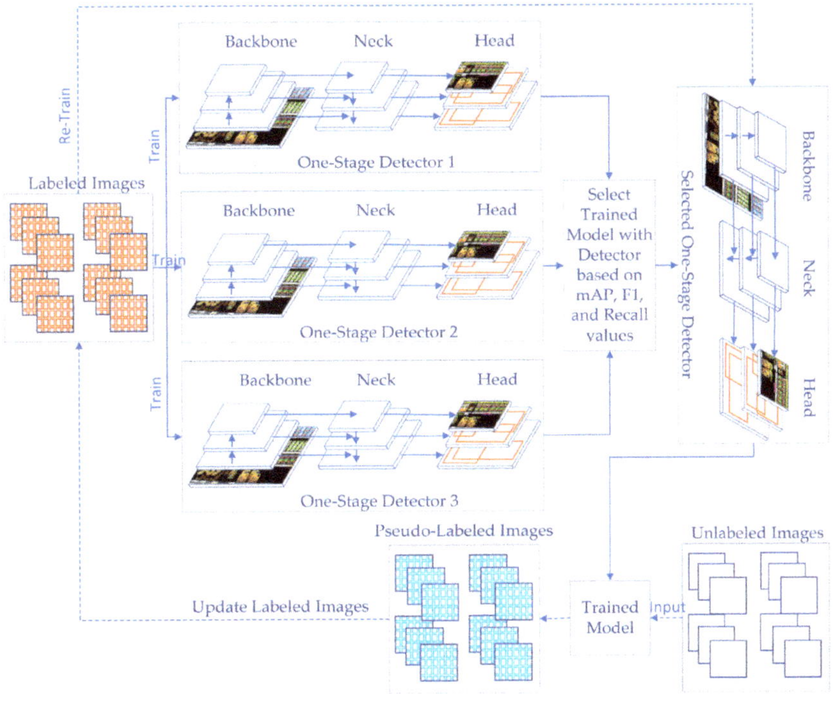

Figure 2. The general structure of the proposed SOSA approach.

In the preprocessing phase of the SOSA method, the images, which are taken from different perspectives in front of the shelves, are labeled to use for the first training. Images are labeled according to the product type(s) it includes. Besides, empty and almost empty shelves are labeled on the images. If one or two of the related products remained on the shelf, these products' areas are labeled as "Almost Empty".

3.2.2. The Formal Definition of the Proposed SOSA Method

Assume that labeled dataset $D = \{(x_1, y_1), (x_2, y_2), \ldots, (x_n, y_n)\}$ has n images with labeled products. Each component (x, y) is composed of d-dimensional vector (x) from a given input space X, such that $x \in X$, and the output variable (y), where $y \in Y = \{c_1, c_2, \ldots, c_m\}$ has m class labels. Unlabeled dataset $U = \{x_{n+1}, x_{n+2}, \ldots, x_{n+s}\}$ has s unlabeled images. We are especially interested in OSA images where labeling the images is difficult and expensive since it requires human labor. The SOSA method considers both D and U to find a decision function f: $X \mapsto Y$ that can correctly predict the class labels $\hat{y}$ of a given unseen input sample image SI. Z refers to one-stage detectors and $Z = \{z_1, z_2, \ldots, z_k\}$ has k

detectors. In this study, three one-stage detectors are used, where z_1 refers to RetinaNet, z_2 refers to YOLOv3, and z_3 refers to YOLOv4, and so k is set to 3.

Definition 1. *Semi-supervised learning on on-shelf availability* (SOSA) refers to a machine learning approach that builds a model to correctly detect "Product", "Empty Shelf", and "Almost Empty Shelf" regions from an input image by using both labeled and unlabeled image data, which are taken from different perspectives in front of the shelves.

Algorithm 1 gives the general framework of the proposed SOSA approach. The algorithm consists of five steps. In the first step, labeled dataset D is split as D_{Train}, and D_{Test} based on the given percentage. In the first loop, a model z_c is trained for each one-stage detector. Trained models are added to Z. In the second step, the constructed models are tested using labeled test dataset D_{Test} and obtained prediction results are compared to select the best one-stage detector that has maximum success rates. The selected-one stage detector is assigned to SD. In the third step, a query instance $x_i \in U$ is submitted to the selected detector SD, and its estimation $\hat{y}$ is assigned to x_i as a pseudo-label. This process is repeated for each instance in the unlabeled dataset U to generate all pseudo labels. At the end of this labeling operation, labeled dataset D is augmented with pseudo-labeled images. In the next step, the classifier is re-trained by using the new dataset D which contains both labeled and pseudo-labeled images. Finally, a sample image SI is given to the trained model TM to be classified, and the predicted classes $\hat{y}$ are obtained from the algorithm.

Algorithm 1 SOSA: Semi-Supervised Learning on OSA

Inputs: D: Labeled dataset $D = \{(x_1, y_1), (x_2, y_2), \ldots, (x_n, y_n)\}$ with n instances
U: Unlabeled dataset $U = \{x_{n+1}, x_{n+2}, \ldots, x_{n+s}\}$ with s instances
Z: One-stage detectors $Z = \{z_1, z_2, \ldots, z_k\}$ with k detectors
Y: Class labels $Y = \{c_1, c_2, \ldots, c_m\}$ with m classes
SI: Sample image
Outputs: TM: Trained model
$\hat{y}$: Predicted class labels for the products included in the sample image
Begin:
$D_{Train} = Split(D, n * percentage)$
$D_{Test} = Split(D, (n - (n * percentage)))$
//Step 1—Training with labeled data
for $c = 1$ **to** k **do**
foreach epoch
foreach (x_i, y_i) **in** D_{Train}
z_c = Train(x_i, y_i)
end foreach
end foreach
$Z = Z \cup z_c$
end for
//Step 2—Testing one-stage detectors and selecting the best one
for $c = 1$ **to** k **do**
foreach (x_i, y_i) in D_{Test}
$Prediction = z_c(x_i)$
$PredictionResult_c = PredictionResult_c \cup Prediction$
end foreach
end for
$SD = MAX\ (PredictionResult_c)$ //SD: Selected detector
//Step 3—Labeling unlabeled image data and generating pseudo-labels
foreach x_i **in** U
$\hat{y} = SD(x_i)$
$D.Add(x_i, \hat{y})$
end foreach
//Step 4—Re-training the model with pseudo-labeled data
$TM = Train(D)$
//Step 5—Classifying a sample image
$\hat{y} = TM(SI)$
End Algorithm

After the preprocessing phase, the algorithm splits the labeled data into training and test sets according to a given ratio. For each one-stage detector, a model is built by using the training set. At the end of the training phase, three trained models are tested using the test set. As test results, the average precision (*AP*), mean average precision (*mAP*), F1-score, and recall values are calculated based on intersection over union (IoU) threshold value. IoU computes the intersection over the union of the given bounding box and the predicted bounding box. The formulas are given in Equations (1) to (5):

$$Precision = \frac{True\ Positive\ (TP)}{True\ Positive\ (TP) + False\ Positive\ (FP)}, \tag{1}$$

$$Recall = \frac{True\ Positive\ (TP)}{True\ Positive\ (TP) + False\ Negative\ (FN)}, \tag{2}$$

$$F1Score = 2 \cdot \frac{Precision \cdot Recall}{Precision + Recall}, \tag{3}$$

$$AP = \frac{1}{N} \cdot \sum_{i=0}^{N} Precision_i, \tag{4}$$

$$mAP = \frac{1}{|T|} \cdot \sum_{t \in T} AP_t, \tag{5}$$

where $|T|$ is total of all *AP*s calculated for each class and *N* is the number of instances in the test set.

After the testing phase, the obtained test results are compared with each other and the detector that has the highest accuracy is selected as the final representative model. In the following phase, semi-supervised learning is performed by labeling the unlabeled data using the trained model of the selected detector. After the labeling operation is completed, the pseudo-labeled data is combined with the labeled data, and then the training process is restarted on the selected detector using the extended dataset. At the end of the re-training operation, the newly trained model is saved for further use for the detection of empty and almost shelves based on sections. To understand which section has empty or almost empty shelves, the relative frequency (*RF*) formula is used as given in Equation (6):

$$RF = \frac{Frequency\ of\ One\ Product\ Class}{Total\ Frequencies\ of\ All\ Products'\ Classes}. \tag{6}$$

Each product is labeled at the end of the detection phase on the shelf image and these labels denote the section of products. For each section, *RF* is calculated and the highest *RF* value gives the final section info. For instance, if the highest *RF* value is obtained for the breakfast products, the section is recognized as a breakfast product. Besides, if three empty and two almost empty shelves are detected on the image, the SOSA algorithm gives the following information: "3 empty and 2 almost empty shelves are existing on the breakfast section."

3.2.3. The Advantages of the Proposed SOSA Method

The proposed method (SOSA) has a number of advantages that can be summarized as follows:

- The traditional OSA applications are limited to using only labeled image data to build a model. However, labeling shelf images are a time-consuming, tedious, expensive, and difficult job because of existing so many products on a one-shelf image, and for this reason, so many human laborers are needed. This is especially true for the OSA applications that include learning from a large number of class labels and distinguishing similar classes. The main advantage of the SOSA method is that it solves OSA problems using a small number of labeled shelf images. The existing

labeled dataset is extended by using unlabeled data with automatically assigned labels, and hence, high accuracy results are taken with the SOSA approach in an efficient way.
- Another advantage is that it includes three different one-stage detector models and the model with the highest accuracy is selected at the beginning of the semi-supervised learning. Hence, satisfactory results can be achieved by the selection of the best model.
- The SOSA approach uses three different deep learning techniques (RetinaNet, YOLOv3, and YOLOv4) without any modification or development of the methods. Therefore, SOSA has advantages in terms of easy implementation. It is possible to implement it in Python by using open-source codes available in the related machine learning libraries.
- The main idea behind the SOSA method is to take advantage of a huge amount of unlabeled image data when building a classifier. In addition to labeled data, the SOSA method also exploits unlabeled data to improve classification performance. Thanks to the SOSA method, the unlabeled data instances provide additional knowledge, and they can be successfully used to improve the generalization ability of the learning system.
- Another advantage is that the SOSA method can be applied to any OSA image data without any prior information about the given dataset. It does not make any specific assumptions for the given data.
- Since a large amount of OSA data generated in real-life is unlabeled, the SOSA method will expand the application field of machine learning in grocery stores.

3.3. Explainable AI for SOSA

Explainable artificial intelligence (XAI) is a growing research field in recent years. The aim of XAI is to make machine learning and AI applications more understandable to users who are not experts in these fields. From rule-based systems to deep learning systems, the transparency of systems is decreasing. Especially, it is hard to understand and interpret the outputs of deep learning applications by users. Therefore, users see this kind of AI application as a black box. They think that inputs are given to these applications; afterward, something happens magically inside the box, and outputs are generated by AI applications [24]. On the other hand, AI-based applications are increasingly being adopted by different sectors, including retail. Therefore, business stakeholders and users of AI systems should be able to understand their systems to trust outputs and manage these applications to their needs without getting help from AI experts or engineers [25].

There are different concepts of XAI to provide transparency for AI models in the literature. This transparency can be provided for different parts of the model based on requirements. One of them is post-hoc explainability. The purpose of post-hoc is to explain decisions of AI models using different approaches such as text explanation, visualization, explanation by example [8]. The results of the model can be explained using text definitions, graphics, and images for users in this way. Besides, the dataset can be interpreted during the training phase and the dataset can be extended for better accuracy without the need for AI experts.

In this study, we present the first demonstration of XAI on OSA using post-hoc techniques, and therefore, users can understand and trust the constructed OSA AI model in this way. Moreover, they can interpret the detection results of the model and enhance the dataset in the training phase to adapt the application for changes. In this study, for the OSA XAI demonstration, we used the following post-hoc techniques: text explanations, visualizations, and explanations by example.

In this study, a new software application, called SOSA XAI, was developed to provide understandability for the users. SOSA XAI consists of four main screens. These are "Train", "Test", "Monitoring", and "Metrics" screens. The main screen is shown in Figure 3. Our proposed SOSA method can be managed using the application. After the first initialization, users can re-train the model using the training screen as shown in Figure 4. In the training screen, the creation dates of the models and the obtained results (mAP, F1-score, and Recall values) are given in an easily understandable format. Furthermore, for each model, the status of the model is shown based on accuracy values using the visualization technique. If one of these three accuracy results under 80% (our threshold value for this study), color is changed for the related accuracy metric. Moreover, during the training process, training progress steps are given in a more understandable way using the text explanation technique, and at the end of the training operation, the meaning of accuracy values are interpreted by the application. The last trained model has activated automatically. On the other hand, users can activate one of the previous models from the list.

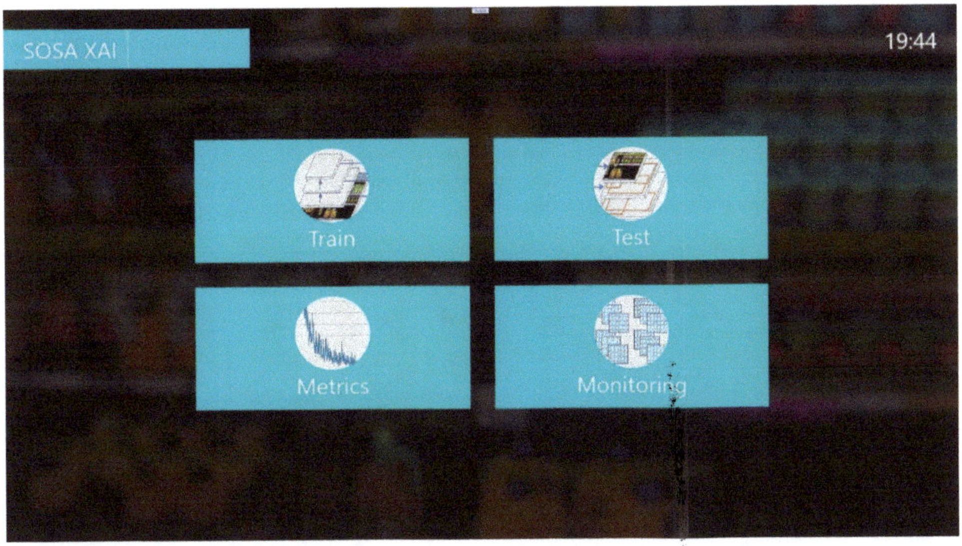

Figure 3. The main screen of SOSA XAI.

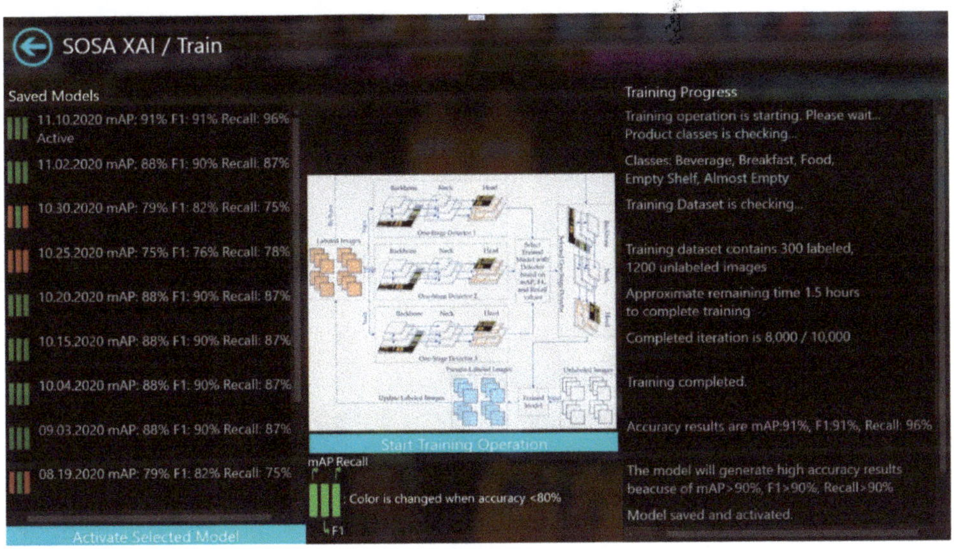

Figure 4. Train screen of SOSA XAI.

Thanks to our SOSA XAI application, the active trained model can be tested using a test screen. Test images are listed on the screen and also new test images can be uploaded. Accuracy values and creation date of the trained model are shown in a straightforward manner. For selected or uploaded test images, detection operation can be started and results are shown both visually and textually. In the textual explanation section, the decision of the model is given in an understandable way and explained how this decision is made. Figure 5 shows a screenshot of the object detection results. Besides, when some of the classes are detected with low accuracy, the application gives suggestions to extend the existing dataset to increase accuracy. For instance, when empty-shelf accuracy is less than 80%, the system suggests adding more images that contain more empty shelves from different perspectives and re-training the model. In the metrics screen of the software, the system shows sample images containing empty shelves from different perspectives, and users can expand the training set by adding new images among these samples. Moreover, the monitoring screen has also similar functionalities, but here, the user can design the test set, instead of the training set. From the monitoring screen, real-time results can be tracked the same as the test screen.

Figure 5. Test screen of SOSA XAI.

Finally, from the metrics screen (Figure 6), the training accuracy graph for the active trained model, and dataset distribution based on classes can be seen easily. In addition, the dataset can be extended using the "Extend Dataset" part with unlabeled images. Sample images are shown to give an idea of how to create a new dataset and suggestions are given as text for this purpose. After the upload operation is completed, the training process can be started from the training screen for the extended dataset. Text explanation, visualization, and explanation by example post-hoc techniques are used in this screen.

In a real-time application, shelf images can be taken approximately every hour during peak times and every three hours during non-peak times to be analyzed by the OSA system. Since this period is parametric in the developed system, it can be easily modified. In order to protect the privacy of customers, when an image is taken, an object detection approach [26] is firstly applied to check the presence of people on the image. If people are detected, this image is not processed by the system, and a new image is taken after a period of time until the image does not contain a person (i.e., shopper or employee). After this step, the trained model is used to classify the new image, and the image is stored in the system. When empty or almost-empty shelves are detected, a notification is sent to the responsible store clerk to check the related area. When a notification is received, a responsible clerk can put new products on the shelf to prevent customer and profit loss.

Figure 6. Metrics screen of SOSA XAI.

SOSA XAI gives an opportunity to understand, trust, and manage AI applications to increase OSA for users who are not AI experts and engineers in grocery stores. Results can be interpreted easily and the dataset can be extended with unlabeled images for requirements changes. Furthermore, the results of the previous models can be compared to the current one and the desired model can be selected and activated. It combines our proposed SOSA approach and XAI.

4. Experimental Studies and Results

In the experiments, the proposed approach SOSA was tested on a real-world dataset. A labeled dataset is needed for the test operation but the largest and well-known computer vision datasets do not provide annotation for grocery store products. In the literature, few datasets have been collected from grocery stores. On the other hand, all of these datasets have unlabeled images for the store products, and these datasets cannot be used directly without labeling some of them.

We used the WebMarket dataset [27] in this study. The dataset contains 3153 unlabeled images that were taken from in front of shelves using three digital cameras that were standing from about 1 m away from shelves. The images were taken without any special illumination changes and without any viewpoint constraint but most of the images are frontal views. The images in the dataset were collected from 18 shelves in a retail store, each of length 30 m and each of which approximately has six levels. Each image generally covers an area of about 2 m in width and 1.5 m in height on shelves, including all the items within three or four shelf levels in range. The dataset contains images of 100 different items. The dataset also includes fine-grained product categories having minor variations in packages [28]. Since the model is trained on such data, it has the capability of dealing with packages. The format of each image is Jpeg and its resolution is either 2592 × 1944 or 2272 × 1704. High resolution holds sufficient information for each object appearing in the image, and thus, the trained model can deal with packages of multiple items and damaged packages, at least with lower accuracy. Until now, the dataset was used for object retrieval studies [29,30]. Namely, our study is the first study that uses the WebMarket dataset for monitoring on-shelf availability.

In order to label the images, LabelImg [31] open-source tool was used and 300 images were labeled based on five classes. Three of them are for product categories that are "Beverage Products", "Breakfast Products", and "Food Products", and two of them are for shelves' areas that are "Empty Shelf" and "Almost Empty Shelf". "Beverage Products" and "Breakfast Products" were selected from fast-moving consumer goods (FMCG) which are nondurable products, consumed at a fast pace. A total of 13,835 products were grouped under three categories, and 818 shelves' areas grouped under two categories. The labeled dataset was divided into a training set (90%) and a test set (10%). The distribution of labeled products and shelves' areas are shown in Figure 7.

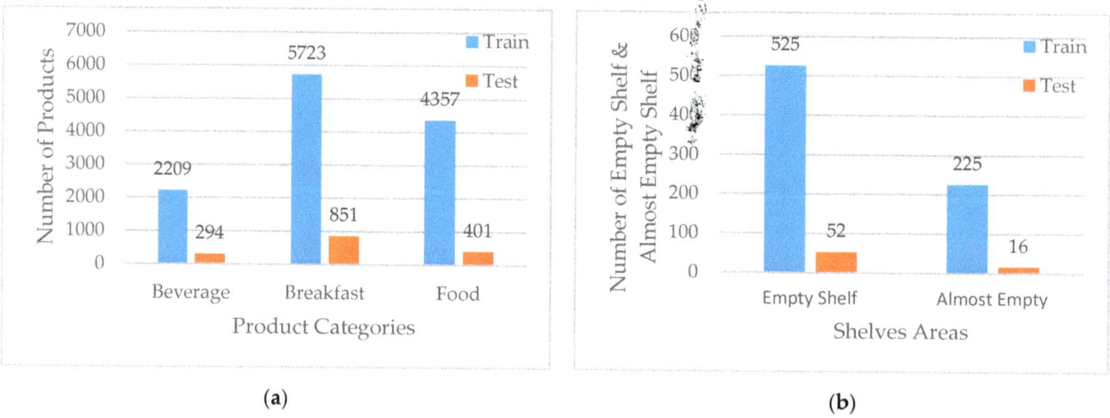

Figure 7. (**a**) Distribution of labeled products; (**b**) Distribution of labeled shelves' areas.

In this study, a file was created to store annotation info for each image, and Pascal visual object classes (VOC) was used as an annotation file, which contains all information about the images such as bounding box coordinates and class names. A sample Pascal VOC format is shown in Figure 8. Since it is in an XML file format, its format can be easily converted to other formats. Different object detection algorithms were used for different file formats to store labeling information of items on the images. Since products and shelf areas are annotated in a Pascal VOC format, we developed several conversion tools in C-Sharp to convert its format to the required formats.

For RetinaNet, each file extension was converted from XML to TXT. Sample file structure and conversion details are shown in Figure 9. The class labels of the products were inserted in a CSV file. Finally, CSV files for the training and test sets were generated from the Pascal VOC file.

For YOLOv3 and YOLOv4, each file extension was converted from XML to TXT, and conversion was applied from Pascal VOC to YOLO format. A sample file structure and conversion details are shown in Figure 10. The class labels of the products were inserted in a NAMES file. Finally, TXT files for the training and test sets were generated to store the locations of images.

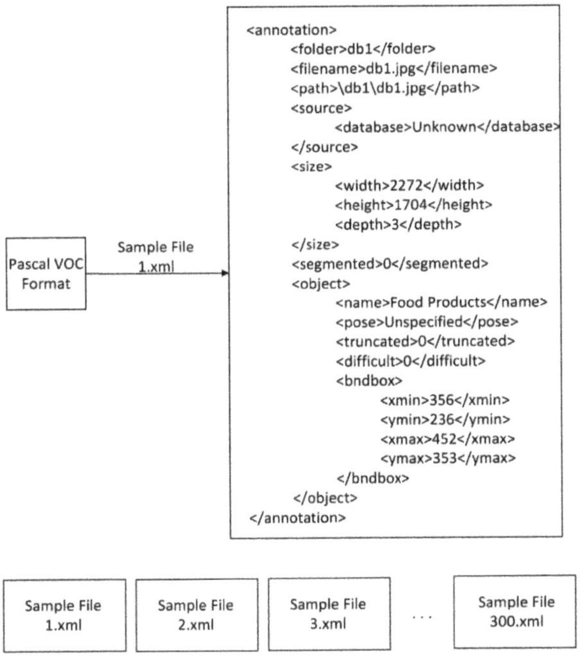

Figure 8. Structure of Pascal visual object classes (VOC) format for a sample file.

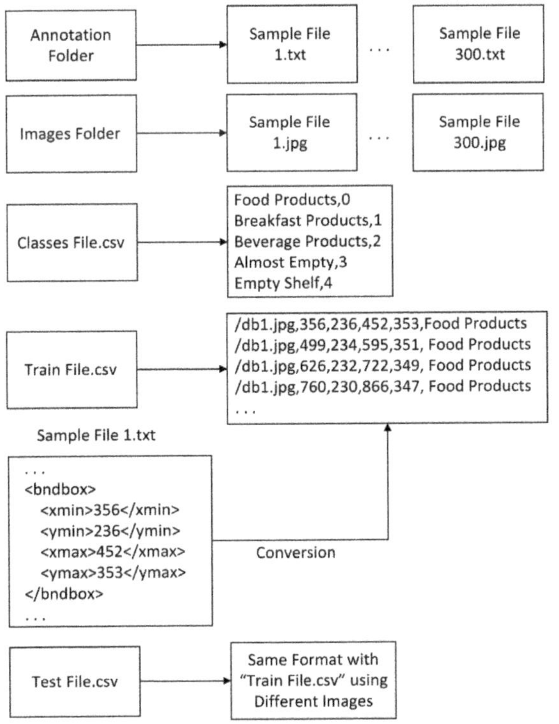

Figure 9. RetinaNet sample file structure and conversion details.

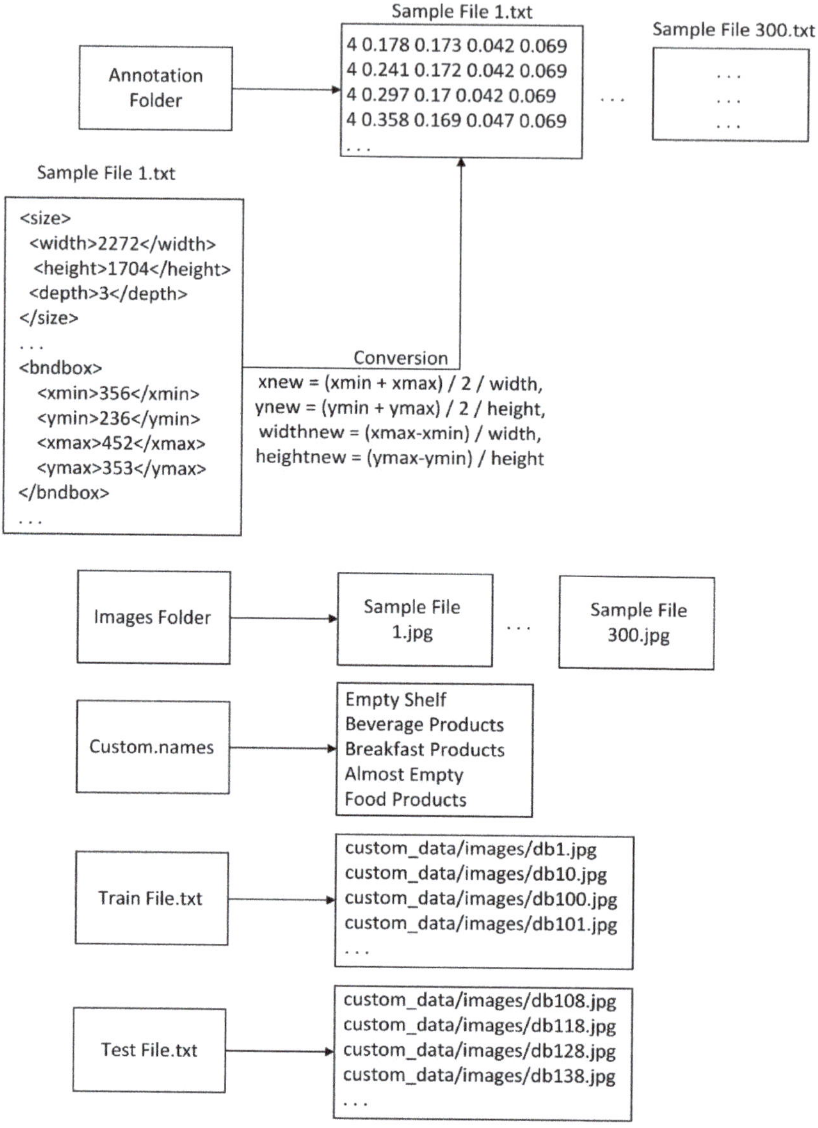

Figure 10. YOLOv3 and YOLOv4 sample file structure and conversion details.

For evaluating our proposed approach, first of all, a training process was started on the selected three one-stage detectors (RetinaNet, YOLOv3, and YOLOv4) using labeled images. The experimental environment was created on a computer that has the following specifications: Ubuntu 20.04 operating system, GeForce RTX 2080 TI GAMING X TRIO 11 GB graphical processing unit (GPU), AMD Ryzen 7 3700X 3.6 GHz/4.4 GHz processor, and 32 GB 3600 MHz DDR4 memory. The training operation is processed on the GPU for each detector.

The following open-source frameworks were used in this study: Keras-based framework for RetinaNet [32], Darknet-based framework for YOLOv3 [33], and YOLOv4 [34]. For each detector, the following parameters were determined: input image size as 512 × 512, learning rate as 0.001, and the number of iterations as 10,000. RetinaNet was trained using two different convolutional neural networks (CNN)-based backbones: ResNet50 and ResNet100. Darknet53, which is a new feature extraction network, was used as the backbone of YOLOv3. For YOLOv4, the CSPDarknet53 backbone was used. This backbone uses cross stage partial network (CSPNet) approach for partitioning and merging feature maps based on a cross-stage hierarchy [22]. Besides, for each backbone, pre-trained models were used, which are "model50.h5" and "model101.h5" for RetinaNet, "yolov3.weights" for YOLOv3, and "yolov4.conv.137" for YOLOv4.

At the end of the first training process, YOLOv4 with CSPDarknet53 had the highest success rates; 0.9187 for mAP, 0.9100 for F1-score, and 0.9600 for recall. These results were followed by RetinaNet with ResNet101 and ResNet50 backbones, and YOLOv3 with Darknet53 backbone. More details of the training results are shown in Table 2. From the results, it can be clearly seen that a significant performance improvement (>20%) was achieved by the proposed method on average. For example; YOLOv4 (0.9616) is significantly better than RetinaNet (0.7245) in the detection of breakfast products. YOLOv4 remarkably outperformed the rest for the "Empty Shelf" and "Almost Empty Shelf" classes with 0.9136 and 0.8125 of average precision values respectively.

Table 2. Comparison of success rates of the methods on the labeled images for three different one-stage detectors with four different backbones.

Classes	RetinaNet (Backbone: ResNet50)	RetinaNet (Backbone: ResNet101)	YOLOv3 (Backbone: Darknet53)	YOLOv4 (Backbone: CSPDarkNet53)
		AP		
Beverage Product	0.9469	0.9625	0.8636	0.9808
Breakfast Product	0.6975	0.7245	0.8003	0.9616
Food Product	0.8886	0.8697	0.6321	0.9252
Empty Shelf	0.8481	0.8387	0.8189	0.9136
Almost Empty Shelf	0.2386	0.1561	0.5884	0.8125
mAP	0.7239	0.7103	0.7406	0.9187
F1-score	0.7333	0.7430	0.6600	0.9100
Recall	0.8105	0.8228	0.6600	0.9600

In this study, we propose semi-supervised learning on OSA, named SOSA, for the first time to take the advantage of unlabeled data because labeling products on image data is an expensive, tedious, difficult, and time-consuming process. For this reason, products and shelf areas were labeled on a small number of images to compare our SOSA approach with the standard supervised OSA approach. The one-stage detector that had the highest success rate was selected with their trained model from the first training phase to use for the semi-supervised learning approach. The proposed SOSA method was evaluated on image datasets, with different ratios of labeled samples varying from 20% to 80%. The distribution of samples is shown in Table 3. For example, in the first experiment, 300 labeled images and 1200 unlabeled images were used to build a classifier. Firstly, unlabeled images were labeled using the selected one-stage detector with its model constructed in the previous training phase. In this way, a pseudo-labeled image dataset is obtained from an unlabeled image dataset. A total of 49,573 products were grouped under three categories and 2145 shelf areas were grouped under two categories for both labeled and pseudo-labeled images.

Table 3. Distribution of labeled samples on image datasets.

Dataset ID	Percentage of Labeled Images	Number of Labeled Images	Number of Unlabeled Images	Number of Total Images
D1	20%	300	1200	1500
D2	40%	300	450	750
D3	60%	300	200	500
D4	80%	300	75	375

Training processes were started for each dataset given in Table 3 to verify the effectiveness of the proposed SOSA method. Here, this experiment is performed to test the different ratios of labeled images, varying from 20% to 80% with an increment of 20%. The comparison of loss values obtained during the training phases is shown in Figure 11. From the figure, it can be seen that loss values decreased dramatically during the training phase after two thousand iterations. Very close loss values were obtained for different ratios. The results indicate that a small amount of labeled data is enough to receive similar loss values since the algorithm benefits from unlabeled data. From the results, it can be concluded that labeled data that contain shelf images taken from a grocery store can be extended with unlabeled images to achieve high accuracy.

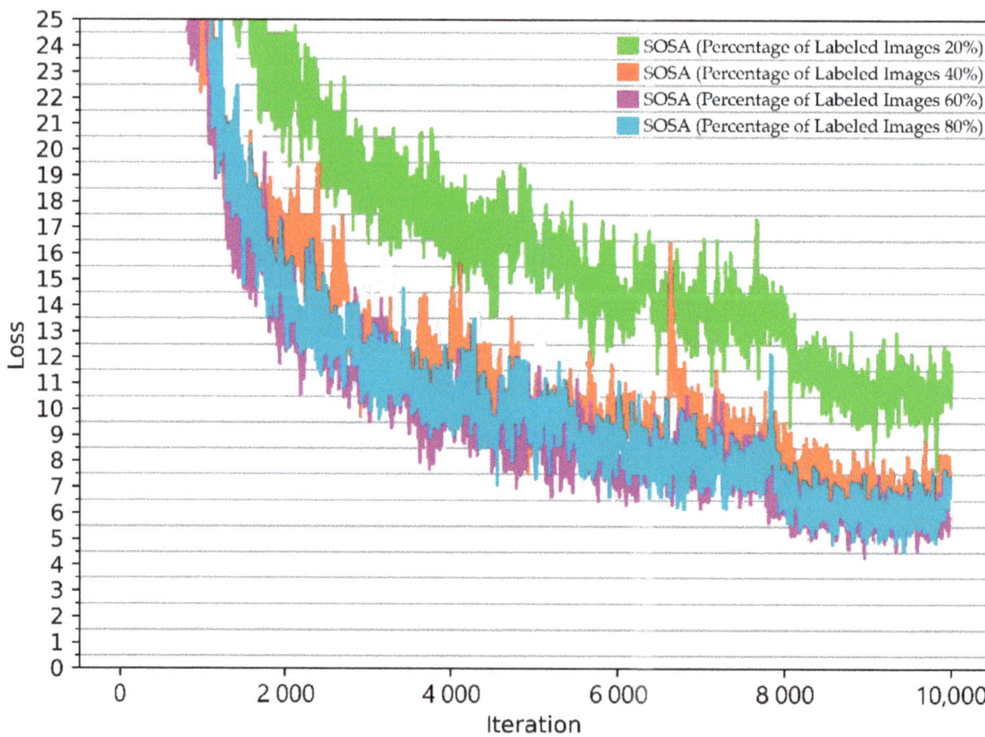

Figure 11. Comparison of loss value changes during the training phase.

As shown in Figure 12, satisfactory results were achieved by the proposed SOSA approach. When the percentage of labeled images was 20%, mAP, F1-score, and recall values were obtained as 0.7209, 0.8100, and 0.8300, respectively. When the percentage of labeled images was 40%, mAP was calculated as 0.7257, F1-score was achieved by 0.7800, and recall was measured as 0.7800. When using a 60–40% labeled-unlabeled ratio, mAP was 0.8622, F1-score was 0.8700, and recall was 0.9000. When the labeled image rate was 80%, mAP was 0.8927, F1-score was 0.9000, and recall was 0.9300. As has been observed in Figure 12, it is possible to provide good generalization ability for the OSA problem by applying a small number of labeled image data. It also can be seen from the results that an improvement could be expected from our method in circumstances where the ratio of labeled data grows. For example, after adding 20% labeled data, we found that the accuracy of the SOSA method increased from 0.7257 to 0.8622, with approximately 14% improvement. The results indicate that an initial amount of labeled ordinal data can be sufficient enough to learn the model; but additional labeled data can improve the classification performance.

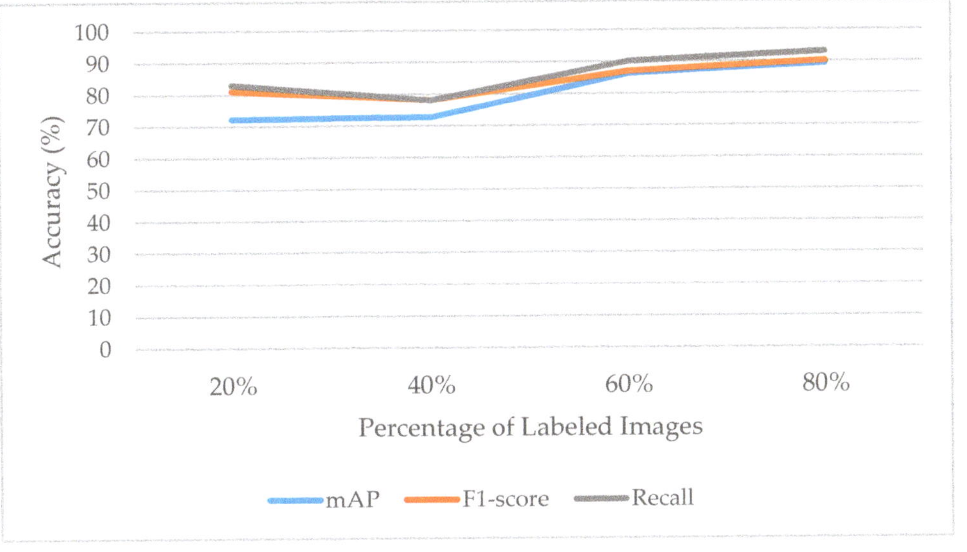

Figure 12. Success rates when different ratios of labeled data are considered.

The sample object detection results related to the breakfast section are shown in Figure 13. Each subfigure shows the results obtained by the models that were trained using labeled samples varying from 20% to 80%. While some almost-empty areas, which are highlighted in red color, could not be detected with a 20–80% labeled-unlabeled ratio, these areas could be detected with higher labeled data ratios. The detection accuracy of almost-empty areas was increased as the labeled data ratio was increased. Besides, from the SOSA XAI perspective, the results were interpreted in a more understandable way for users. SOSA XAI output message is as follows: "8 almost-empty and 4 empty shelves are existing on the breakfast section. In order to determine the section, the products on the picture are counted and the category that has the most products is selected".

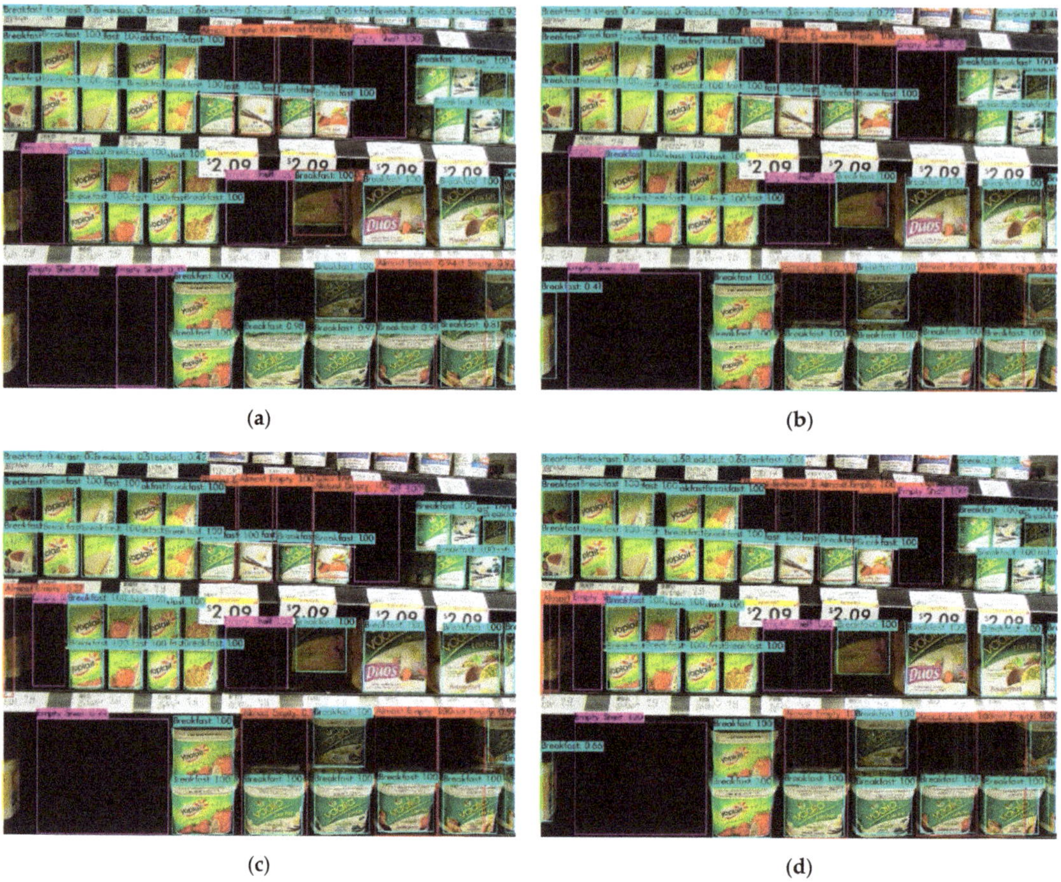

Figure 13. Sample object detection results of the breakfast section. It shows detection results using the model that was trained with (**a**) 20% of labeled images; (**b**) 40% of labeled images; (**c**) 60% of labeled images; and (**d**) 80% of labeled images.

For the beverage section, the sample object detection results are shown in Figure 14. The sample shelf image has eight empty and one almost-empty shelves. All trained models detected empty shelves (highlighted in violet color) and almost-empty shelf (highlighted in red color) with high probabilities. For example, the model that was trained with 20% labeled images detected one of the empty shelves with 99% probability and the almost-empty shelf with 100% probability. The model that was trained with 80% labeled images detected one of the empty shelves with 100% probability and the almost empty shelf with 100% probability. Besides, all models detected beverage products with high probabilities.

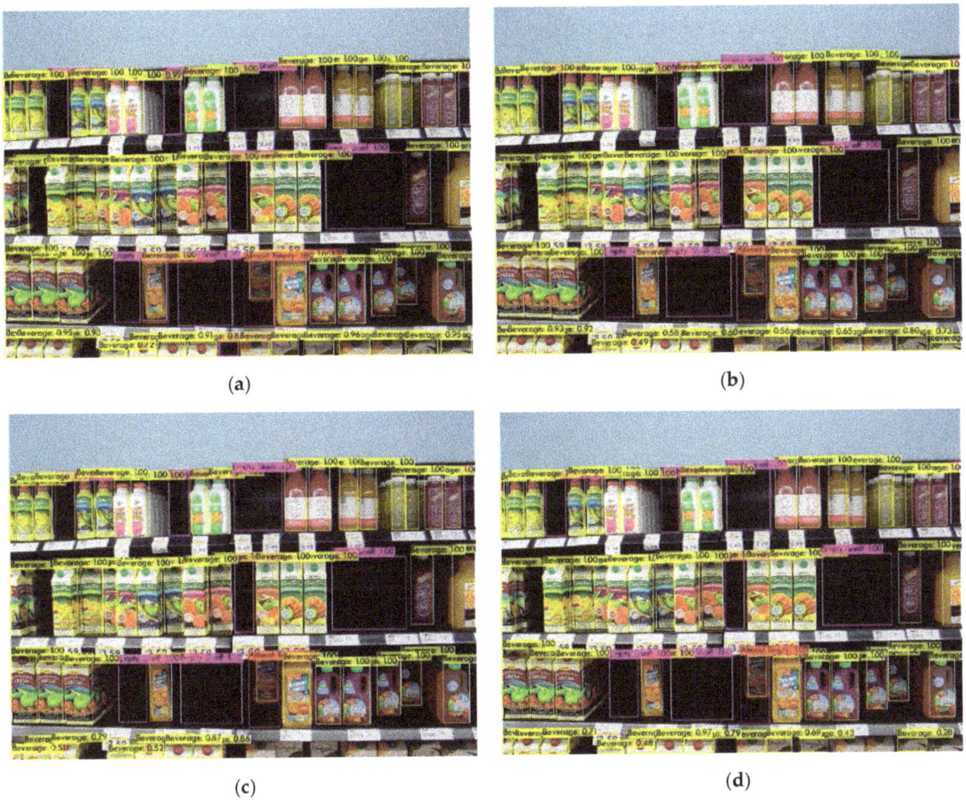

Figure 14. Sample object detection results of the beverage section. It shows detection results using the model that was trained with (**a**) 20% of labeled images; (**b**) 40% of labeled images; (**c**) 60% of labeled images; and (**d**) 80% of labeled images.

In this study, we focused on detecting empty and almost-empty areas on the breakfast and the beverage sections. The images that do not include beverage and breakfast products are labeled as "Food Product". The SOSA approach was also tested using shelf images that were taken from the food sections. Figure 15 shows the sample object detection results for the rice and pasta section. The results showed that products were correctly labeled as "Food Product" with high probabilities. Most of the images in the dataset are frontal views and products in front of the shelves are fundamentally analyzed to be detected in this study. When a product is on the backside of a shelf, the system generates an almost-empty shelf notification. However, if the depth of a shelf is too deep, the system may detect the products at the backside with low probability. In the cases of empty and almost-empty shelf notifications, a responsible clerk can put new items on the shelf to prevent customer and profit loss. It can also be noted that a package may contain multiple items but it may look like a single object from the outside, and in this case, the package is detected as a single object by the system, as expected. Besides, some products can be damaged. In this case, these products can also be detected by the system, but with lower probability due to the change in their appearance.

Figure 15. Sample object detection results of the rice and pasta section. It shows detection results using the model that was trained with (**a**) 20% of labeled images; (**b**) 40% of labeled images; (**c**) 60% of labeled images; and (**d**) 80% of labeled images.

The comparison of the proposed SOSA approach with the existing approaches in terms of accuracy is shown in Table 4. The experimental results showed that the proposed approach outperformed the existing approaches (RetinaNet and YOLOv3) in terms of accuracy. The best performance was achieved for the breakfast products. When the percentage of labeled images varied between 80% and 20%, the AP values were obtained as 0.9467, 0.9626, 0.9253, and 0.8414, respectively. At most, all the algorithms had difficulty in distinguishing the "Almost Empty" class. Compared to the existing methods, approximately 25% improvement was achieved for the almost-empty class by the proposed method. Besides, this achievement of the SOSA method can be increased by extending the dataset with unlabeled images that contain more almost-empty areas. From the empty-shelf class perspective, satisfactory AP results (>0.8) were obtained by the proposed method. In addition, when AP results were evaluated based on product classes, it was seen that AP values increased compared to the previous methods on average. Based on the results, it can be concluded that the proposed SOSA method has good generalization ability in distinguishing all the classes, so it can be effectively used to recognize them well.

Table 4. Comparison of accuracies between SOSA and existing approaches.

Classes	RetinaNet (Backbone: ResNet50)	RetinaNet (Backbone: ResNet101)	YOLOv3 (Backbone: Darknet53)	SOSA (80% Labeled Images)	SOSA (60% Labeled Images)	SOSA (40% Labeled Images)	SOSA (20% Labeled Images)
			AP				
Beverage Product	0.9469	0.9625	0.8636	0.9586	0.8797	0.7477	0.8224
Breakfast Product	0.6975	0.7245	0.8003	0.9467	0.9626	0.9253	0.8414
Food Product	0.8886	0.8697	0.6321	0.9303	0.8680	0.7308	0.8718
Empty Shelf	0.8481	0.8387	0.8189	0.8410	0.8736	0.7601	0.7216
Almost Empty Shelf	0.2386	0.1561	0.5884	0.7866	0.7269	0.4646	0.3471
mAP	0.7239	0.7103	0.7406	0.8927	0.8622	0.7257	0.7209
F1-score	0.7333	0.7430	0.6600	0.9000	0.8700	0.7800	0.8100
Recall	0.8105	0.8228	0.6600	0.9300	0.9000	0.7800	0.8300

5. Discussion

Providing high on-shelf availability is a key factor to increase profits in grocery stores. For this purpose, this study is the first attempt to combine two concepts "semi-supervised learning" and "on-shelf availability" (SOSA). The proposed SOSA method aims to decrease out-of-stock by automatically detecting and classifying products using shelf images based on the semi-supervised learning principle. The main purposes of the SOSA method are to decrease the need for manual image labeling, to obtain satisfactory results using a small amount of manually labeled images, and to provide additional knowledge present in unlabeled image data.

SOSA builds a classification model that allows the detection of "Product", "Empty Shelf", and "Almost-Empty Shelf" classes by using both labeled and unlabeled image data. When building the model, three different deep learning techniques are used: RetinaNet, YOLOv3, andYOLOv4. This study shows that the proposed approach improves accuracy when monitoring OSA.

In the experimental studies, the effectiveness of the proposed SOSA method was demonstrated on a real-world image dataset, with different ratios of labeled samples varying from 20% to 80%. As can be seen from Table 4, the proposed SOSA approach outperformed the previous approaches on average. For instance, according to the experimental results, 15–18% improvement is achieved on average accuracy.

When developing an OSA system, some challenges could be encountered and overcome. The first one is that some products can lie on the shelf or their packages can be damaged. In these cases, these kinds of products can be detected with low probability due to the change in their appearance. To increase detection probability, the images containing this type of product can be included in the training set. In this way, the model can learn from these kinds of objects, and therefore these products can be detected with higher accuracies. Another difficulty is that some objects can stand in front of the shelf and they can prevent detecting products on the shelf. These objects can be shoppers, employees, grocery store trolleys, or something else. In this study, we implemented a solution to check the presence of people on the image. If a person is detected, this image is discarded by the system, and then a new image is taken after a period of time. This issue is also important to protect the privacy of customers. Similar solutions can also be implemented for other objects.

The SOSA method is powered by the SOSA XAI application. SOSA XAI was developed to provide transparency and understandability of the OSA model for the users. Thanks to this tool, detection results can be interpreted more easily. Furthermore, it provides insight into what the OSA model is paying attention to. In addition, it provides evidence for the results, and validity of the process. Moreover, the training set can be enhanced in a proper manner to adopt the application for changes without an expert in the AI field.

The main findings of this study can be listed as follows:

(1) It was observed that "semi-supervised learning" provides many advantages for monitoring OSA, including improving efficiency, reducing labeling cost, providing additional knowledge present in unlabeled data, and increasing the applicability of machine learning in the retail sector.
(2) The combination of three deep learning techniques (RetinaNet, YOLOv3, YOLOv4) improves accuracy when monitoring OSA.
(3) Explainable AI is a powerful tool in monitoring OSA since it provides users with an explanation of individual decisions and enables users to manage, understand, and trust the OSA model.
(4) The proposed SOSA method has the potential to expand the application of machine learning in grocery stores, thanks to its advantages.

6. Conclusions and Future Work

Providing high on-shelf availability is a key factor to increase profits in grocery stores. For this purpose, the traditional OSA applications use the labeled image data when building a classifier. However, a large amount of data generated in real-life is unlabeled, and manually labeling products on the images is an expensive process. To overcome this problem, this paper proposes a new approach, called SOSA, which combines two concepts "semi-supervised learning" and "on-shelf availability" for the first time. The proposed SOSA method addresses the problem by automatically allowing the model to integrate the available unlabeled image data with very little cost. Our proposed method detects empty and almost empty shelves based on each product category using a small amount of labeled data and a large amount of unlabeled data.

It is the first time that YOLOv4 object detection architecture is used for monitoring OSA. This study is also original in that it compares three deep learning approaches (RetinaNet, YOLOv3, YOLOv4) for monitoring OSA in the retail sector.

The experiments were conducted on image datasets with different ratios of labeled samples varying from 20% to 80%. The experimental results show that the proposed approach outperforms the existing approaches (RetinaNet and YOLOv3) in terms of accuracy.

Moreover, this paper presents the first demonstration of explainable artificial intelligence (XAI) on OSA, called SOSA XAI. Thanks to the SOSA XAI, users can understand, trust, and manage the proposed deep learning model and modify the dataset to adapt the deep learning model for changes.

Currently, the developed SOSA XAI software application uses rule-based interpretation to explain the outputs of the proposed approach. Instead of rule-based interpretation, a deep learning-based interpretation can be developed in the future. Furthermore, augmented reality technology can be implemented to render some virtual objects related to the products detected on the shelves. Moreover, more than three one-stage detectors can be integrated into the SOSA approach as future work.

Author Contributions: Conceptualization, R.Y.; Data curation, R.Y.; Formal analysis, R.Y.; Investigation, R.Y.; Methodology, R.Y.; Project administration, D.B.; Resources, R.Y.; Software, R.Y.; Supervision, D.B.; Validation, R.Y.; Visualization, R.Y.; Writing—original draft, R.Y.; Writing—review & editing, D.B. All authors have read and agreed to the published version of the manuscript.

Funding: This research received no external funding.

Institutional Review Board Statement: Not applicable.

Informed Consent Statement: Not applicable.

Data Availability Statement: Publicly available dataset was used in this study. This data can be found here: http://yuhang.rsise.anu.edu.au/.

Conflicts of Interest: The authors declare no conflict of interest.

References

1. Corsten, D.; Gruen, T. Desperately seeking shelf availability: An examination of the extent, the causes, and the efforts to address retail out-of-stocks. *Int. J. Retail Distrib. Manag.* **2003**, *31*, 605–617. [CrossRef]
2. Musalem, A.; Olivares, M.; Bradlow, E.T.; Terwiesch, C.; Corsten, D. Structural estimation of the effect of out-of-stocks. *Manage. Sci.* **2010**, *56*, 1180–1197. [CrossRef]
3. Bottani, E.; Bertolini, M.; Rizzi, A.; Romagnoli, G. Monitoring on-shelf availability, out-of-stock and product freshness through RFID in the fresh food supply chain. *Int. J. RF Technol. Res. Appl.* **2017**, *8*, 33–55. [CrossRef]
4. Michael, K.; McCathie, L. The Pros and Cons of RFID in Supply Chain Management. In Proceedings of the 4th Annual 4th International Conference on Mobile Business, ICMB, Sydney, NSW, Australia, 11–13 July 2005; pp. 623–629. [CrossRef]
5. Moorthy, R.; Behera, S.; Verma, S.; Bhargave, S.; Ramanathan, P. Applying image processing for detecting on-shelf availability and product positioning in retail stores. In Proceedings of the ACM International Conference Proceeding Ser., Kochi, India, 10–13 August 2015; pp. 451–457. [CrossRef]
6. Higa, K.; Iwamoto, K. Robust estimation of product amount on store shelves from a surveillance camera for improving on-shelf availability. In Proceedings of the IST 2018—IEEE International Conference Imaging Systems and Techniques Proceeding, Kraków, Poland, 16–18 October 2018; pp. 1–6. [CrossRef]
7. Zhu, X. *Semi-Supervised Learning, Encyclopedia of Machine Learning and Data Mining*; Springer: Berlin/Heidelberg, Germany, 2017; Volume 3, ISBN 9781489976871. [CrossRef]
8. Barredo Arrieta, A.; Díaz-Rodríguez, N.; Del Ser, J.; Bennetot, A.; Tabik, S.; Barbado, A.; Garcia, S.; Gil-Lopez, S.; Molina, D.; Benjamins, R.; et al. Explainable Explainable Artificial Intelligence (XAI): Concepts, taxonomies, opportunities and challenges toward responsible AI. *Inf. Fusion* **2020**, *58*, 82–115. [CrossRef]
9. Moorthy, R.; Behera, S.; Verma, S. On-Shelf Availability in Retailing. *Int. J. Comput. Appl.* **2015**, *115*, 47–51. [CrossRef]
10. Satapathy, R.; Prahlad, S.; Kaulgud, V. Smart Shelfie-Internet of shelves: For higher on-shelf availability. In Proceedings of the 2015 IEEE Region 10 Symposium TENSYMP, Ahmedabad, India, 13–15 May 2015; pp. 70–73. [CrossRef]
11. Rosado, L.; Goncalves, J.; Costa, J.; Ribeiro, D.; Soares, F. Supervised Learning for Out-of-Stock Detection in Panoramas of Retail Shelves. In Proceedings of the IST 2016 IEEE International Conference on Imaging Systems and Techniques, Chania, Greece, 4–6 October 2016; pp. 406–411. [CrossRef]
12. Kejriwal, N.; Garg, S.; Kumar, S. Product counting using images with application to robot-based retail stock assessment. In Proceedings of the IEEE Conference on Technologies for Practical Robot Applications, Woburn, MA, USA, 11–12 May 2015; pp. 1–6. [CrossRef]
13. Higa, K.; Iwamoto, K. Robust shelf monitoring using supervised learning for improving on-shelf availability in retail stores. *Sensors* **2019**, *19*, 2722. [CrossRef] [PubMed]
14. Liu, S.; Tian, H. Planogram Compliance Checking Using Recurring Patterns. In Proceedings of the 2015 IEEE International Symposium on Multimedia (ISM), Miami, FL, USA, 14–16 December 2015; pp. 27–32. [CrossRef]
15. Liu, S.; Li, W.; Davis, S.; Ritz, C.; Tian, H. Planogram compliance checking based on detection of recurring patterns. *IEEE Multimed.* **2016**, *23*, 54–63. [CrossRef]
16. Saran, A.; Hassan, E.; Maurya, A.K. Robust visual analysis for planogram compliance problem. In Proceedings of the 14th IAPR International Conference on Machine Vision Applications MVA, Tokyo, Japan, 18–22 May 2015; pp. 576–579. [CrossRef]
17. Girshick, R.; Donahue, J.; Darrell, T.; Malik, J. Rich feature hierarchies for accurate object detection and semantic segmentation. In Proceedings of the IEEE Computer Society Conference on Computer Vision Pattern Recognition, Columbus, OH, USA, 23–28 June 2014; pp. 580–587. [CrossRef]
18. Ren, S.; He, K.; Girshick, R.; Sun, J. Faster R-CNN: Towards Real-Time Object Detection with Region Proposal Networks. *IEEE Trans. Pattern Anal. Mach. Intell.* **2017**, *39*, 1137–1149. [CrossRef] [PubMed]
19. Lin, T.Y.; Dollár, P.; Girshick, R.; He, K.; Hariharan, B.; Belongie, S. Feature pyramid networks for object detection. In Proceedings of the 30th IEEE Conference on Computer Vision Pattern Recognition (CVPR 2017), Honolulu, HI, USA, 21–26 July 2017; pp. 936–944. [CrossRef]
20. Lin, T.Y.; Goyal, P.; Girshick, R.; He, K.; Dollar, P. Focal Loss for Dense Object Detection. *IEEE Trans. Pattern Anal. Mach. Intell.* **2018**, *42*, 318–327. [CrossRef] [PubMed]
21. Redmon, J.; Farhadi, A. YOLOv3: An Incremental Improvement. *arXiv* **2018**, arXiv:1804.02767.
22. Bochkovskiy, A.; Wang, C.-Y.; Liao, H.-Y.M. YOLOv4: Optimal Speed and Accuracy of Object Detection. *arXiv* **2020**, arXiv:2004.10934.
23. Krizhevsky, A.; Sutskever, I.; Hinton, G.E. ImageNet Classification with Deep Convolutional Neural Networks. In Proceedings of the 25th International Conference on Neural Information Processing Systems, Lake Tahoe, NV, USA, 3–6 December 2012; 2012; pp. 1097–1105.
24. Emmert-Streib, F.; Yli-Harja, O.; Dehmer, M. Explainable artificial intelligence and machine learning: A reality rooted perspective. *Wiley Interdiscip. Rev. Data Min. Knowl. Discov.* **2020**, *10*, 1–8. [CrossRef]

25. Confalonieri, R.; Coba, L.; Wagner, B.; Besold, T.R. A historical perspective of explainable Artificial Intelligence. *Wiley Interdiscip. Rev. Data Min. Knowl. Discov.* **2020**, 1–21. [CrossRef]
26. Kajabad, E.N.; Ivanov, S.V. People detection and finding attractive areas by the use of movement detection analysis and deep learning approach. *Procedia Comput. Sci.* **2019**, *156*, 327–337. [CrossRef]
27. WebMarket Is An Image Database Built for Computer Vision Research. Available online: http://yuhang.rsise.anu.edu.au (accessed on 11 November 2020).
28. Santra, B.; Mukherjee, D.P. A comprehensive survey on computer vision based approaches for automatic identification of products in retail store. *Image Vision Comput.* **2019**, *86*, 45–63. [CrossRef]
29. Zhang, Y.; Wang, L.; Hartley, R.; Li, H. Where's the Weet-Bix? In Proceedings of the ACCV'07: 8th Asian Conference on Computer Vision, Tokyo, Japan, 18–22 November 2007; Part 1. pp. 800–810. [CrossRef]
30. Zhang, Y.; Wang, L.; Hartley, R.; Li, H. Handling significant scale difference for object retrieval in a supermarket. *DICTA 2009 Digit. Image Comput. Tech. Appl.* **2009**, 468–475. [CrossRef]
31. LabelImg. Available online: https://github.com/tzutalin/labelImg (accessed on 11 November 2020).
32. Keras RetinaNet. Available online: https://github.com/fizyr/keras-retinanet (accessed on 11 November 2020).
33. YOLOv3 Framework. Available online: https://pjreddie.com/darknet/yolo (accessed on 11 November 2020).
34. YOLOv4 Framework. Available online: https://github.com/AlexeyAB/darknet (accessed on 11 November 2020).

sensors

MDPI

Article

Estimating the Growing Stem Volume of Coniferous Plantations Based on Random Forest Using an Optimized Variable Selection Method

Fugen Jiang [1,2,3], Mykola Kutia [4], Arbi J. Sarkissian [4], Hui Lin [1,2,3], Jiangping Long [1,2,3], Hua Sun [1,2,3,*] and Guangxing Wang [1,5]

1 Research Center of Forestry Remote Sensing and Information Engineering, Central South University of Forestry and Technology, Changsha 410004, China; jiangkriging@csuft.edu.cn (F.J.); linhui@csuft.edu.cn (H.L.); longjiangping@csuft.edu.cn (J.L.); gxwang@siu.edu (G.W.)

2 Key Laboratory of Forestry Remote Sensing Based Big Data and Ecological Security for Hunan Province, Changsha 410004, China

3 Key Laboratory of National Forestry and Grassland Administration on Forest Resources Management and Monitoring in Southern Area, Changsha 410004, China

4 Bangor College China, Bangor University, 498 Shaoshan Rd., Changsha 410004, China; m.kutia@bangor.ac.uk (M.K.); a.sarkissian@bangor.ac.uk (A.J.S.)

5 Department of Geography and Environmental Resources, Southern Illinois University, Carbondale, IL 62901, USA

* Correspondence: sunhua@csuft.edu.cn; Tel.: +86-138-758-821-84

Received: 29 September 2020; Accepted: 14 December 2020; Published: 17 December 2020

Abstract: Forest growing stem volume (GSV) reflects the richness of forest resources as well as the quality of forest ecosystems. Remote sensing technology enables robust and efficient GSV estimation as it greatly reduces the survey time and cost while facilitating periodic monitoring. Given its red edge bands and a short revisit time period, Sentinel-2 images were selected for the GSV estimation in Wangyedian forest farm, Inner Mongolia, China. The variable combination was shown to significantly affect the accuracy of the estimation model. After extracting spectral variables, texture features, and topographic factors, a stepwise random forest (SRF) method was proposed to select variable combinations and establish random forest regressions (RFR) for GSV estimation. The linear stepwise regression (LSR), Boruta, Variable Selection Using Random Forests (VSURF), and random forest (RF) methods were then used as references for comparison with the proposed SRF for selection of predictors and GSV estimation. Combined with the observed GSV data and the Sentinel-2 images, the distributions of GSV were generated by the RFR models with the variable combinations determined by the LSR, RF, Boruta, VSURF, and SRF. The results show that the texture features of Sentinel-2's red edge bands can significantly improve the accuracy of GSV estimation. The SRF method can effectively select the optimal variable combination, and the SRF-based model results in the highest estimation accuracy with the decreases of relative root mean square error by 16.4%, 14.4%, 16.3%, and 10.6% compared with those from the LSR-, RF-, Boruta-, and VSURF-based models, respectively. The GSV distribution generated by the SRF-based model matched that of the field observations well. The results of this study are expected to provide a reference for GSV estimation of coniferous plantations.

Keywords: forest growing stem volume; coniferous plantations; variable selection; texture feature; random forest; red-edge band

1. Introduction

Forest plays a crucial role in the global carbon cycle as one of the largest carbon sinks in the biosphere [1,2]. Estimating forest growth and productivity is, therefore, essential for informing climate change research and forest management efforts globally [3]. Forest growing stem volume (GSV) refers to the total volume of tree trunks in a given forested area that reflects the richness of forest resources and the health of the forest ecosystem. Accurate estimation of the GSV, therefore, plays an important role in both forest resource management and the monitoring of ecosystem dynamics [4–6]. Compared with traditional field surveys, remote sensing technology allows for a more efficient approach to the monitoring and management of forest resources in real-time [7–9]. However, the accuracy of GSV estimation using remotely sensed data is determined together by data sources, feature variable selection methods, and estimation models.

Extracting information on vegetation growth from remote sensing data is the basis of forest parameter estimation. The common remote sensing data sources for GSV estimation are varied in terms of active and passive remote sensing data. Optical data is a typical passive remote sensing data that is derived from multi-spatial resolution and multi-spectral sensors, which provides ground feature information as a range of different wavelength bands. For example, MODIS, SPOT-VEGETATION, NOAA/AVHRR, FY-3/MERSI [10–13] are common optical datasets used for GSV estimation at global and regional scales. Hyper-spectral data, such as those from Hyperion [14], AVIRIS [15] and HYDICE [16], can effectively detect the characteristics of ground objects. Hyper-spectral data contains nearly continuous spectral information, which greatly improves the recognition of different objects on the land surface. Low-resolution sensor data having high temporal resolution and wide spatial coverage enables the analysis of time-series characteristics of vegetation at large scales. However, a large number of spectral bands and information redundancy of hyper-spectral data lead to huge computational loads. With the enhancement of remote sensing technology, a large number of datasets with higher spatial resolution have been recently used for GSV estimation. Medium and high-resolution data, such as those provided by Landsat [17] and GaoFen (GF) [18] systems, can obtain more accurate vegetation attributes. Using these data, the mixed pixels are reduced, thus significantly improving the mapping of vegetation parameters. However, due to cloud coverage, data stripe, long satellite revisit periods, and high data costs in acquiring data, it is often difficult to obtain sufficient satellite data covering a given target area.

Active remote sensing refers to the transmitting of electromagnetic waves to a detected target area and receiving the echo signal of an object [19], mainly including light detection and ranging (LiDAR) and synthetic aperture radar (SAR). LiDAR can provide three-dimensional structural information efficiently and has unique advantages in estimating tree height and spatial structure. Among LiDAR datasets, airborne LiDAR is the most effective but also the most expensive. SAR can provide observations under all weather conditions and is not affected by atmospheric propagation [20,21]. However, SAR signals are influenced by the terrain, and the terrain effects cannot be completely corrected [22,23]. LiDAR and SAR offer three-dimensional point cloud data to record the location and characteristics of a target object in detail. But due to massive data processing requirements, there are limitations in using LiDAR and SAR for acquiring vegetation information at large scales.

The Sentinel-2 carries a multi-spectral imager (MSI) for land monitoring, providing vegetation, soil, and water cover. The red-edge bands have priority over other spectral variables in modeling. The vegetation indices composed of reflectance of the red-edge bands are highly correlated with GSV, which can effectively improve the estimation accuracy of the structural parameters of the planted forests [24–26]. Currently, Sentinel-2 is the only source of freely available optical data that exceeds three red-edge bands. Sentinel-2 data can, therefore, provide a more cost-effective potential for accurately mapping GSV of forest plantations [27]. Compared with MODIS and Landsat, the Sentinel-2 data has a higher spatial resolution, which can obtain more accurate vegetation information. Its four 10 m resolution bands and six 20 m resolution bands get much more accurate vegetation growth status in forests. In addition, Sentinel-2 data characterizes higher temporal resolution (revisit period of five days using 2A and 2B). The time-series data provided by the satellite system allow us to obtain high-quality records of seasonal forest changes for forest resources monitoring [27].

Feature variable combination is extremely important for model development and prediction of forest GSV. In the process of modeling, using a large number of independent variables or predictors often leads to complex models and overfitting. Moreover, as the number of predictors increases, the prediction accuracy of GSV from the models may not necessarily increase and, in turn, may decrease due to the increased noise and errors of the input variables. Before modeling, it is thus critical to select the predictors that significantly contribute to improving the estimation accuracy, increasing the interpretability of models, and reducing the running time of the models [28,29]. Pearson correlation coefficient combined with linear stepwise regression (LSR) is commonly used for selecting feature variables, which are linearly related [30]. However, due to the complexity of forest ecosystems, the linear relationship may limit the estimation accuracy of models [27]. The importance assessment of random forest (RF) provides a metric for judging the contributions of feature variables to increasing model prediction accuracy, which can offer an evaluation of non-linear relationships between feature variables and GSV. RF can provide the importance weights of individual variables, but it lacks the ability to directly determine the optimal combination of feature variables. The same feature variable in different combinations of predictors may have different contributions to reducing the errors of models. There have been a few studies on variable selection using random forest associated importance, such as Boruta and Variable Selection Using Random Forests (VSURF), but the effect of these methods in forest estimation is limited [31]. Moreover, different combinations of feature variables may lead to different overall contributions to the improvement of models. The existing methods for the selection of feature variables lack the ability to determine an optimal combination of feature variables. Furthermore, the existing methods may take much time to complete in the large data sets available. In fact, the primary reason why many existing methods are established is to improve accuracy, and the runtime is often hard to balance [31]. There is a strong need to improve the existing methods or to develop a new method that can be used to determine an optimal and stable combination of feature variables for modeling [32].

Presently, remote sensing images or laser point clouds are usually combined with the field measurements to establish models for regional GSV estimation by parametric or non-parametric methods [33,34]. The parametric methods are easy to understand and fast to learn, but they cannot quickly retrieve the correct objective function form in complex conditions [35]. Non-parametric methods do not make strong assumptions about the form of objective functions and are more suitable for the prediction of complex data [33–38]. As a representative of non-parametric methods, RF has become popular for forest GSV estimation [36,37]. RF is an ensemble learning method that is insensitive to noise data and does not require any assumptions about the distribution of input datasets. It estimates forest GSV by swiftly constructing a large number of regression trees. During training, the regression trees are independent of each other, and the training speed is rapid. In addition, RF can evaluate the importance of each feature variable in the model, which can effectively judge the contribution of individual variables to the model [38].

This study aims to establish a random forest regression (RFR) model for GSV estimation of coniferous plantations through developing a novel feature variable selection method based on importance evaluation and analyzing its accuracy and effectiveness. In order to verify the application of the modified feature variable selection method, Sentinel-2 and the observed GSV data of Wangyedian Forest Farm were combined with mapping the GSV in the study area. Moreover, four widely used feature variable selection methods (LSR, RF, Boruta, and VSURF) were used and analyzed for comparison. In addition, the effect of texture features in the red-edge bands on the improved GSV mapping was also studied.

2. Materials and Methods

2.1. General Description of the Study Area

The Wangyedian forest farm is located in Harqin (longitude 118°09′–118°30′ E, latitude 41°21′–41°39′ N), Inner Mongolia Autonomous Region, China (Figure 1). The altitude ranges from 500 to 1890 m. The forest farm has a mid-temperate continental monsoon climate with an annual average temperature of 4.2 °C, a frost-free period of 117 days, average annual sunshine of 2913.3 h, and average annual precipitation of 400 mm. The total area of this region is 25,958 ha, with a forested area of 23,118 ha. The total volume of living trees is 1.28 million m^3. The dominant tree species are Chinese pine (*Pinus tabuliformis*) and larch (*Larix principis-rupprechtii* and *Larix ologensis*).

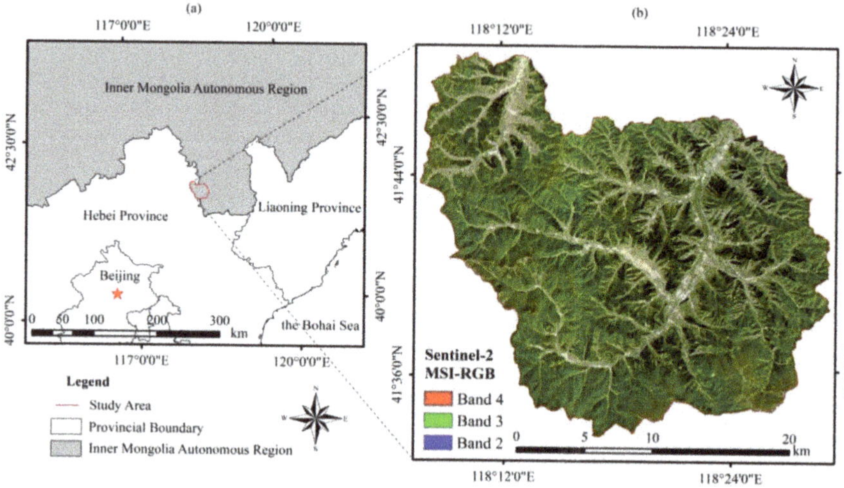

Figure 1. (**a**) Location and boundary of the Wangyedian forest farm and (**b**) Sentinel-2 image covering the study area.

2.2. Sampling Design and GSV Survey

According to the forest survey data of the Wangyedian forest farm in 2017, the land types were divided into (1) coniferous forest; (2) other forest types (broadleaf forests, coniferous, and broad-leaved mixed forests); and (3) non-forest land (farmland, buildings, water, and unused land). The species, distribution, and coverage range of coniferous forests were analyzed and sorted. The statistical results show that the planted coniferous forests consisted of mainly Chinese pine and Larch. The boundary between the two tree species was determined, and random sampling was used to select 81 sample plots of 25 m × 25 m from the coniferous forests in the study area (Figure 2).

The field survey was carried out from 20 September to 15 October 2017. Trimble Geo 7x Global Positioning System (GPS) was used to record the central coordinates of each sample plot. The forest compass was used to determine the boundaries of the sample plots. Tree height, diameter at breast height (DBH), and environmental factors (air temperature and soil moisture) in each sample plot were measured. Trees in the sample plots with DBH greater than 5 cm were selected and examined. The GSV of each tree was determined using tree height and DBH based volume formula stated by the National Forestry and Grassland Administration of China (http://www.forestry.gov.cn/). The plot-level GSV value of each sample plot was obtained by summing the tree volumes within each plot and then converted to the hectare-level. The GSV values in the study area range from 86.17 to 514.96 m^3/ha. The mean value, standard deviation, and coefficient of variation of all the sample plots are 209.01 m^3/ha, 119.87 m^3/ha, and 44.2%, respectively. At the significant level of 0.05, the confidence interval for GSV is from 209.55 to 254.92 m^3/ha (Table 1).

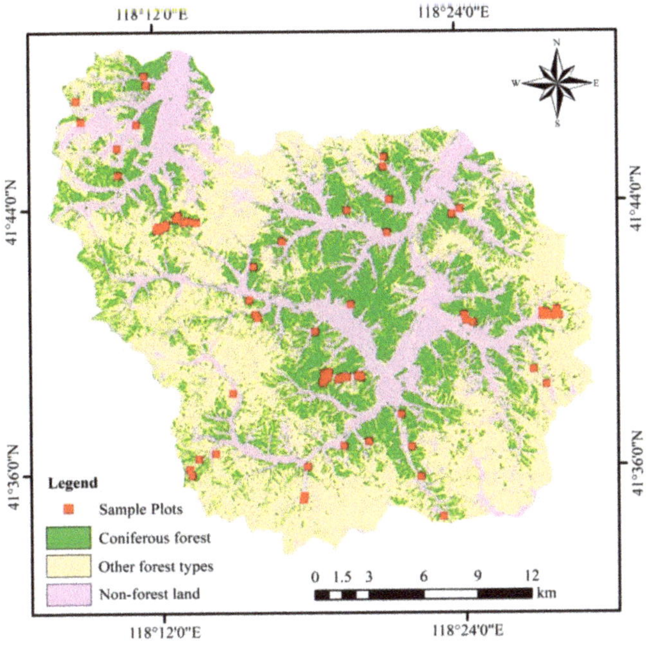

Figure 2. Land-cover types in the Wangyedian forest farm and the spatial distribution of sampled plots.

Table 1. Summary statistics of per unit growing stem volume (GSV) values (m^3/ha).

Tree Specie	Plot Number	Range of Values	Mean	Standard Deviation	Coefficient of Variation (%)
Chinese pine	43	91.97–514.96	245.52	110.59	45.0
Larch	38	86.17–466.23	217.21	91.84	42.3
Total	81	86.17–514.96	232.24	102.59	44.2

2.3. Remote Sensing Data and Preprocessing

Two Sentinel-2 multi-spectral images were downloaded as the Level-1C product covering the whole study area from the scientific data hub (https://scihub.copernicus.eu/). These images were acquired during the field investigation time on 22 September, 2017 (Table 2). The official Sen2cor module version 2.5.5 was used to transform the Level-1C product into the Level-2A product [27]. The Level-2A product is the bottom-of-atmosphere corrected reflectance after radiometric calibration and atmospheric correction. Sentinel-2 has 13 bands with three resolutions [26]. Four 10 m spatial resolution and six 20 m spatial resolution bands were used in this study to extract spectral information (Table 3). The cubic convolution interpolation method was used to resample the selected Sentinel-2 bands in order to match the pixel size with the sample plot size and acquire accurate vegetation information.

Table 2. Information of the acquired Sentinel-2 data.

Image Identification	Product Level	Acquisition Date
S2A_MSIL1C_20170922T025541_N0205_R032_T50TNM_20170922T 030440	L1C	22 September, 2017
S2A_MSIL1C_20170922T025541_N0205_R032_T50TPM_20170922T 030440	L1C	22 September, 2017

L1C: Level-1C

Table 3. Bands of the Sentinel-2 images used in the study.

Sentinel-2 Bands	Central Wavelength (nm)	Bandwidth (nm)	Spatial Resolution (m)
Band 2-Blue	496.6	98	10
Band 3-Green	560.0	45	10
Band 4-Red	664.5	38	10
Band 5-Vegetation Red Edge	703.9	19	20
Band 6-Vegetation Red Edge	740.2	18	20
Band 7-Vegetation Red Edge	782.5	28	20
Band 8-NIR	835.1	145	10
Band 8A-Vegetation Red Edge	864.8	33	20
Band 11-SWIR	1613.7	143	20
Band 12-SWIR	2202.4	242	20

2.4. Extraction and Selection of Spectral Variables and Topographic Factors

Spectral variables are the basis of GSV modeling and mapping. An appropriate spectral variable combination can markedly improve the accuracy and efficiency of modeling [39,40]. Each forest parameter has different reflection characteristics in different bands. Vegetation indices (VIs) were obtained by combining the bands, which can be used to quantitatively describe the vegetation condition. VIs have been widely used in vegetation coverage monitoring and forest parameter estimation. Additionally, the red-edge vegetation index can accurately reveal vegetation health, which correlates closely with the GSV [41–43]. Topographic factors are significantly correlated with forest cover in mountainous regions. Slope, aspect, and elevation are topographic factors commonly used. These factors are extracted from the Digital Elevation Model (DEM) raster data in the study area. Twenty feature variables were derived from the Sentinel-2 images and the DEM covering the Wangyedian forest farm, including ten multi-spectral bands, seven VIs (four common VIs and three red-edge VIs), and three topographic factors (Table 4).

Table 4. Spectral variables and topographic factors used in this study.

Feature Variable	Description	Reference
Band Reflectivity	B2-BLUE, B3-GREEN, B4-RED, B5-Red Edge1, B6-Red Edge2, B7-Red Edge3, B8-NIR, B8A-Red Edge4, B11-SWIR1, B12-SWIR2	[39]
Vegetation Index	Normalized Difference Vegetation index (NDVI)	[39]
	Enhanced Vegetation index (EVI)	[39]
	Red-green vegetation index (RGVI)	[39]
	Atmospherically resistant vegetation index (ARVI)	[39]
	Red Edge Normalized Difference Vegetation index (RENDVI)	[44]
	Red Edge Chlorophyll Index (RECI)	[45]
	Red Edge Simple Ratio (RESR)	[45]
Topographic Factor	Elevation	[46]
	Slope	[46]
	Aspect	[46]

The correlation coefficient reflects the strength of the relationship between variables. Pearson correlation coefficient was used to measure the relationship between GSV and the 20 feature variables (Figure 3). Eighteen feature variables significantly related to the GSV were selected for subsequent modeling. Overall, the Pearson correlation coefficients between VIs and GSV were greater than those of the original bands and topographic factors. The feature variables with higher correlation coefficients mainly come from the combinations of red-edge and near-infrared bands.

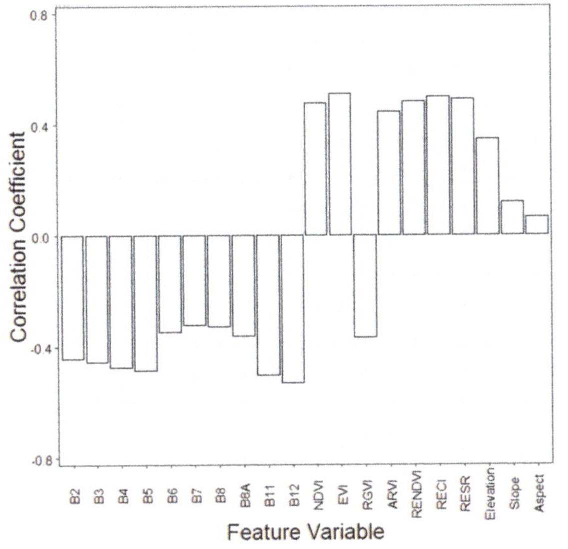

Figure 3. Correlation coefficients of the extracted feature variables with GSV.

A complex combination of feature variables leads to low modeling efficiency, resulting in exceeded calculations and suboptimal accuracy of estimation. Therefore, the feature variables that contribute most to reducing the error of a model are first selected to form variable combinations, which are then used to build the model. The appropriate combination of variables can greatly shorten the running time of the model, improve the accuracy of estimation and the interpretability of the model [38]. In order to explore the influence of combinations of feature variables on modeling, feature variables were selected to form the combinations by five methods, including LSR, RF, Boruta, VSURF, and stepwise random forest (SRF) that was proposed in this study. The feature variable combinations selected by the five methods were then used for developing GSV estimation models.

The LSR requires feature variables with a high correlation with GSV to be used in the screening process. The feature variables were introduced into the model, and the significance test was carried out one by one. Statistically significant variables (p-value < 0.05) in line with the range were retained to form the final variable combination in the LSR [47]. For each variable combination, the collinearity between the feature variables was detected to avoid the estimation distortion of the model. The variance inflation factor (VIF) was used to measure the collinearity between feature variables, and the threshold value of the VIF was set at 5. The feature variable combination formed by LSR consisted of B7, Elevation, Red-green vegetation index (RGVI), and Red Edge Normalized Difference Vegetation index (RENDVI). The statistical results of the LSR showed that all the selected variables had significant correlations with GSV, and there was no obvious collinearity between the variables (Table 5).

Table 5. Statistical results of the linear stepwise regression (LSR) method.

Variable	Coefficient	Significance	VIF
Constant	−125.48	0.03	-
B7	−1162.73	0.00	1.10
Elevation	0.29	0.00	1.07
RGVI	−111.09	0.04	1.60
RENDVI	379.13	0.00	1.65

VIF: variance inflation factor.

The random forest is based on the importance assessment of feature variables and error balance [48,49]. The importance is determined by their error contribution to the model. The Mean Decrease in Accuracy (MDA) is regarded as the indicator for measuring the importance of these feature variables [38,44]. The importance of all the feature variables was calculated and converted to percentages and by which the feature variables were ranked (Figure 4a). Multiple random forest regressions were established by gradually increasing the number of feature variables. When the relative root means square error (rRMSE) of the model reached the lowest, the variable combination formed by the number of variables was selected as the final result. Figure 4b shows that rRMSE reaches the minimum when the top 13 variables in the importance ranking form a variable combination.

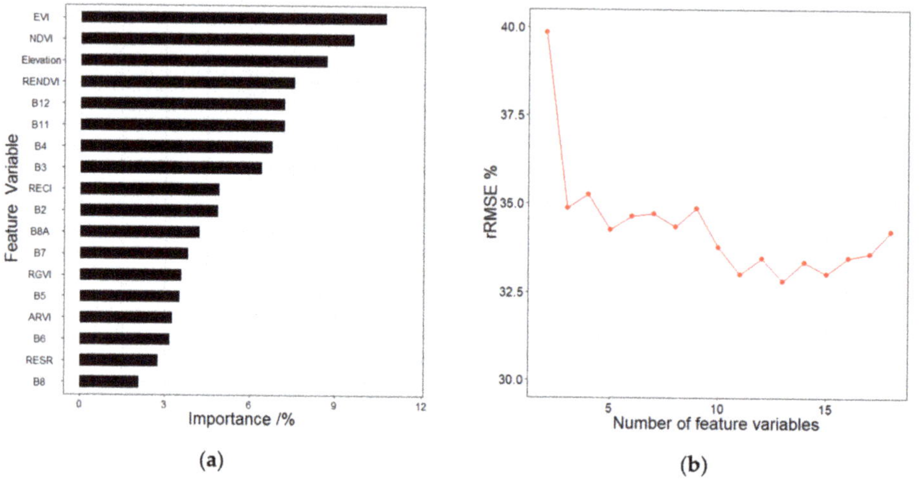

Figure 4. (**a**) The importance ranking of feature variables and (**b**) the random forest (RF) regression results (rRMSE) under different numbers of feature variables.

The RF method can evaluate the importance of a single feature variable but cannot directly provide the appropriate feature variable combinations. Different combinations of feature variables bring different rRMSE of RF regression, which can significantly affect the estimation accuracy. In order to select the appropriate combination of feature variables that can cause the smallest rRMSE, the stepwise random forest (SRF) was used to construct multiple random forest regression (RFR) models according to the importance ranking. Firstly, the feature variable with the highest importance was selected as a determining variable. Through the importance ranking, the variables were selected one by one to form the combinations with previously determined variables, and the RFR models were then respectively established. The corresponding variable combination with the smallest rRMSE was regarded as the newly determined variable combination, and the rRMSE was considered as the threshold. With the arrival of the next variable combination, if the new rRMSE was smaller than the threshold, the variable combination and threshold were then updated simultaneously. Until the variable combination and threshold were no longer updated, the variable combination with the minimum rRMSE was taken as the final selection result (Figure 5). Figure 6 shows that the lowest rRMSE of the feature variable combination consisting of EVI, B2, Elevation, B11, and B7 was 30.3%. The final feature variable combination was then used for GSV estimation. In addition, Boruta and VSURF are also used for variable selection. Finally, 16 and 3 variables are obtained for modeling and GSV estimation, respectively.

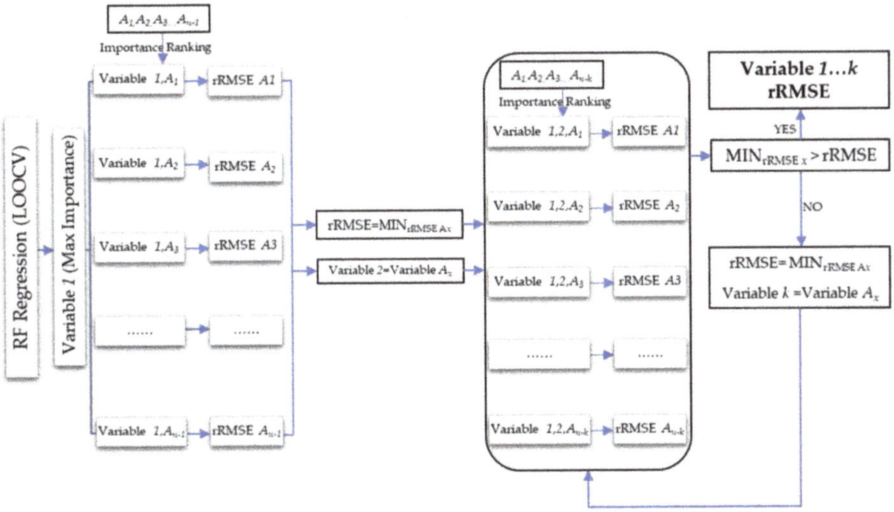

Figure 5. Feature variable selection process of the stepwise random forest (SRF) method.

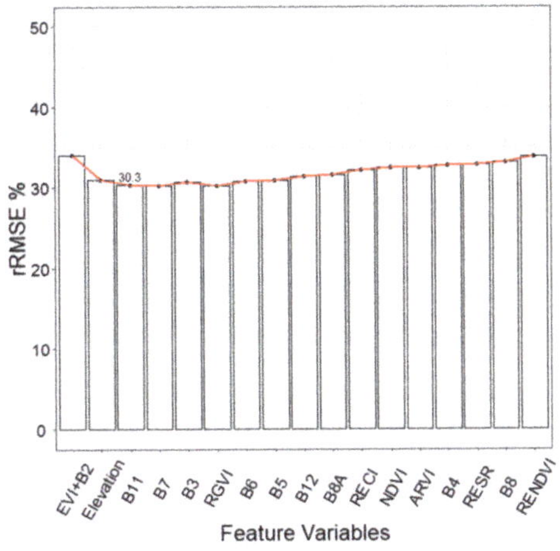

Figure 6. The SRF method applied to spectral variables for GSV estimation.

2.5. Extraction of the Texture Feature of Red-Edge Bands

The texture features can reveal the homogenous phenomenon in the image and reflect the arrangement characteristics of the surface features and structural organization of an object. Texture features have huge potential advantages for complex forest parameter estimation [47]. Due to the availability of multiple red-edge bands, the spectral information of Sentinel-2 images has proven to be very sensitive to vegetation parameters, but the effect of texture features on GSV estimation would still need to be verified. Gray-scale co-occurrence matrix (GLCM) has been widely used to extract texture information of remote sensing images [50]. In order to reduce the effect of different texture windows on texture feature values, eight texture features of red-edge bands were extracted through five texture windows (3 × 3, 5 × 5, 7 × 7, 9 × 9 and 11 × 11) [51–54]. Details of the extracted texture features are shown in Table 6.

Table 6. Texture features extracted under different texture windows.

Texture Window	Red Edge Band	Texture Metric
3 × 3, 5 × 5, 7 × 7, 9 × 9, 11 × 11	Band 5-Vegetation Red Edge 1 (RE1), Band 6-Vegetation Red Edge 2 (RE2), Band 7-Vegetation Red Edge 3 (RE3), Band 8A-Vegetation Red Edge 4 (RE4)	Mean (Men) Variance (Var) Homogeneity (Hom) Contrast (Con) Dissimilarity (Dis) Entropy (Ent) Second moment (Sec) Correlation (Cor)

Each of the texture features was defined using the corresponding window size plus the red edge band and the texture metric. For example, 5 × 5_RE1_Men meant the mean texture measure from a window of 5 × 5 and band 5-Vegetation Red Edge 1. Correlation coefficients of all extracted texture features with GSV were calculated. It was found that the correlations of some texture features (e.g., 5 × 5_RE1_Men, 7 × 7_RE1_Men, 9 × 9_RE1_Men, and 11 × 11_RE1_Men) were higher than those of all the spectral variables. The distribution trends of correlation coefficients of the texture features extracted from the four red edge bands were similar. The overall correlation coefficient of the texture features in RE1 was relatively high, and the mean derived texture features under different texture windows provided the highest correlation coefficient (Figure 7).

Together with the spectral variables and topographic factors, the importance of the texture features was evaluated. Figure 8 shows the partial importance ranking and the changing trend of errors under the RF method. Finally, a combination of eight variables with the minimum rRMSE was selected as the result. The SRF method was also used for the variable selection after adding the texture features. As the rRMSE started to decrease with the increase of the number of predictors, it finally reached the minimum when nine predictors were utilized and then continued to increase (Figure 9). With the participation of the texture features, the feature variable combinations were then selected by the LSR, RF, Boruta, VSURF, and SRF methods to establish the RFR models for estimating the GSV in the study area.

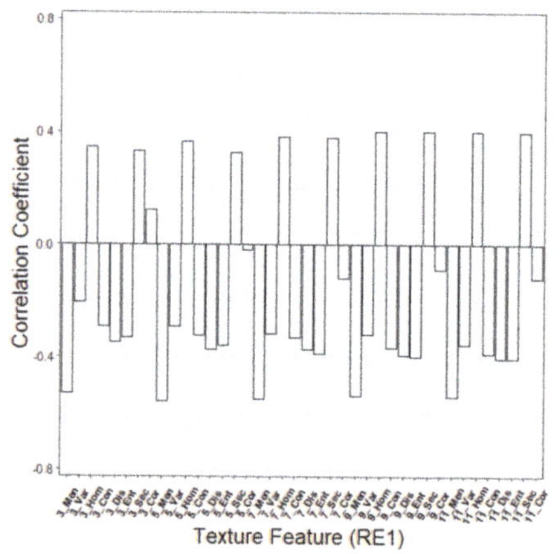

(a)

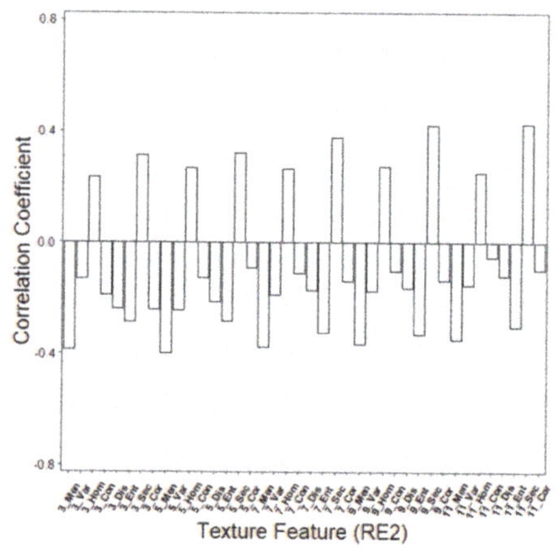

(b)

Figure 7. *Cont.*

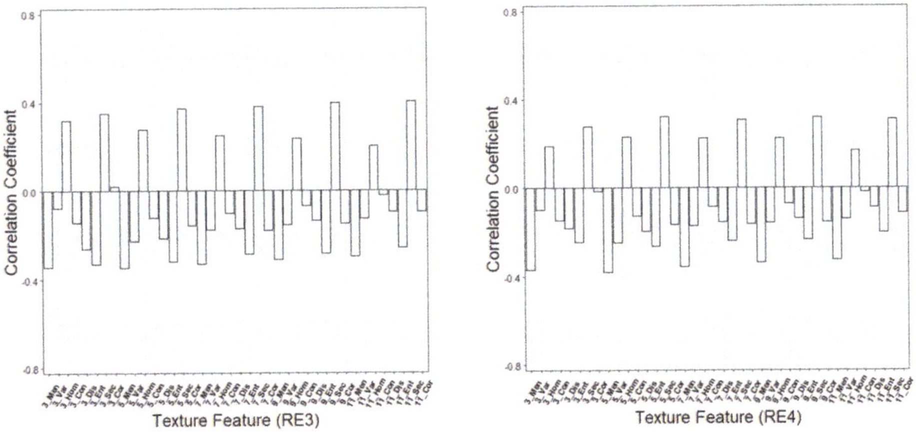

(c) (d)

Figure 7. Correlation coefficients of texture features extracted from (**a**) RE1, (**b**) RE2, (**c**) RE3 and (**d**) RE4.

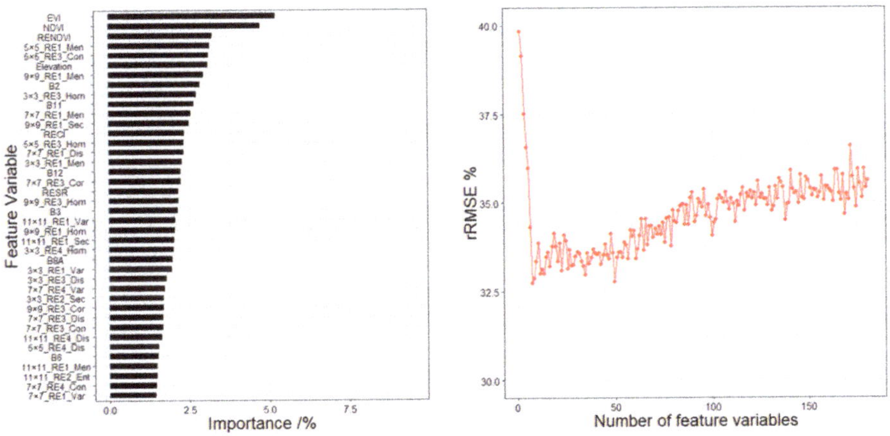

(a) (b)

Figure 8. (**a**) The partial importance ranking of spectral variables, topographic factors, and texture features, (**b**) the change trend of rRMSE under different numbers of the feature variables.

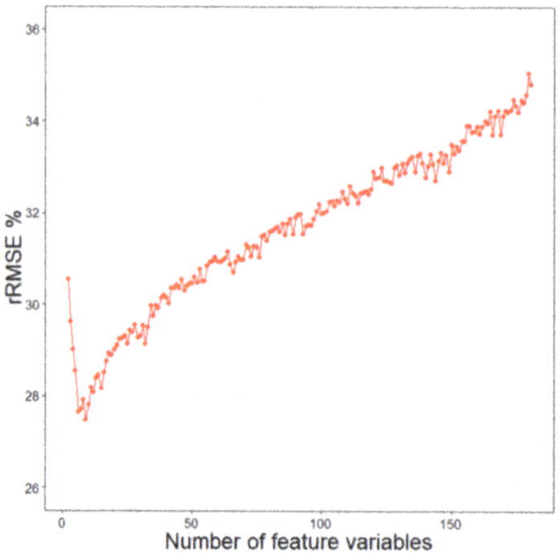

Figure 9. The change trend of rRMSE under the stepwise random forest method.

2.6. Random Forest Regression for GSV Estimation

As a non-parametric algorithm that is less sensitive to noise data, RF has been widely used in image classification and vegetation parameter estimation because it does not need the assumption of data distribution. Random forest regression can randomly generate a large number of regression trees used for estimation based on the used datasets and does not need to consider the collinearity between feature variables. It can effectively process large datasets and does not require reducing the dimension of high-dimensional data, which provides the potential for significantly improving its applicability [38,55]. Substantial research has shown that as a non-parametric algorithm, RF often performs better than the widely used parametric methods and thus has become more popular for estimation of vegetation parameters [36,37,55]. The *mtry* and *ntrees* are two important parameters that affect the model estimation accuracy. The *mtry* refers to the number of sample predictors of decision tree nodes, of which the default does not exceed the number of feature variables. And *ntrees* is the number of decision trees constructed by RF. Excessive *ntrees* will reduce the computing efficiency, and the final number can be determined based on the error change during the model construction [38]. In this study, RF modeling was therefore divided into participation with and without texture features, and then the parameter group results of the five variable combinations were determined respectively according to the error changes. The final prediction of each pixel depends on the average of the individual results of all the regression trees. The leave-one-out cross-validation (LOOCV) method [47] was applied in the RF modeling to assess the accuracy of all the predicted GSV values.

When no texture features were involved, the *mtry* of the five feature variable combinations were set as 2, 3, 1, 3, 2 and the *ntrees* were 150, 200, 200, 200 and 150, respectively. Similarly, the parameter groups (*mtry*, *ntrees*) with texture feature participation by five methods were finally determined as (2, 200), (2, 200), (7, 200), (2, 250) and (3, 200).

2.7. Accuracy Assessment of GSV Estimation

Using the GSV plot data and the feature variable combinations selected by the LSR, RF, Boruta, VSURF, and SRF methods, ten RFR models were developed to predict and map the GSV. These models were denoted as LSR-RFR, RF-RFR, and SRF-RFR. A LOOCV was used to assess the ten models [47]. The absolute residuals between the estimated values and the observations for the models were tested using the Student's *t*-test. The estimation accuracy of the models was then evaluated by R^2; Root Mean

Squared Error (RMSE), rRMSE, the mean absolute error (MAE), and standard deviation of estimation (SD_e) [56] as indicated in the formulae below

$$R^2 = 1 - \frac{\sum_{i=1}^{n}(y_i - \hat{y}_i)^2}{\sum_{i=1}^{n}(y_i - \overline{y})^2} \tag{1}$$

$$RMSE = \sqrt{\frac{\sum_{i=1}^{n}(\hat{y}_i - y_i)^2}{n}} \tag{2}$$

$$rRMSE = \frac{\sqrt{\frac{\sum_{i=1}^{n}(\hat{y}_i - y_i)^2}{n}}}{\overline{y}} \times 100\% \tag{3}$$

$$MAE = \frac{1}{n}\sum_{i=1}^{N}\left|\hat{y}_i - y_i\right| \tag{4}$$

$$SD_e = \sqrt{\frac{\sum_{i=1}^{n}(\hat{y}_i - \overline{y})^2}{n}} \tag{5}$$

where y_i is the observed GSV, $\hat{y}_i$ is the estimated GSV based on RFR, $\overline{y}$ is the mean of the observed GSV, and n is the number of sample plots. All the models and calculations were done with the R 3.5.5 software.

3. Results

3.1. GSV Estimation and Mapping

Ten models (five RFR models with and without the texture features) were developed using the observed GSV combined with the feature variable combinations selected by the LSR, RF Boruta, VSURF, and SRF. The variable numbers without texture features in models were four, 13, 16, three, and five, while the numbers of variables with texture features were five, eight, 16, nine, and nine, respectively (Table 7). The estimated results represented in Table 7 shows that there was no significant difference between the estimation accuracy of LSR-RFR and RF-RFR. However, the SRF-RFR always achieved the best estimation performance whether there were texture features or not, which attained the highest determinant coefficient (R^2 = 0.53 and 0.62) and the lowest RMSE, rRMSE, and MAE. The results implied that the SRF method achieved the best estimation effect, remaining statistically significant (Table 8). In addition, after the texture features participated in the modeling of the optimal SRF method, RMSE greatly decreased by 7.4%. In addition, the 95% confidence interval of SRF-RFR' RMSE ranges from 46.64 to 56.06 m^3/ha, and the RMSEs of other models do not fall within the range.

The scatter plots in Figure 10 show the fitting between the observed GSV and the GSV predicted by the ten models. The fitting effect of the ten models is similar, having both overestimation and underestimation values. Among them, the SRF-RFR shows to be the best fit. While the estimated values are basically distributed on both sides of the fitting line, there are also some overestimation values. Compared with the one without the texture features, the model SRF-RFR with the texture features greatly reduced the overestimations and underestimations.

Table 7. Accuracy comparison of the GSV estimates (m^3/ha) from the RFR models with various combinations selected by the LSR, RF, Boruta, VSURF, and SRF variable selection methods.

Model	Method	Variables Combination	R^2	RMSE	rRMSE (%)	MAE	SDe
Without texture features	LSR	B7, Elevation, RENDVI, RGVI	0.41	78.84	33.9	62.17	70.95
	RF	EVI, NDVI, Elevation, RENDVI, B12, B11, B4, B3, RECI, B2, B8A, B7, RGVI	0.43	77.09	33.2	61.13	72.27
	Boruta	EVI, NDVI, Elevation, RENDVI, B12, B11, B4, B3, RECI, B2, B8A, B7, RGVI, B5, B6, RESR	0.41	78.72	33.9	63.06	73.43
	VSURF	EVI, NDVI, B2	0.44	77.54	33.4	59.90	80.73
	SRF	EVI, B2, Elevation, B11, B7	0.53	70.26	30.3	56.06	71.35
With texture features	LSR	5 × 5_RE1_Men, Elevation, 11 × 11_RE2_Sec, 11 × 11_RE2_Ent, 9 × 9_RE2_Con, B7, 9 × 9_RE3_Cor	0.42	77.82	33.5	59.04	58.18
	Boruta	NDVI, EVI, B11, B12, B2, B3, B4, Elevation, RECI, RENDVI, 3 × 3_RE1_Men, 5 × 5_RE1_Men, 7 × 7_RE1_Men, 9 × 9_RE1_Men, 11 × 11_RE1_Men, 3 × 3_RE3_Hom	0.43	77.74	33.5	61.34	72.34
	RF	EVI, NDVI, RENDVI, 5 × 5_RE1_Men, 5 × 5_RE3_Con, Elevation, 9 × 9_RE1_Men, B2	0.45	76.04	32.7	59.67	73.96
	VSURF	EVI, NDVI, 5 × 5_RE1_Men, 9 × 9_RE1_Men, B12, Elevation, B2, B3, 9 × 9_RE1_Sec	0.49	72.73	31.3	58.26	73.59
	SRF	EVI, 11 × 11_RE1_Sec, B2, 5 × 5_RE1_Sec, Elevation, 11 × 11_RE3_Ent, 3 × 3_RE3_Var, 3 × 3_RE1_Hom, 9 × 9_RE4_Var	0.62	65.05	28.0	52.69	65.04

Table 8. The test results (p-values) of significant differences among the models based on the absolute residuals between the estimated and observed GSV values using student's t test.

Model	Variable Selection Method	LSR	RF	Boruta	VSURF
Without texture features	LSR				
	RF	−7.86 (0.00)			
	Boruta	−0.42 (0.67)	7.84 (0.00)		
	VSURF	0.51 (0.61)	8.38 (0.00)	0.77 (0.44)	
	SRF	2.01 (0.04)	8.63 (0.00)	2.63 (0.01)	2.13 (0.00)
With texture features	LSR				
	RF	0.08 (0.93)			
	Boruta	−0.33 (0.74)	−0.82 (0.41)		
	VSURF	0.42 (0.66)	0.65 (0.52)	1.82 (0.07)	
	SRF	2.01 (0.04)	2.40 (0.01)	2.66 (0.00)	2.00 (0.04)

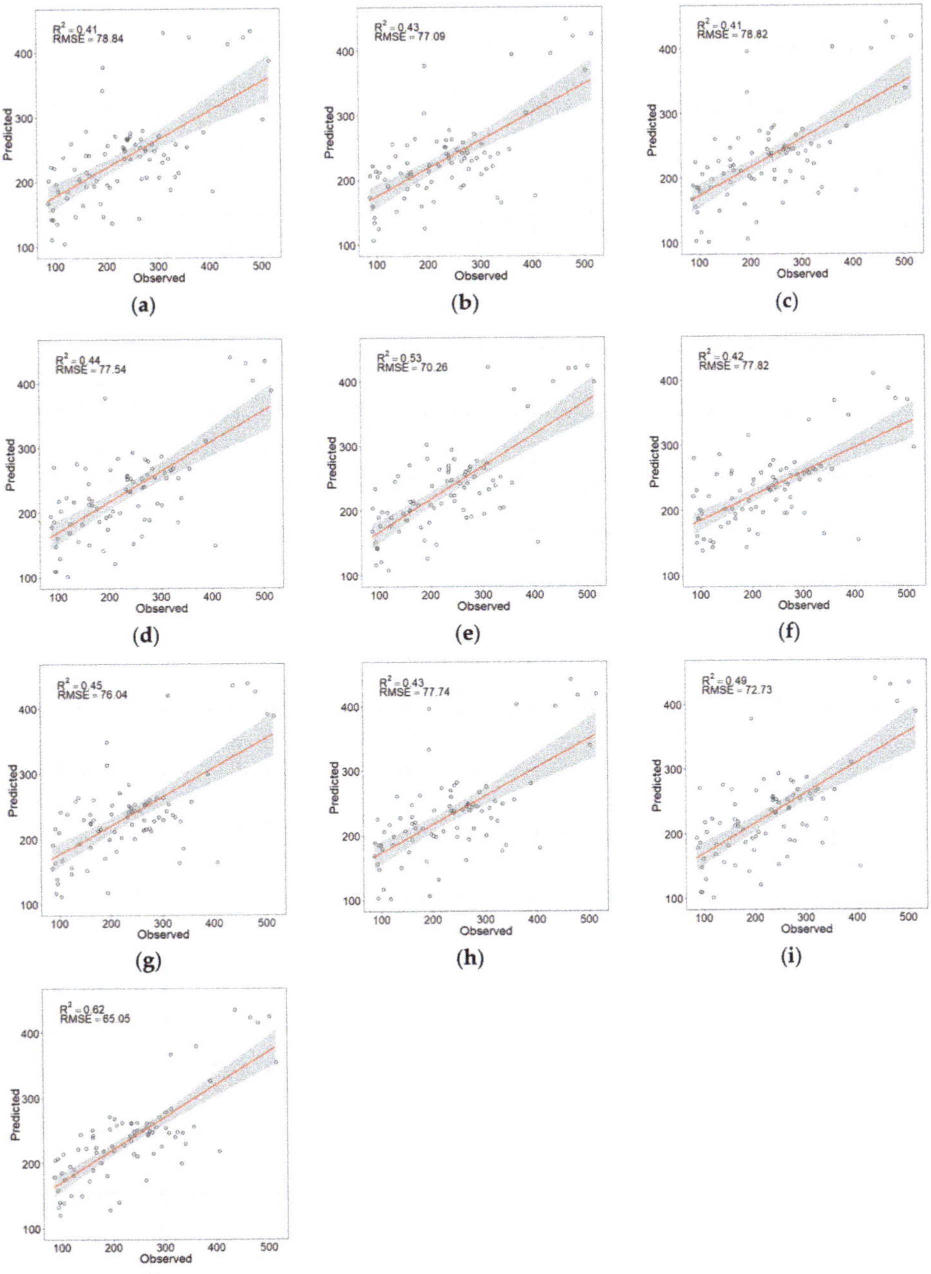

Figure 10. Scatter plots of the observed and predicted GSV values by (**a**) LSR-RFR without texture features, (**b**) RF-RFR without texture features, (**c**) Boruta-RFR without texture features, (**d**) VSURF-RFR without texture features, (**e**) SRF-RFR without texture features, (**f**) LSR-RFR with texture features, (**g**) RF-RFR with texture features, (**h**) Boruta-RFR with texture features, (**i**) VSURF-RFR with texture features, and (**j**) SRF-RFR with texture features.

The ten RFR models were separately used for mapping the GSV of the Wangyedian forest farm and led to similar spatial distributions of GSV (Figure 11). The GSV spatial distribution by the SRF-RFR model with the texture features best agrees with the actual forest distribution. The GSV values of 300–500 m^3/ha are mainly distributed in the northwest and west of the forest farm. The eastern and central areas are mainly farmlands and built areas where the distribution of GSV is negligible.

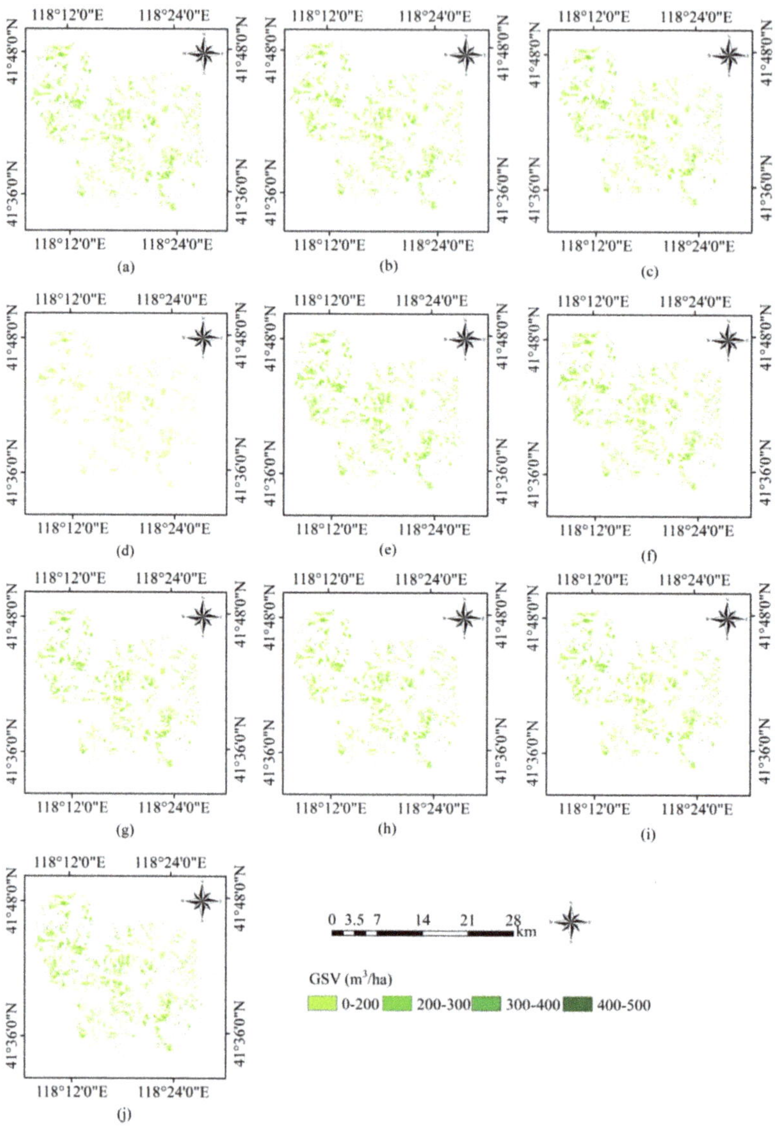

Figure 11. Spatial distributions of GSV estimates in the Wangyedian forest farm predicted by (**a**) LSR-RFR without texture features, (**b**) RF-RFR without texture features, (**c**) Boruta-RFR without texture features, (**d**) VSURF-RFR without texture features, (**e**) SRF-RFR without texture features, (**f**) LSR-RFR with texture features, (**g**) RF-RFR with texture features, (**h**) Boruta-RFR with texture features, (**i**) VSURF-RFR with texture features, and (**j**) SRF-RFR with texture features.

3.2. Uncertainty Analysis

In order to evaluate the adaptability of the SRF-RFR, the following two methods were used to analyze the uncertainty of the residuals generated by the SRF-RFR: (1) the significance of the relationship between the feature variables and the residuals; and (2) the comparison of estimates between the separate sample plot sets by tree species and the pooled sample plot set.

Our results of the correlation analysis show that there were positive and negative correlations found between the feature variables and residuals. The feature variable that possesses the highest correlation with residuals is 'Elevation' with a Pearson correlation coefficient of −0.17 ($p > 0.05$). The absolute values of all the correlations were similarly low, and not statistically significant (Table 9). In addition, the VIF values of variables determined by SRF were very low, which further improves the reliability of the model and SRF method.

Table 9. The Pearson correlation coefficients of the residuals with the associated feature variables.

Study Area	Feature Variable	Residual	VIF
Wangyedian Forest Farm	EVI	−0.09	3.222
	11 × 11_RE1_Sec	−0.04	6.270
	B2	0.12	3.244
	5 × 5_RE1_Sec	0.01	5.308
	Elevation	−0.17	1.064
	11 × 11_RE3_Ent	−0.04	6.115
	3 × 3_RE3_Var	−0.13	1.568
	3 × 3_RE1_Hom	−0.08	2.768
	9 × 9_RE4_Var	0.04	2.430

The feature variables show dissimilar importance in different sample sizes. Ranking of the feature variables was separately conducted under the Chinese pine and larch sample plot sets, and the SRF method was used to determine the final feature variable combination to establish the RFR for GSV estimation. The results show that the selected feature variables of the Chinese pine were basically the same as those for the total or pooled sample plot set; however, the results for larch were quite different from those for the total sample plot set. The correlation coefficients between the predicted GSV values from the pooled sample plot and the separate sample plot sets were 0.807 ($p < 0.01$) for the overall, 0.879 ($p < 0.01$) for Chinese pine, and 0.616 ($p < 0.01$) for larch (Figure 12). There were significant correlations for the three groups, which indicate that it is acceptable to estimate GSV whether or not the separation of the sample plots by tree species is made. The estimation accuracies from the pooled sample plot set and the separate sample plot sets by tree species were compared in Table 10. Chinese pine provided more accurate estimations with smaller rRMSE values of 25.4% and 24.9% compared to the corresponding values of larch, 31.3%, and 29.8% for the pooled sample plot set and the species separate sample plot sets, respectively.

Table 10. Feature variable results based on SRF method under two tree species (m^3/ha).

Sample Size	Tree Specie	R^2	RMSE	rRMSE (%)	MAE
Total or pooled sample plot set	Chinese pine	0.73	62.40	25.4	51.05
	Larch	0.44	67.92	31.3	54.53
	Total	0.62	65.05	28.0	56.69
Separate sample plot sets by species	Chinese pine	0.68	63.20	24.9	50.06
	Larch	0.53	65.67	29.8	49.45
	Total	0.62	64.37	27.7	49.77

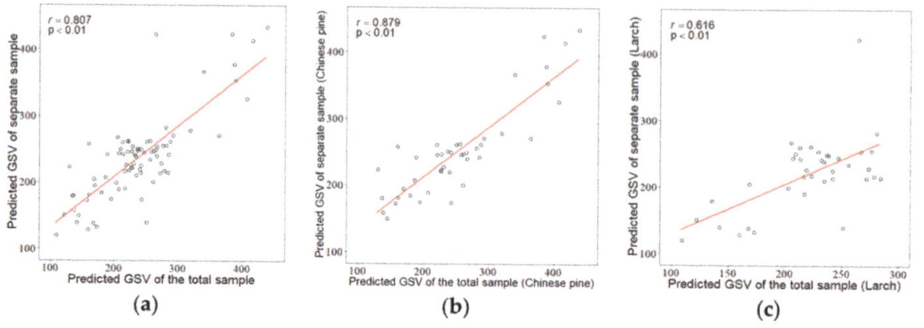

Figure 12. Scatter plots of the GSV predictions by the total or pooled sample plot set against the GSV predictions by separate sample plot sets of tree species: (**a**) all the tree species, (**b**) Chinese pine, and (**c**) Larch.

4. Discussion

4.1. Feature Variable Selection Method

Different combinations of feature variables will lead to different modeling accuracies. Choosing the appropriate variable selection method can significantly improve the GSV estimation. Original spectral bands, VIs, and topographic factors are commonly used feature variables for GSV estimation [39,57]. The predictors significantly correlated with GSV were extracted in our study and formed the feature variable combinations by five selection methods. It was found that the LSR method provided the best linear combination of feature variables, but it requires variables without collinearity. However, the relationships between GSV and feature variables may not be linear, likely due to the complexity of forest ecosystems, which limits the estimation accuracy of the model [46]. The RF method can evaluate the importance of feature variables based on non-linear relationships, allowing for the selection of feature variables with high importance for modeling [28,38]. Xie et al. [58] used an RF algorithm to measure the importance of all candidate feature variables and then selected the predictors for regional GSV prediction and mapping, and the obtained results were acceptable (R^2 = 0.618). This implies that RF algorithms can be used to select robust and stable predictors based on importance measures. The Boruta and VSURF methods [31] were compared to the SRF method. And the RMSEs obtained by the two methods were 77.74 m^3/ha and 72.73 m^3/ha, respectively, which were increased by 16.3% and 10.6% compared with SRF ($p < 0.001$). In addition, the two methods were tested in time costing, which spent 78s and 297s in 181 variables (with texture features), respectively. However, the SRF method with higher accuracy takes 89s. Compared with Boruta, the time is not significantly increased, but the accuracy of 16.3% is improved, which makes the time cost worth it.

While different from the LSR method based on linear correlation, the RF method selects variables based on their importance rankings, thus improving the selection process. However, the importance assessment is relative, where the importance of a single variable in different combinations of feature variables can also vary [32,38]. In response, we proposed the SRF method for the selection of appropriate feature variable combinations. In this method, we first used the importance ranking determined by the whole sample as a reference and then selected the appropriate combination according to the error change as the number of feature variables increased. Finally, we found that the RFR with the variable combination determined by the SRF method had the highest estimation accuracy. As a result, SRF can effectively reduce the estimation error by establishing multiple RFR models based on importance ranking and comparing the results from different variable combinations.

4.2. Uncertainty Analysis

Through the comparison of five feature variable selection methods combined with the RF model, a map showing the most accurate estimates of GSV for the planted coniferous forests in Wangyedian

Forest Farm was produced. However, we found that the estimates were associated with uncertainties. The uncertainty of GSV estimates often results from atmospheric effects, sensor effects, sample plot GSV measurement errors, feature variable selection, and estimation model [27,59]. At present, the atmospheric effects and sensor errors cannot be completely eliminated [57,59]. Also, there was a substantial variation of the GSV among the training sample plots. This suggests that the sampling design and sample size could be better improved in prospective studies.

Collecting tree-by-tree measurements requires a sampling design to represent the average level of forest cover in a given study area. Accurate measurements and the selection of appropriate volume equations can greatly determine the acquisition of true volume values [59]. Deforestation, forest fires, and natural regeneration can all affect the applicability of the volume formula [27]. It is therefore crucial for the forest survey program in Wangyedian Forest Farm to obtain updated forest resource data annually while checking and calibrating volume equations periodically [9,27]. This would help to substantially reduce the error of the plot level GSV estimates.

It was found that the combination of feature variables directly affects the accuracy of model estimation [38,57]. Too many variables tend to increase model errors and program running time. Therefore, selecting the appropriate feature variables can significantly improve the efficiency of modeling and prediction [32,38]. The commonly used LSR method is sensitive to the linear model, but the model itself is limited in GSV estimation given the complex forest structure [51]. The RF can evaluate the importance of feature variables, which helps to efficiently select the appropriate feature variables [38,49,51]. Compared with the LSR, the RF showed significant improvement to the model given that feature variables were selected according to their importance ranking. While RF can only provide the importance of a single feature variable, the importance of the same feature variable may change in different combinations of the variables [32]. On the basis of primary importance ranking, variable combinations should be considered to extract significant combinations of the variables. The SRF provides the potential to select the combinations of feature variables and can lead to the smallest error in GSV estimation. All the variable combinations selected by SRF can be tested for their significance with absolute residuals produced by the models. The variable combinations that have no significant contribution to the reduction of residual error can be eliminated. Thus, SRF provides a powerful tool for the selection of predictors and improvement of estimation accuracy.

Linear models are easy to realize and understand. However, their estimation accuracy of forest parameters in complex forest ecosystems is limited [51]. In contrast, non-parametric methods can model non-linear relationships and result in more accurate results in GSV estimation [27,35,57–59]. In our analysis, the SRF-RFR achieved the minimum rRMSE of 28.0% in the GSV estimation of Wangyedian forests, while the LSR-RFR, RF-RFR, Boruta-RFR, and VSURF-RFR resulted in the rRMSE values of 33.5%, 32.7%, 33.5%, and 31.3%, respectively. The RMSE obtained by SRF-RFR was 16.4%, 14.4%, 16.3%, and 10.6% smaller than those by LSR-RFR, RF-RFR, Boruta-RFR, and VSURF-RFR. Compared with the pooled sample plot set moreover, using the separate sample plot sets by species for the development of SRF-RFR models slightly increased the estimation accuracy of the GSV. The reason might be because separating the sample plots by tree species decreased the variation of the within model GSV. Due to the limited space, this study only dealt with the improvement of RF. In future studies, other non-parametric methods such as *k*-nearest neighbors (*k*NN), support vector machine (SVM), and artificial neural network (ANN) should be considered [59,60].

5. Conclusions

Accurate estimation of GSV is crucial for regional and global forest resource assessment and ecosystem dynamic monitoring. This study proposed the SRF, an improved RF selection method of feature variables, and compared its results with those from four widely used methods (LSR, RF, Boruta, and VSURF) for GSV estimation using Sentinel-2 and observed GSV data for the Wangyedian forest farm. The following conclusions were drawn: (1) The red-edge bands of Sentinel-2 images have more significant correlations with GSV than other used feature variables, and the red-edge bands derived feature variables have priority in terms of their contribution to the reduction of model errors. Introducing the red edge features into the GSV models greatly improves the estimation accuracy of the GSV; and (2) Compared with the LSR, RF, Boruta, and VSURF methods, the SRF performed best in the selection of feature variables, and the SRF-RFR led to the smallest rRMSE of the GSV predictions. Compared with the LSR-RFR, RF-RFR, Boruta-RFR and VSURF-RFR, the SRF-RFR model reduced the RMSE by 16.4%, 14.4%, 16.3% and 10.6%, respectively. Thus, the SRF-RFR method offers the potential for improving the estimation accuracy of the GSV and provides a reference for forest dynamic monitoring.

Author Contributions: All authors read the manuscript; Conceptualization, H.S. and G.W.; methodology, H.S. and G.W.; validation, F.J. and H.L.; formal analysis, F.J.; investigation, F.J., J.L., and H.S., and H.L.; draft, F.J., H.S., and M.K.; supervision, H.S.; review, editing and revision, H.S., G.W., and A.J.S.; funding acquisition, H.S., G.W. All authors have read and agreed to the published version of the manuscript.

Funding: This research was funded by the National Key Research and Development Program of China project "Research of Key Technologies for Monitoring Forest Plantation Resources", grant number 2017YFD0600900; National Natural Science Foundation of China, grant number 31971578; Scientific Research Fund of Hunan Provincial Education Department, grant number 17A225; Forestry Administration of Hunan Province, grant number XLK201986; Training Fund of Young Professors from Hunan Provincial Education Department, grant number 90102-7070220090001.

Conflicts of Interest: The authors declare no conflict of interest.

References

1. Gower, S.T. Patterns and mechanisms of the forest carbon cycle. *Ann. Rev. Environ. Resour.* **2003**, *28*, 169–204. [CrossRef]
2. Pedro, M.S.; Rammer, W.; Seidl, R. Tree species diversity mitigates disturbance impacts on the forest carbon cycle. *Oecologia* **2015**, *177*, 619–630. [CrossRef]
3. Zhang, H.; Zhu, J.; Wang, C.; Lin, H.; Long, J.; Zhao, L.; Fu, H.; Liu, Z. Forest Growing Stock Volume Estimation in Subtropical Mountain Areas Using PALSAR-2 L-Band PolSAR Data. *Forests* **2019**, *10*, 276. [CrossRef]
4. Jung, J.; Kim, S.; Hong, S.; Kim, K.; Kim, E.; Im, J.; Heo, J. Effects of national forest inventory plot location error on forest carbon stock estimation using k-nearest neighbor algorithm. *ISPRS J. Photogramm. Remote Sens.* **2013**, *81*, 82–92. [CrossRef]
5. Condés, S.; McRoberts, R.E. Updating national forest inventory estimates of growing stock volume using hybrid inference. *For. Ecol. Manag.* **2017**, *400*, 48–57. [CrossRef]
6. Di Cosmo, L.; Gasparini, P.; Tabacchi, G. A national-scale, stand-level model to predict total above-ground tree biomass from growing stock volume. *For. Ecol. Manag.* **2016**, *361*, 269–276. [CrossRef]
7. Brigot, G.; Colin-Koeniguer, E.; Plyer, A.; Janez, F. Adaptation and Evaluation of an Optical Flow Method Applied to Coregistration of Forest Remote Sensing Images. *IEEE J. Sel. Top. Appl. Earth Obs. Remote Sens.* **2016**, *9*, 2923–2939. [CrossRef]
8. Klinge, M.; Böhner, J.; Erasmi, S. Modelling forest lines and forest distribution patterns with remote sensing data in a mountainous region of semi-arid Central Asia. *Biogeosci. Discuss.* **2014**, *11*, 14667–14698. [CrossRef]
9. Tang, X.; Fehrmann, L.; Guan, F.; Forrester, D.I.; Guisasola, R.; Kleinn, C. Inventory-based estimation of forest biomass in Shitai County, China: A comparison of five methods. *Ann. For. Res.* **2016**, *59*, 269–280. [CrossRef]
10. Isaacson, B.N.; SerbiniD, S.; Townsend, P.A. Detection of relative differences in phenology of forest species using Landsat and MODIS. *Landsc. Ecol.* **2012**, *27*, 529–543. [CrossRef]
11. Amiro, B.D.; Chen, J.M. Forest-fire-scar aging using SPOT-VEGETATION for Canadian ecoregions. *Can. J. For. Res.* **2003**, *33*, 1116–1125. [CrossRef]
12. Guo, X.-Y.; Zhang, H.-Y.; Wu, Z.; Zhao, J.-J.; Zhang, Z.-X. Comparison and Evaluation of Annual NDVI Time Series in China Derived from the NOAA AVHRR LTDR and Terra MODIS MOD13C1 Products. *Sensors* **2017**, *17*, 1298. [CrossRef]
13. Liu, Q.S.; Tang, Y.; Qin, L.; Dai, H.S.; Zhao, Q.C.; Yang, Y. On-board radiometric calibration for thermal emission band of FY-3C/MERSI. *Int. J. Remote Sens.* **2019**, *40*, 1–14. [CrossRef]

14. Pu, R.; Gong, P.; Yu, Q. Comparative Analysis of EO-1 ALI and Hyperion, and Landsat ETM+ Data for Mapping Forest Crown Closure and Leaf Area Index. *Sensors* **2008**, *8*, 3744–3766. [CrossRef] [PubMed]
15. Jacquemoud, S. Inversion of the PROSPECT + SAIL canopy reflectance model from AVIRIS equivalent spectra: Theoretical study. *Remote Sens. Environ.* **1993**, *44*, 281–292. [CrossRef]
16. Zhong, Y.; Zhao, J.; Zhang, L. A Hybrid Object-Oriented Conditional Random Field Classification Framework for High Spatial Resolution Remote Sensing Imagery. *IEEE Trans. Geosci. Remote Sens.* **2014**, *52*, 7023–7037. [CrossRef]
17. Chrysafis, I.; Mallinis, G.; Gitas, I.Z.; Tsakiri-Strati, M. Estimating Mediterranean forest parameters using multi seasonal Landsat 8 OLI imagery and an ensemble learning method. *Remote Sens. Environ.* **2017**, *199*, 154–166. [CrossRef]
18. Chaozong, X.; Yuxing, Z.; Wei, W. A relief-based forest cover change extraction using GF-1 images. *IEEE Geosci. Remote Sens. Symp.* **2014**, 4212–4215. [CrossRef]
19. Inman, D.; Khosla, R.; Reich, R.M.; Westfall, D.G. Active remote sensing and grain yield in irrigated maize. *Precis. Agric.* **2007**, *8*, 241–252. [CrossRef]
20. Lin, C.; Thomson, G.; Popescu, S. An IPCC-Compliant Technique for Forest Carbon Stock Assessment Using Airborne LiDAR-Derived Tree Metrics and Competition Index. *Remote Sens.* **2016**, *8*, 528. [CrossRef]
21. Hakkenberg, C.R.; Zhu, K.; Peet, R.K.; Song, C. Mapping multi-scale vascular plant richness in a forest landscape with integrated LiDAR and hyperspectral remote-sensing. *Ecology* **2018**, *99*, 474–487. [CrossRef] [PubMed]
22. Thiel, C.; Schmullius, C. The potential of ALOS PALSAR backscatter and InSAR coherence for forest growing stock volume estimation in Central Siberia. *Remote Sens. Environ.* **2016**, *173*, 258–273. [CrossRef]
23. Mahdianpari, M.; Salehi, B.; Mohammadimanesh, F.; Motagh, M. Random forest wetland classification using ALOS-2 L-band, RADARSAT-2 C-band, and TerraSAR-X imagery. *ISPRS J. Photogramm. Remote Sens.* **2017**, *130*, 13–31. [CrossRef]
24. Majasalmi, T.; Rautiainen, M. The potential of Sentinel-2 data for estimating biophysical variables in a boreal forest: A simulation study. *Remote Sens. Lett.* **2016**, *7*, 427–436. [CrossRef]
25. Mura, M.; Bottalico, F.; Giannetti, F.; Bertani, R.; Giannini, R.; Mancini, M.; Orlandini, S.; Travaglini, D.; Chirici, G. Exploiting the capabilities of the Sentinel-2 multi spectral instrument for predicting growing stock volume in forest ecosystems. *Int. J. Appl. Earth Obs. Geoinf.* **2018**, *66*, 126–134. [CrossRef]
26. Chrysafis, I.; Mallinis, G.; Siachalou, S.; Patias, P. Assessing the relationships between growing stock volume and Sentinel-2 imagery in a Mediterranean forest ecosystem. *Remote Sens. Lett.* **2017**, *8*, 508–517. [CrossRef]
27. Hu, Y.; Xu, X.; Wu, F.; Sun, Z.; Xia, H.; Meng, Q.; Huang, W.; Zhou, H.; Gao, J.; Li, W.; et al. Estimating Forest Stock Volume in Hunan Province, China, by Integrating in Situ Plot Data, Sentinel-2 Images, and Linear and Machine Learning Regression Models. *Remote Sens.* **2020**, *12*, 186. [CrossRef]
28. Chen, Y.; Li, L.; Lu, D.; Li, D. Exploring Bamboo Forest Aboveground Biomass Estimation Using Sentinel-2 Data. *Remote Sens.* **2018**, *11*, 7. [CrossRef]
29. Gao, Y.; Lu, D.; Li, G.; Wang, G.; Chen, Q.; Liu, L.; Li, D. Comparative Analysis of Modeling Algorithms for Forest Aboveground Biomass Estimation in a Subtropical Region. *Remote Sens.* **2018**, *10*, 627. [CrossRef]
30. Yu, X.; Ge, H.; Lu, D.; Zhang, M.; Lai, Z.; Yao, R. Comparative Study on Variable Selection Approaches in Establishment of Remote Sensing Model for Forest Biomass Estimation. *Remote Sens.* **2019**, *11*, 1437. [CrossRef]
31. Speiser, J.L.; Miller, M.E.; Tooze, J.A.; Ip, E.H. A comparison of random forest variable selection methods for classification prediction modeling. *Expert Syst. Appl.* **2019**, *134*, 93–101. [CrossRef] [PubMed]
32. Jiang, F.; Smith, A.R.; Kutia, M.; Wang, G.; Liu, H.; Sun, H. A Modified KNN Method for Mapping the Leaf Area Index in Arid and Semi-Arid Areas of China. *Remote Sens.* **2020**, *12*, 1884. [CrossRef]
33. McRoberts, R.E.; Liknes, G.C.; Domke, G.M. Using a remote sensing-based, percent tree cover map to enhance forest inventory estimation. *For. Ecol. Manag.* **2014**, *331*, 12–18. [CrossRef]
34. Marselis, S.M.; Yebra, M.; Jovanovic, T.; Van Dijk, A.I. Deriving comprehensive forest structure information from mobile laser scanning observations using automated point cloud classification. *Environ. Model. Softw.* **2016**, *82*, 142–151. [CrossRef]
35. García-Gutiérrez, J.; Martínez-álvarez, F.; Troncoso, A.; Riquelme, J.C. A comparison of machine learning regression techniques for lidar-derived estimation of forest variables. *Neurocomputing* **2015**, *167*, 24–31. [CrossRef]

36. Chirici, G.; Barbati, A.; Corona, P.; Marchetti, M.; Travaglini, D.; Maselli, F.; Bertini, R. Non-parametric and parametric methods using satellite images for estimating growing stock volume in alpine and Mediterranean forest ecosystems. *Remote Sens. Environ.* **2008**, *112*, 2686–2700. [CrossRef]
37. Wu, J.; Yao, W.; Choi, S.; Park, T.; Myneni, R.B. A Comparative Study of Predicting DBH and Stem Volume of Individual Trees in a Temperate Forest Using Airborne Waveform LiDAR. *IEEE Geosci. Remote Sens. Lett.* **2015**, *12*, 2267–2271. [CrossRef]
38. Breiman, L. Random forests. *Mach. Learn.* **2001**, *45*, 5–32. [CrossRef]
39. Haapanen, R.; Tuominen, S. Data Combination and Feature Selection for Multi-source Forest Inventory. *Photogramm. Eng. Remote Sens.* **2008**, *74*, 869–880. [CrossRef]
40. Eitel, J.U.H.; Vierling, L.A.; Litvak, M.E.; Long, D.S.; Schulthess, U.; Ager, A.A.; Krofcheck, D.J.; Stoscheck, L. Broadband, red-edge information from satellites improves early stress detection in a New Mexico conifer woodland. *Remote Sens. Environ.* **2011**, *115*, 3640–3646. [CrossRef]
41. Candra, E.D.; Hartono; Wicaksono, P. *Above Ground Carbon Stock Estimates of Mangrove Forest Using Worldview-2 Imagery in Teluk Benoa, Bali*; IOP Publishing: Bristol, UK, 2016; Volume 47, p. 12014.
42. Zarco-Tejada, P.; Hornero, A.; Hernández-Clemente, R.; Beck, P. Understanding the temporal dimension of the red-edge spectral region for forest decline detection using high-resolution hyperspectral and Sentinel-2a imagery. *ISPRS J. Photogramm. Remote Sens.* **2018**, *137*, 134–148. [CrossRef] [PubMed]
43. A Sims, D.; Gamon, J.A. Relationships between leaf pigment content and spectral reflectance across a wide range of species, leaf structures and developmental stages. *Remote Sens. Environ.* **2002**, *81*, 337–354. [CrossRef]
44. Gitelson, A.A.; Viña, A.; Arkebauer, T.J.; Rundquist, D.; Keydan, G.; Leavitt, B. Remote estimation of leaf area index and green leaf biomass in maize canopies. *Geophys. Res. Lett.* **2003**, *30*. [CrossRef]
45. Wallis, C.I.; Homeier, J.; Peña, J.; Brandl, R.; Farwig, N.; Bendix, J. Modeling tropical montane forest biomass, productivity and canopy traits with multispectral remote sensing data. *Remote Sens. Environ.* **2019**, *225*, 77–92. [CrossRef]
46. Li, Y.; Li, C.; Li, M.; Liu, Z. Influence of Variable Selection and Forest Type on Forest Aboveground Biomass Estimation Using Machine Learning Algorithms. *Forest* **2019**, *10*, 1073. [CrossRef]
47. Zheng, S.; Cao, C.; Dang, Y.; Xiang, H.; Zhao, J.; Zhang, Y.; Wang, X.; Guo, H. Retrieval of forest growing stock volume by two different methods using Landsat TM images. *Int. J. Remote Sens.* **2014**, *35*, 29–43. [CrossRef]
48. Myroniuk, V.; Kutia, M.; Sarkissian, A.J.; Bilous, A.M.; Liu, S. Regional-Scale Forest Mapping over Fragmented Landscapes Using Global Forest Products and Landsat Time Series Classification. *Remote Sens.* **2020**, *12*, 187. [CrossRef]
49. Chelgani, S.C.; Matin, S.; Makaremi, S. Modeling of free swelling index based on variable importance measurements of parent coal properties by random forest method. *Measurement* **2016**, *94*, 416–422. [CrossRef]
50. Cutler, M.E.J.; Boyd, D.S.; Foody, G.M.; Vetrivel, A. Estimating tropical forest biomass with a combination of SAR image texture and Landsat TM data: An assessment of predictions between regions. *ISPRS J. Photogramm. Remote Sens.* **2012**, *70*, 66–77. [CrossRef]
51. Li, Y.; Han, N.; Li, X.; Du, H.; Mao, F.; Cui, L.; Liu, T.; Xing, L. Spatiotemporal Estimation of Bamboo Forest Aboveground Carbon Storage Based on Landsat Data in Zhejiang, China. *Remote Sens.* **2018**, *10*, 898. [CrossRef]
52. Liu, J.; Pattey, E.; Jégo, G. Assessment of vegetation indices for regional crop green LAI estimation from Landsat images over multiple growing seasons. *Remote Sens. Environ.* **2012**, *123*, 347–358. [CrossRef]
53. Calvão, T.; Palmeirim, J.M. Mapping Mediterranean scrub with satellite imagery: Biomass estimation and spectral behaviour. *Int. J. Remote Sens.* **2004**, *25*, 3113–3126. [CrossRef]
54. Zhang, M.; Du, H.; Zhou, G.; Li, X.; Mao, F.; Dong, L.; Zheng, J.; Liu, H.; Huang, Z.; He, S. Estimating Forest Aboveground Carbon Storage in Hang-Jia-Hu Using Landsat TM/OLI Data and Random Forest Model. *Forest* **2019**, *10*, 1004. [CrossRef]
55. Li, C.; Li, Y.; Li, M. Improving Forest Aboveground Biomass (AGB) Estimation by Incorporating Crown Density and Using Landsat 8 OLI Images of a Subtropical Forest in Western Hunan in Central China. *Forest* **2019**, *10*, 104. [CrossRef]
56. Willmott, C.J.; Ackleson, S.G.; Davis, R.E.; Feddema, J.J.; Klink, K.M.; LeGates, D.R.; O'Donnell, J.; Rowe, C.M. Statistics for the evaluation and comparison of models. *J. Geophys. Res. Space Phys.* **1985**, *90*, 8995–9005. [CrossRef]

57. Li, X.; Liu, Z.; Lin, H.; Wang, G.; Sun, H.; Long, J.; Zhang, M. Estimating the Growing Stem Volume of Chinese Pine and Larch Plantations based on Fused Optical Data Using an Improved Variable Screening Method and Stacking Algorithm. *Remote Sens.* **2020**, *12*, 871. [CrossRef]
58. Xie, B.; Cao, C.; Xu, M.; Bashir, B.; Singh, R.P.; Huang, Z.; Lin, X. Regional Forest Volume Estimation by Expanding LiDAR Samples Using Multi-Sensor Satellite Data. *Remote Sens.* **2020**, *12*, 360. [CrossRef]
59. Lu, D.; Chen, Q.; Wang, G.; Liu, L.; Li, G.; Moran, E.F. A survey of remote sensing-based aboveground biomass estimation methods in forest ecosystems. *Int. J. Digit. Earth* **2016**, *9*, 63–105. [CrossRef]
60. Zhou, J.-J.; Zhou, Z.; Zhao, Q.; Han, Z.; Wang, P.; Xu, J.; Dian, Y. Evaluation of Different Algorithms for Estimating the Growing Stock Volume of Pinus massoniana Plantations Using Spectral and Spatial Information from a SPOT6 Image. *Forest* **2020**, *11*, 540. [CrossRef]

sensors

MDPI

Article

Classification of Variable Foundation Properties Based on Vehicle–Pavement–Foundation Interaction Dynamics

Robin Eunju Kim

Department of Civil and Environmental Engineering, Hanyang University, Seoul 04763, Korea; robinekim@hanyang.ac.kr; Tel.: +82-2220-0413

Received: 15 October 2020; Accepted: 31 October 2020; Published: 3 November 2020

Abstract: The dynamic interaction between vehicle, roughness, and foundation is a fundamental problem in road management and also a complex problem, with their coupled and nonlinear behavior. Thus, in this study, the vehicle–pavement–foundation interaction model was formulated to incorporate the mass inertia of the vehicle, stochastic roughness, and non-uniform and deformable foundation. Herein, a quarter-car model was considered, a filtered white noise model was formulated to represent the road roughness, and a two-layered foundation was employed to simulate the road structure. To represent the non-uniform foundation, stiffness and damping coefficients were assumed to vary either in a linear or in a quadratic manner. Subsequently, an augmented state-space representation was formulated for the entire system. The time-varying equation governing the covariance of the response was solved to examine the vehicle response, subject to various foundation properties. Finally, a linear discriminant analysis method was employed for classifying the foundation types. The performance of the classifier was validated by test sets, which contained 100 cases for each foundation type. The results showed an accuracy of over 90%, indicating that the machine learning-based classification of the foundation had the potential of using vehicle responses in road managements.

Keywords: machine learning-based classification; non-uniform foundation; stochastic analysis; vehicle–pavement–foundation interaction

---

1. Introduction

Road infrastructure forms a basic component in transportation, providing connectivity between local, regional and global value chains. Despite the impacts of road's serviceability on the economy and public safety, maintenance is inadequate, due to its extensive nature. For example, the Federal Highway Administration reported that 26 percent of major urban roads in the U.S. are in a poor condition [1]. FHWA also reported that a capital of 182 billion dollars was spent in 2008 on improvements and maintenance of federal highway, while they are still in shortage [1]. Thus, research that develops tools and methods for assessing road conditions assume greater importance.

Typically, a pavement's condition is assessed by measured information, such as ride comfort, surface defects, and structural adequacy [2]. For example, the Pavement Condition Index (PCI), developed by the U.S. Army Corps of Engineers rates the surface operational conditions including rutting, potholes, crackings, etc. [3]. Recent advances in image processing and deep learning technologies demonstrated that an automatic rating of PCI using a visual platform is available. For example, the automated pavement management system equipped with visual inspecting tools were developed to evaluate the pavement deteriorations and cracking [4,5]. Multimetric sensors including wireless sensing modules are utilized as well to examine the surface condition [6,7], fatigue [8], and certain anomalies on road structures [9]. However, unlike visible surface defects, structural adequacy is associated with the load transfer capability of the subgrade layers.

The falling weight deflectometer (FWD) is an essential non-destructive tool that is widely used for evaluating the structural adequacy of the pavement [10]. In FWD, as the weight falls on the pavement to be examined at some height, the response of the pavement, including deflection, is measured. Then, the responses are related to the strain and elasticity of the pavement, to examine the adequacy of the sublayer [11]. However, due to complexities in the testing method, which is usually performed during post-processing of the collected data, and due to required technical expertise, significant time, and costs, the network-level application is limited [12]. With the advancement of the wireless module's sensing and calculating capability, the inspection and monitoring fields are in the transition from human-oriented inspection to machine-based inspection [13]. The following literature shows some successful examples of monitoring foundation noise excitation [14], decentralized road networks [15], which are known to be complicated, compared to other applications. Thus, so far, predicting the capability of road structure with rather portable and automated devices are of interest, but a challenging task, due to its complicated mechanisms.

Within a road structure, the main excitation source is a moving vehicle. To understand the responses of the moving loads, various foundation types were examined, based on analytical models. One of the simplest model was developed to examine vehicle response due to road roughness on a non-deformable foundation. Roughness was first modeled as Gaussian random signals [13]. Then, the models improved to contain a more realistic input, such as a stationary zero-mean process with a certain power spectral density (PSD) [16–18]. Among various research works, Wedig derived a closed-form expression of the covariance response of a vehicle model, by integrating the PSD of road roughness [19]. These models examined the impact of road roughness on vehicle responses, while the interactions due to pavement deflection were neglected.

To consider the deformable foundation, an Euler-Bernoulli beam resting on the viscoelastic foundation was investigated. Hardy and Cebon (1993) developed a quarter-car model on a smooth beam on a uniform Winkler foundation, to examine the vehicle–pavement–foundation interaction [20]. Similar approaches were adopted by other researchers and they used the models to understand the impact of vehicle parameters (including speed) on foundation responses [18,21]. Kelvin foundation under the Bernoulli beam was also adopted by authors in [22]. In their model, the interaction responses were examined by coupling the solutions of two systems—(1) vehicle on rough road and (2) elastic foundation subject to a single load. To eliminate the boundary condition effects, the frequency domain analysis of the interaction problem was performed on an infinite length beam [23]. Instead of handling infinite length, Kim et al. (2019) formulated the interaction system, based on a moving coordinate system, and examined the second-order stationary response of the interaction problem [24]. In the aforementioned studies, the foundation properties such as stiffness and density were assumed to be uniform, while in reality, those quantities might vary along the length of the road.

The non-uniform foundation on a beam was investigated by several groups of researchers. Early efforts focused on formulating a closed-form equation for varying foundation modulus, targeting statistical analyses. The linearly varying solutions were presented by Franklin and Scott (1979) [25] and higher-order variations were solved by the authors in [26,27]. The free vibration of the beams on the non-uniform foundation was studied by the following authors [28–30]. The authors in [30] compared the impact of nonlinear foundation on the deflection shapes and natural frequencies of the beam. Then, dynamic responses of a beam on the variable Winkler foundation, subject to a moving load, were studied by [31–33], and a moving mass was investigated by [34]. Although previous studies captured the effect of variable foundation on the pavement system, due to computational complexities, studies mostly neglected the inertial force effect from the moving vehicles.

In this study, the impact of the non-uniform foundation on vehicle responses was solved by developing the vehicle–pavement–(non-uniform)-foundation interaction model. In the model, the vehicle was represented with a moving-oscillator (a quarter-car). The pavement roughness was described with a filtered white noise model. The rigid foundation was modeled to have a finite-length Euler–Bernoulli beam on a deformable foundation. The top layer was modeled using the assumed modes method. The subgrade was modeled with a Winkler-type foundation, in which the stiffness and damping properties varied along the length. The interaction model was then formulated in an augmented state-space representation. To effectively examine the response of the vehicle, the covariance of the response was then solved for the time-varying Lyapunov equation. Then, the equations were solved for various pavement roughness and foundation cases, to construct the covariance responses. Based on the estimated responses, six features that could distinguish the foundation types were selected and employed on a classifier. Subsequently, noise-added responses were employed on a linear classifier and demonstrated that the measured dynamics of a vehicle due to interaction could distinguish the foundation types and variations with an accuracy of over 90%.

2. Model Formulation

2.1. Overview

The vehicle–pavement–foundation interaction model considered herein is shown in Figure 1. The rigid pavement was modeled with an Euler–Bernoulli beam that had constant material properties. Elastic modulus (E), the moment of inertia (I), thickness (t_b), cross-sectional area (A), density (ρ), and length (L). The vertical displacement of the beam due to interaction was defined as $u_B(x,t)$. The elastic foundation was taken as a Winkler-type foundation with varying stiffness ($k_f(x)$) and viscous damping ($c_f(x)$), along the length. The roughness of the pavement was modeled as a profile $\xi(x)$ and superimposed on top of the beam.

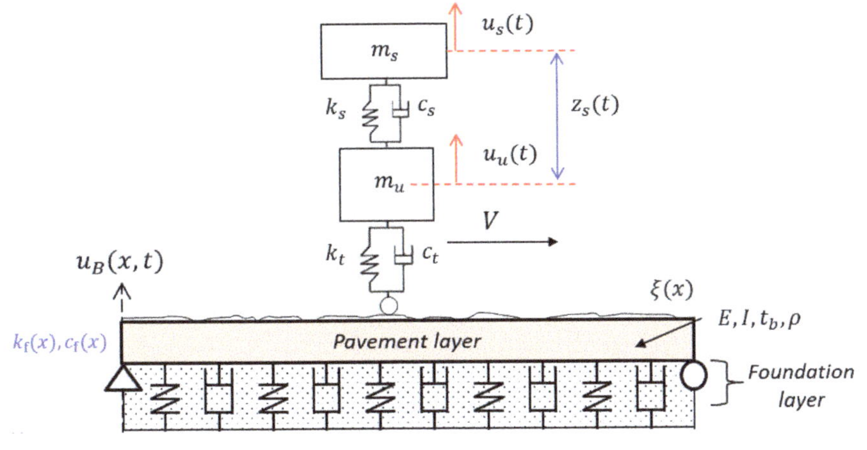

Figure 1. Vehicle–pavement–foundation system.

The vehicle was represented with a quarter-car model, consisting of sprung mass (m_s) and unsprung mass (m_u). Their vertical movement was defined as u_s and u_u, while spring stiffness and damping coefficients at suspension and tire were denoted with k_s, k_t, c_s, and c_t, respectively. The vehicle was assumed to have constant velocity (V), as it traveled along the length. In the subsequent sections, a model formulation of the interaction system was introduced. Note that some derivations such as a state-space representation of the roughness were briefly discussed, while more detailed formulations could be found in the related literature [24,35,36].

2.2. Basic Equations

Employing the assumed modes method, the vertical deflection of the Euler–Bernoulli beam $u_B(x,t)$ could be defined in a series of sine functions, assuming a simple support boundary condition:

$$\phi(n) = \sin\left(\frac{n\pi(x+L)}{L}\right),\ n = 1,\ 2,\ 3,\ \ldots N \tag{1}$$

where N is the total number of modes in the shape function. Then, the deflection of the beam could be rewritten as $u_B(x,t) = \mathbf{N}_B(x)q_B(t)$. $\mathbf{N}_B(x)$ is a mode shape vector containing defined mode shapes $(\phi(n))$ and $q_B(t)$ is a time-dependent generalized displacement of the beam. The relationships for the first- and second-time derivative are $\dot{u}_B(x,t) = \mathbf{N}_B(x)\dot{q}_B(t)$ and $\ddot{u}_B(\eta,t) = \mathbf{N}_B(x)\ddot{q}_B(t)$.

Then, defining the displacement vector as $x_c(t) = \begin{bmatrix} q_B(t) & u_u & z_s \end{bmatrix}^T$, the equations of motion for the vehicle–pavement–foundation interaction system could be formulated as follows:

$$\mathbf{M}_c\ddot{x}_c(t) + \mathbf{C}_c\dot{x}_c(t) + \mathbf{K}_c x_c(t) = -\boldsymbol{P}_g + \mathbf{P}_c\xi(t) \tag{2}$$

where

$$\mathbf{M}_c = \begin{bmatrix} \left[\int_0^L \rho A\mathbf{N}_B^T\mathbf{N}_B dx\right]_{N\times N} & [\mathbf{0}]_{N\times 1} & [\mathbf{0}]_{N\times 1} \\ [\mathbf{0}]_{1\times N} & m_s + m_u & m_s \\ [\mathbf{0}]_{1\times N} & m_s & m_s \end{bmatrix} \tag{3}$$

$$\mathbf{C}_c = \begin{bmatrix} c_f(Vt)\int_0^L \mathbf{N}_B^T\mathbf{N}_B dx + c_t\mathbf{N}_B^T(Vt)\mathbf{N}_B(Vt) & -c_t\mathbf{N}_B^T(Vt) & [\mathbf{0}]_{N\times 1} \\ -c_t\mathbf{N}_B(Vt) & c_t & 0 \\ [\mathbf{0}]_{1\times N} & 0 & c_s \end{bmatrix} \tag{4}$$

$$\mathbf{K}_c = \begin{bmatrix} \int_0^L \mathrm{EI}\mathbf{N}_B''^T\mathbf{N}_B'' dx + k_f(Vt)\int_0^L \mathbf{N}_B^T\mathbf{N}_B dx + k_t\mathbf{N}_B^T(Vt)\mathbf{N}_B(Vt) & -k_t\mathbf{N}_B^T(Vt) & [\mathbf{0}]_{N\times 1} \\ -k_t\mathbf{N}_B(Vt) & k_t & 0 \\ [\mathbf{0}]_{1\times N} & 0 & k_s \end{bmatrix} \tag{5}$$

$$P_g = \begin{bmatrix} [\mathbf{0}]_{N\times 1}, & -(m_s + m_u)g, & -m_s g \end{bmatrix}^T \tag{6}$$

$$\mathbf{P}_c = \begin{bmatrix} -k_t\mathbf{N}_B^T(Vt) & -c_t\mathbf{N}_B^T(Vt) \\ k_t & c_t \\ 0 & 0 \end{bmatrix} \tag{7}$$

$$\xi(t) = \begin{bmatrix} \hat{\xi}(\mathrm{t}) & \dot{\hat{\xi}}(\mathrm{t}) \end{bmatrix}^T \tag{8}$$

Note that $\mathbf{N}_B(x)$ is short noted as $\mathbf{N}_B$, except for the case when evaluated at $x = Vt$. Additionally, $\hat{\xi}(\mathrm{t}) = \xi(x)\big|_{x=Vt}$, $k_f(Vt) = k_f(x)\big|_{x=Vt}$, $c_f(Vt) = c_f(x)\big|_{x=Vt}$ and g is the gravity term.

Then, defining a state vector $x_T = \begin{bmatrix} x_c & \dot{x}_c \end{bmatrix}^T$ and organizing Equation (2) in a state-space representation, yields:

$$\dot{x}_T(t) = \mathbf{A}_T x_T(t) + \mathbf{B}_T\xi(t) = \begin{bmatrix} [\mathbf{0}]_{N_T\times N_T} & [\mathbf{I}]_{N_T\times N_T} \\ -\mathbf{M}_c^{-1}\mathbf{K}_c & -\mathbf{M}_c^{-1}\mathbf{C}_c \end{bmatrix} x_T(t) + \begin{bmatrix} [\mathbf{0}]_{N_T\times 2} \\ \mathbf{M}_c^{-1}\mathbf{P}_c \end{bmatrix}\xi(t) \tag{9}$$

where N_T is $N+2$. The output vector $y_T(t)$ could be defined to contain arbitrary information about the system. In this study, the vehicle responses including displacement and velocity of the unsprung and sprung masses were considered as the output, i.e., $y_T(t) = \begin{bmatrix} u_u & z_s & \dot{u}_u & \dot{z}_s \end{bmatrix}$. Then, the observation and feedthrough matrices yield:

$$y_T(t) = \mathbf{C}_T x_T(t) + \mathbf{D}_T\xi(t) = \begin{bmatrix} [\mathbf{0}]_{1\times N} & 1 & 0 & [\mathbf{0}]_{1\times N} & 1 & 0 \\ [\mathbf{0}]_{1\times N} & 0 & 1 & [\mathbf{0}]_{1\times N} & 0 & 1 \end{bmatrix} x_T(t) + [\mathbf{0}]_{2N_T}\xi(t) \tag{10}$$

2.3. Augmented Equations of Motion

This section further arranges the equations derived for the interaction problem in Equation (9) to yield an augmented system in which the primary input is white noise. The white noise input allows a much simpler calculation and the use of stochastic analyses, when compared with manually inputting the roughness profile to the system. Authors in [35,36] constructed a state-space model for the stochastic roughness, when it follows a specified PSD, ($S_{\zeta\zeta}(\omega)$). In their approach, the transfer function ($H_{\zeta w}(\omega)$) was approximated using polynomial representation, as below:

$$S_{\zeta\zeta}(\omega) = H^2_{\zeta w}(\omega) S_0 \tag{11}$$

where S_0 is the degree of unevenness and ω is radian per second.

Then, $H_{\zeta w}(\omega)$ is realized in a state-space model as below to have output vector $y_f = \begin{bmatrix} \hat{\xi}(t) & \dot{\hat{\xi}}(t) \end{bmatrix}^T$, i.e.,

$$\dot{x}_f = \mathbf{A}_f x_f(t) + \mathbf{B}_f w(t) \tag{12}$$

$$y_f(t) = \mathbf{C}_f x_f(t) \tag{13}$$

where x_f is the state vector, $\mathbf{A}_f$, $\mathbf{B}_f$, and $\mathbf{C}_f$ are system, input, and observation matrices, respectively. The output vector is defined to contain the roughness and time derivative term of the roughness, i.e., $y_f(t) = \begin{bmatrix} \widetilde{\xi}(t) & \dot{\widetilde{\xi}}(t) \end{bmatrix}^T$. An example of designing a pavement filter using a second-order low-pass filter and polynomial approximation approaches is discussed in detail [35].

Finally, by combining Equations (9), (10), (12), and (13), the augmented state vector was defined as $x_a = \begin{bmatrix} x_T^T & x_f^T \end{bmatrix}^T$. Then,

$$\dot{x}_a = \mathbf{A}_a x_a + \mathbf{B}_a w(t) = \begin{bmatrix} \mathbf{A}_T & \mathbf{B}_{T1}\mathbf{C}_{f1} + \mathbf{B}_{T2}\mathbf{C}_{f2} \\ 0 & \mathbf{A}_f \end{bmatrix} \begin{bmatrix} x_T \\ x_f \end{bmatrix} + \begin{bmatrix} 0 \\ \mathbf{B}_f \end{bmatrix} w(t) \tag{14}$$

$$y_T = \mathbf{C}_a x_a = \begin{bmatrix} \mathbf{C}_T & \mathbf{D}_{T1}\mathbf{C}_{f1} + \mathbf{D}_{T2}\mathbf{C}_{f2} \end{bmatrix} \begin{bmatrix} x_T \\ x_f \end{bmatrix} \tag{15}$$

Note that $\mathbf{B}_{T1}$ and $\mathbf{B}_{T2}$ indicate the first and the second columns of $\mathbf{B}_T$, respectively. Similarly, $\mathbf{D}_{T1}$ and $\mathbf{D}_{T2}$ correspond to the first and the second columns of $\mathbf{D}_T$. The equations do not contain the feedthrough terms, implying that the system was strictly proper. In addition, $\mathbf{A}_a$ and $\mathbf{C}_a$ were time-dependent matrices, due to the variable foundation coefficients, $k_f(Vt)$ and $c_f(Vt)$.

3. Stochastic Vehicle Response

The covariance of the augmented system, $\mathbf{\Gamma}_{x_a}$, could be determined through a linear differential equation, when the input is a white noise process [37]:

$$\dot{\mathbf{\Gamma}}_{x_a}(t) = \mathbf{A}_a(t)\mathbf{\Gamma}_{x_a} + \mathbf{\Gamma}_{x_a}(t)\mathbf{A}_a^T(t) + 2\pi S_0 \mathbf{B}_a \mathbf{B}_a^T \tag{16}$$

where S_0 is the level of the white noise indicating the level of roughness. Solving Equation (16) is beneficial as it does not contain the principal matrix, in which an explicit format of the matrix is generally unknown in time-varying systems.

In the case of uniform foundation, i.e., $k_f(x) = k_f$, $c_f(x) = c_f$ with an infinite length beam, the system becomes stationary. Then, assuming that the initial conditions could be described by a random vector, $x_a(0) = x_{a0}$ the initial condition of Equation (16) $\Gamma_{x_a}(0) = \mathbf{\Gamma}_0$ becomes:

$$\mathbf{\Gamma}_0 = E[(x_{a0} - \mu_{x_{a0}})(y_{a0} - \mu_{x_{a0}})^T] \tag{17}$$

where $\mu_{x_{a0}}$ and $\boldsymbol{\Gamma}_0$ indicates the mean and the covariance, respectively. If the initial conditions are all deterministic, $\Gamma_0 = 0$. Then, the covariance of the structure responses, Γ_{y}, is given by:

$$\Gamma_{\mathrm{y}} = \mathbf{C}_{\mathrm{a}}\Gamma_{x_{\mathrm{a}}}\mathbf{C}_{\mathrm{a}}^{\mathrm{T}} \tag{18}$$

Further, with a zero-mean white noise being the input to the augmented system in Equation (14), the stationary covariance responses could be obtained by the solution of

$$0 = \mathbf{A}_{\mathrm{a}}\Gamma_{x_{\mathrm{a}}} + \Gamma_{x_{\mathrm{a}}}\mathbf{A}_{\mathrm{a}}^{\mathrm{T}} + 2\pi\mathbf{B}_{\mathrm{a}}S_0B_{\mathrm{a}}^{\mathrm{T}} \tag{19}$$

which is known as the Lyapunov equation [38]. Note that the equation is linear in unknown covariances and can only examine the moments of the responses under the stationary process.

However, the presented study consisted of a non-uniform foundation, in which the quantities varied over the length of the beam. Thus, the basic assumptions made in Equation (20) was no longer valid. Instead of directly integrating Equation (14), the general covariance response in Equation (16) was solved for $\boldsymbol{\Gamma}_{x_a}$, for which the matrix components are described below:

$$\boldsymbol{\Gamma}_{xa} = \begin{bmatrix} \boldsymbol{\Gamma}_{\mathrm{qq}} & \boldsymbol{\Gamma}_{\mathrm{qu}} & \boldsymbol{\Gamma}_{\mathrm{qz}} & \boldsymbol{\Gamma}_{\mathrm{q\dot{q}}} & \boldsymbol{\Gamma}_{\mathrm{q\dot{u}}} & \boldsymbol{\Gamma}_{\mathrm{q\dot{z}}} & \boldsymbol{\Gamma}_{\mathrm{qf}} \\ & \boldsymbol{\Gamma}_{\mathrm{uu}} & \boldsymbol{\Gamma}_{\mathrm{uz}} & \boldsymbol{\Gamma}_{\mathrm{u\dot{q}}} & \boldsymbol{\Gamma}_{\mathrm{u\dot{u}}} & \boldsymbol{\Gamma}_{\mathrm{u\dot{z}}} & \boldsymbol{\Gamma}_{\mathrm{uf}} \\ & & \boldsymbol{\Gamma}_{\mathrm{zz}} & \boldsymbol{\Gamma}_{\mathrm{z\dot{q}}} & \boldsymbol{\Gamma}_{\mathrm{z\dot{u}}} & \boldsymbol{\Gamma}_{\mathrm{z\dot{z}}} & \boldsymbol{\Gamma}_{\mathrm{zf}} \\ & & & \boldsymbol{\Gamma}_{\mathrm{\dot{q}\dot{q}}} & \boldsymbol{\Gamma}_{\mathrm{\dot{q}\dot{u}}} & \boldsymbol{\Gamma}_{\mathrm{\dot{q}\dot{z}}} & \boldsymbol{\Gamma}_{\mathrm{\dot{q}f}} \\ - & sym & - & & \boldsymbol{\Gamma}_{\mathrm{\dot{u}\dot{u}}} & \boldsymbol{\Gamma}_{\mathrm{\dot{u}\dot{z}}} & \boldsymbol{\Gamma}_{\mathrm{\dot{u}f}} \\ & & & & & \boldsymbol{\Gamma}_{\mathrm{\dot{z}\dot{z}}} & \boldsymbol{\Gamma}_{\mathrm{\dot{z}f}} \\ & & & & & & \boldsymbol{\Gamma}_{\mathrm{ff}} \end{bmatrix} \tag{20}$$

where $\mathbf{q_B}$, u_u, and z_s are short noted as $\mathbf{q}$, u, and z, respectively. Then, the symbolic covariance matrix, for which the number of variables is $\mathrm{N_{var}} = (\mathrm{N_a} \times (N_a + 1)/2)$ was plugged into Equation (15) to construct $\mathrm{N_{var}}$ distinct differential relationships.

Finally, the desired time-varying covariance responses of the vehicle, $\boldsymbol{\Gamma}_{y_{\mathrm{T}}}(t) = \begin{bmatrix} \boldsymbol{\Gamma}_{\mathrm{uu}}(t) & \boldsymbol{\Gamma}_{\mathrm{zz}}(t) & \boldsymbol{\Gamma}_{\mathrm{\dot{u}\dot{u}}}(t) & \boldsymbol{\Gamma}_{\mathrm{\dot{z}\dot{z}}}(t) \end{bmatrix}^{\mathrm{T}}$, was obtained by solving Equation (16) via the time-step integration method embedded in Matlab®(e.g., ode45).

4. Illustrative Examples

This section demonstrates the proposed approach by examining the covariance response of a vehicle on a non-uniform foundation. To first validate the solution procedure, steady-state covariance responses were compared by slowing the speed of the vehicle. Then, covariance responses were compared for various pavement scenarios. Finally, covariance response features were selected to classify and examine the foundation properties.

4.1. Vehicle and Pavement Model Properties

The properties of the quarter-car model used in the numerical examples are drawn from [39,40] and summarized in Table 1. Note that k_t is sought using a calibration index to well approximate the empirical model in [41], on a non-deformable rigid foundation with varying roughness [35].

Table 1. Vehicle properties.

Symbol	Components	Value
m_s	Sprung mass	1460 kg
m_u	Unsprung mass	80 kg
c_s	Suspension Damping	8760 Ns/m
c_t	Tire Damping	700 Ns/m
k_s	Suspension Stiffness	29.44 kN/m
k_t	Tire Stiffness	2500 kN/m
V	Velocity	20 km/h (stated otherwise)

The transfer function to approximate the PSD of road roughness, as in Equation (11), is considered as follows:

$$S_{\xi\xi}(\omega) = S_0\left(\frac{\Omega}{\Omega_0}\right)^{-v} \tag{21}$$

where Ω is the spatial circular frequency (ω/V); $\Omega_0 = 1\ \text{rad/m}$, v is the waviness that is taken as 2.45, to match the average roads in the U.S. [42]. Note that S_0 is varied to match the International Roughness Index (IRIs), which is a measure of road roughness on ride comfort [43]. A lower IRI value indicates a smooth pavement, while a higher IRI implies a rough pavement. In this study, IRIs are varied from 1 to 5, and the corresponding S_0's are approximated at those integers, using the golden car approach described in [35].

The typical pavement system was adopted herein, and the uniform properties of the Euler–Bernoulli beam (top layer in Figure 1) are summarized in Table 2. The elasticity of the top-layer used in the study represents the medium soil [44].

Table 2. Euler–Bernoulli beam property.

Symbol	Components	Value
h	Thickness	200 mm
b	Width	1.8 m
E	Elastic modulus	8760 Ns/m
ρ	Density	700 Ns/m
L	Length	5 m

Finally, to accommodate different Winkler type foundations, spring and damping coefficients were varied linearly or quadratically. Thus, the following equations were adopted for each case:

$$k_f(x) = k_{f0} \times z_f(x) c_f(x) = c_{f0} \times z_f(x) \tag{22}$$

$$\text{Linear: } z_f(x) = 1 - \alpha x, \quad 0 \le x \le L/2 z_f(x) = (2\alpha - 1) - 2x(\alpha - 1)/L, \quad L/2 \le x \le L \tag{23}$$

$$\text{Quadratic: } z_f(x) = 1 + 4x(\alpha - 1)/L - (\alpha - 1)4x^2/L^2,\ 0 \le x \le L \tag{24}$$

where the reference parameters for stiffness (k_{f0}) and damping (c_{f0}) are 30 kPa/mm and 2.4×10^7 N·s/m, respectively. The reduction factor $\alpha \le 1$ was selected such that the soil had the most reduced value at the mid-span of the beam (L/2). In this study, α was varied from 0.5, 0.7, and 0.9, which implied 50%, 70%, and 90% of the reference parameters. An illustration of $z_f(x)$ for each α is shown in Figure 2. Herein, the variation profiles are described with L and Q, for a linear and quadratic shape, respectively, followed by two digits describing α. For example, a dashed-dot plot in Figure 2a (linear with α = 0.5) was denoted as 'L50'. Similarly, 'Q90' indicates that $z_f(x)$ varies in a quadratic manner (see a solid line in Figure 2b).

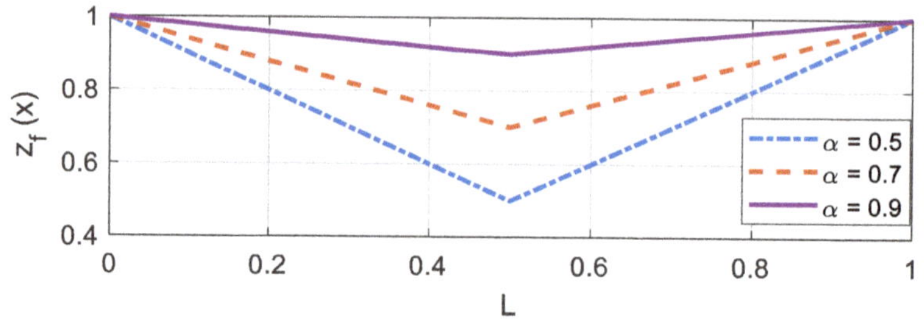

(a)

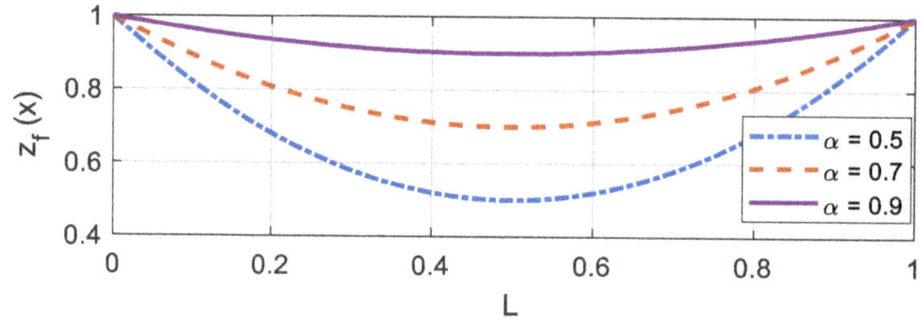

(b)

Figure 2. Nonlinear variation in the Winkler foundation (**a**) linear variation; (**b**) quadratic variation.

4.2. Validation of the Solution Approach

In this section, the steady-state responses of the vehicle were compared with that of near-stationary responses. The purpose of the presented study was to validate the time-varying covariance solutions in Equation (16) by slowing the speed of the vehicle, and also to illustrate the difference in the response due to various profiles of the subgrade and roughness, to be used in the following sections.

Figure 3 shows $\boldsymbol{\Gamma}_{y_T}(t)$ in comparison with steady-state responses. Here, the speed of the vehicle was slowed to 0.5 m/s (1.8 km/h). The total number of modes used in the Euler–Bernoulli beam was 10 sine modes and a roughness of IRI = 3 m/km was used. Regarding subgrade, the uniform foundation was represented by $z_f(x) = 1$ in Equation (19), while $\alpha = 0.5$ was used for linearly and quadratically varying foundations. Steady-state responses were calculated by fixing the location of the vehicle at the mid-span and solving Equation (16) with $\dot{\boldsymbol{\Gamma}}_{x_a}(t) = 0$, where the responses were as small as $\Gamma_{u_u u_u} = 1.39 \times 10^{-5}\ [\mathrm{m/s}]^2$; $\Gamma_{z_s z_s} = 1.40 \times 10^{-5}\ [\mathrm{m/s}]^2$; $\Gamma_{\dot{u}_u \dot{u}_u} = 1.42 \times 10^{-5}\ [\mathrm{m/s}]^2$; $\Gamma_{\dot{z}_s \dot{z}_s} = 0.189 \times 10^{-5}\ [\mathrm{m/s}]^2$. Note that the impact of foundation property change in the case of a stationary process was negligible, as reported by [24]. However, non-stationary covariance responses showed large humps within the first 2 s, due to the dynamic effect of boundary conditions. Peaks on velocity covariance responses were much higher than the displacement responses, rapidly converging to stationary responses.

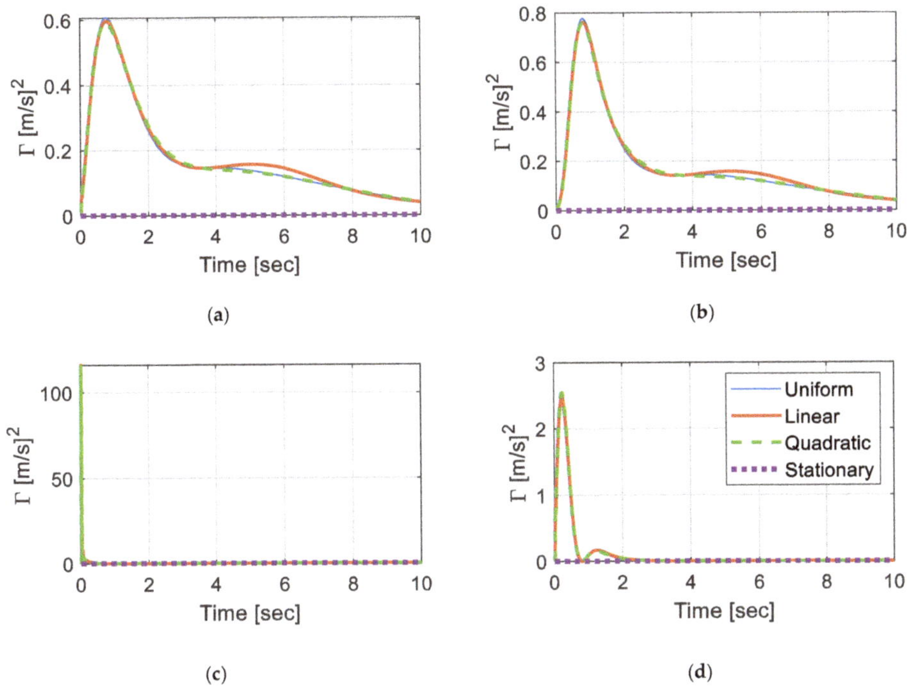

Figure 3. Covariance response of the vehicle over time (**a**) $\Gamma_{u_u u_u}$; (**b**) $\Gamma_{z_s z_s}$; (**c**) $\Gamma_{\dot{u}_u \dot{u}_u}$; and (**d**) $\Gamma_{\dot{z}_s \dot{z}_s}$.

Furthermore, the responses were examined on various road roughness. To compare the results efficiently, Figure 4a plots $\Gamma_{u_u u_u}$ with linearly varying foundation, as the IRIs varied from 1 to 5 m/km. A time-step integration method, ode45 was used as the vehicle crossed over the length with vehicle speed (V = 20 km/h). As could be seen, the effect of surface roughness on the nonstationary covariance responses were negligible. To better visualize the difference in the responses, the differences were plotted in Figure 4b, in percentage. The difference was estimated for each IRI (Γ_{IRI_i}), with respect to the response at IRI = 1 km/m (Γ_{IRI_1}), as below:

$$\Delta\Gamma_{IRI} = \left(\frac{\Gamma_{IRI_i} - \Gamma_{IRI_1}}{\Gamma_{IRI_1}}\right) * 100\ [\%]\,(i = 1, 2, \ldots, 5) \tag{25}$$

$\Delta\Gamma_{IRI}$ tended to diverge as the vehicle moved along the beam, indicating that dynamic responses were accumulated. However, within the domain, the maximum difference was less than 0.025% which was negligible, compared to the governing dynamic responses. Although the impact of change in IRI might increase as the beam length gets longer and the speed of the vehicle increases, the result indicated that the responses were governed more by the non-uniform features of the foundation. This fact emphasized the importance of conducting nonstationary response analyses because the stationary response analyses could not capture such differences, as reported by [24].

Based on the study, non-stationary responses under various subgrades converged to steady-state responses with time, while velocity covariance responses showed a faster rate. Additionally, the subgrade variation types affected the vehicle responses while road roughness had a negligible impact. Therefore, in the subsequent sections, road roughness was fixed to IRI = 3 m/km, to examine $\Gamma_{u_u u_u}(t)$ on various α's.

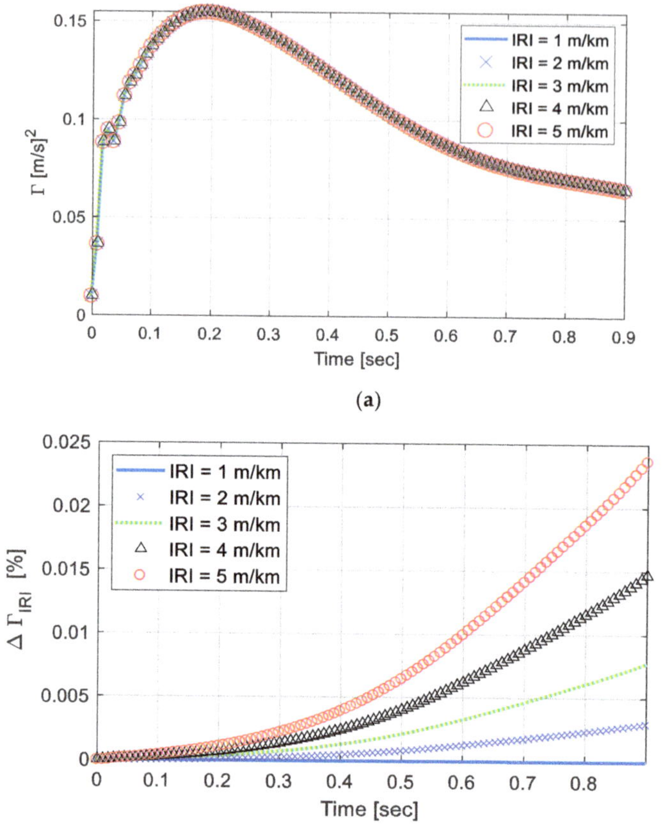

(a)

(b)

Figure 4. (**a**) Covariance response of vehicle under varying IRIs. (**b**) Difference in the response among IRIs in percentage.

4.3. Time-Varying Covariance Responses

Figure 5 shows the time-varying covariance responses, $\Gamma_{u_u u_u}(t)$, as the vehicle runs over the pavement with different types of foundation properties. The simulations were carried out on the basis of combinations of two variables—(1) foundation profile, implying how the subgrade properties change, i.e., either in a linear or in a quadratic manner, and (2) α, which varied from 0.5, 0.7, to 0.9. Then, the variable time-step was used for the integration and then downsampled to present 100 data points within the duration. After the resampling procedure, one could consider that the responses on the vehicle were measured at about 110 Hz. The chosen sampling rate was low enough to be easily realized, yet could capture the key features of the responses. The response shown in Figure 5 are deterministic, as the foundation variation profiles, α, and the speed of the vehicle are known. However, deterministic identification based on Figure 5 was unrealistic, because the measured signals tended to be contaminated with noises.

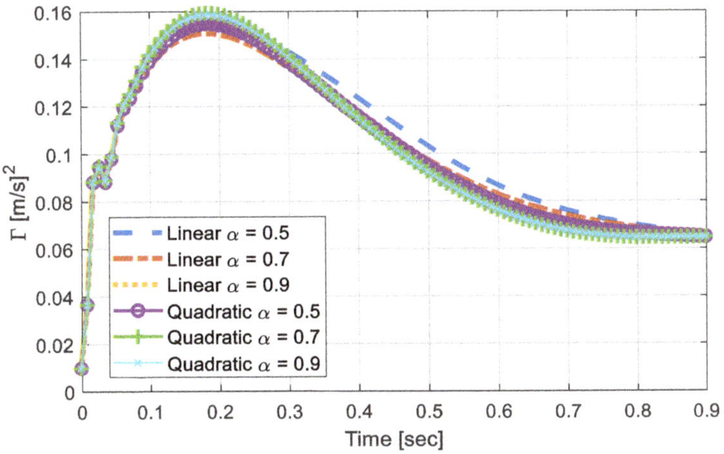

Figure 5. Covariance response on various foundation properties (without noise).

Now, noises were randomly selected to have a signal to noise level (SNR) between 25–50 dB. Note that the noises were added to the raw signals and then resampling was performed to represent the signal noises. Here, only the measured noises were considered because the zero-mean noises in the responses did not affect the covariance responses. For example, Figure 6 illustrates the covariance response with noises added on linearly varying foundation with α = 50. SNR of Figure 6a is about 45 dB and Figure 6b is about 50 dB.

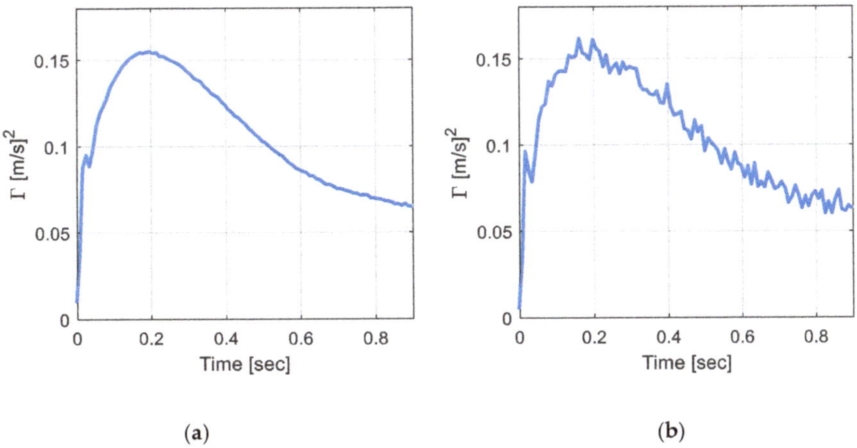

Figure 6. Covariance response with measured noise (**a**) SNR of 45 dB; and (**b**) SNR of 25 dB.

From the outcome of this subsection, the following conclusions could be made:

- Variations in profiles and α differed the maximum response and rate of convergence, while a general shape of the responses was preserved.
- A larger maximum value was obtained in quadratically varying profiles, compared to the linear case when the same α was used.

- Within the same profile, a larger α tended to increase the rate of convergence. However, the responses exhibit highly nonlinear relationships between the variables, making the prediction of subgrade's property change difficult.
- Although the vehicle responses were somewhat deterministic, once the foundation and vehicle parameters were determined, the analytical approach in the prediction was not realistic, due to the noises in the measured signal.

Thus, to resolve the issue, a machine-learning based classification of subgrades based on $\Gamma_{u_u u_u}(t)$ is discussed in the subsequent section.

5. Machine-Learning Based Classification

Machine-learning techniques are recognized in the civil engineering field as a promising component for monitoring and inspecting [13]. Machine-learning tools can provide pattern recognition strategies, when a deterministic model is difficult to be identified [45]. With their highlighted importance and computational advances, the Matlab software incorporated the Statics and Machine Learning toolbox containing considerable machine-learning techniques [46,47].

Among classifiers provided in the Classification Learner App in Matlab R2019b, one of the traditional classifier, linear discriminant analysis (LDA) is implemented for identifying the changes in the foundation properties from vehicle responses. The LDA method assumes that the data are distributed in Gaussian and that each attribute has the same variance. Then, the Bayes' theorem is applied to estimate the posterior probability that the observation belongs to a certain class. Then, the costs are evaluated from the maximal difference between the computed sample covariance and the empirical covariance matrix. In LDA, the cost function is linear with respect to the observation [46,48]. With these assumptions, the LDA model attempts to express one dependent variable in terms of a linear combination of other features or measurements [46]. Thus, to enhance the classification accuracy, features in the covariance responses must be selected carefully.

Based on the previously presented results, the following six features were selected—(1) maximum amplitude, A1; (2) time corresponding to A1, T1; (3) minimum tangent occurring between 0.2 s and 0.8 s, A2; (4) time corresponding to A2, T2; (5) slope of the linear regression between 0.2 s and 0.8 s, A3; and (6) y-intercept of the regression found in (5), A4. Then, to incorporate the measured noise in covariance responses, RMS noises were added in $\Gamma_{u_u u_u}(t)$. The extracted features showed some relationships among them. Figure 7 illustrates the distribution of features over the range of A1. As can be seen, some features, such as A2 and A4, show a higher correlation with A1, while other features are more scattered over the range of A1. In addition, L70, L90, and Q90 seem to overlap with each other (as in Figure 7a–e, making it hard to differentiate with classification models that are based on decision trees, etc.

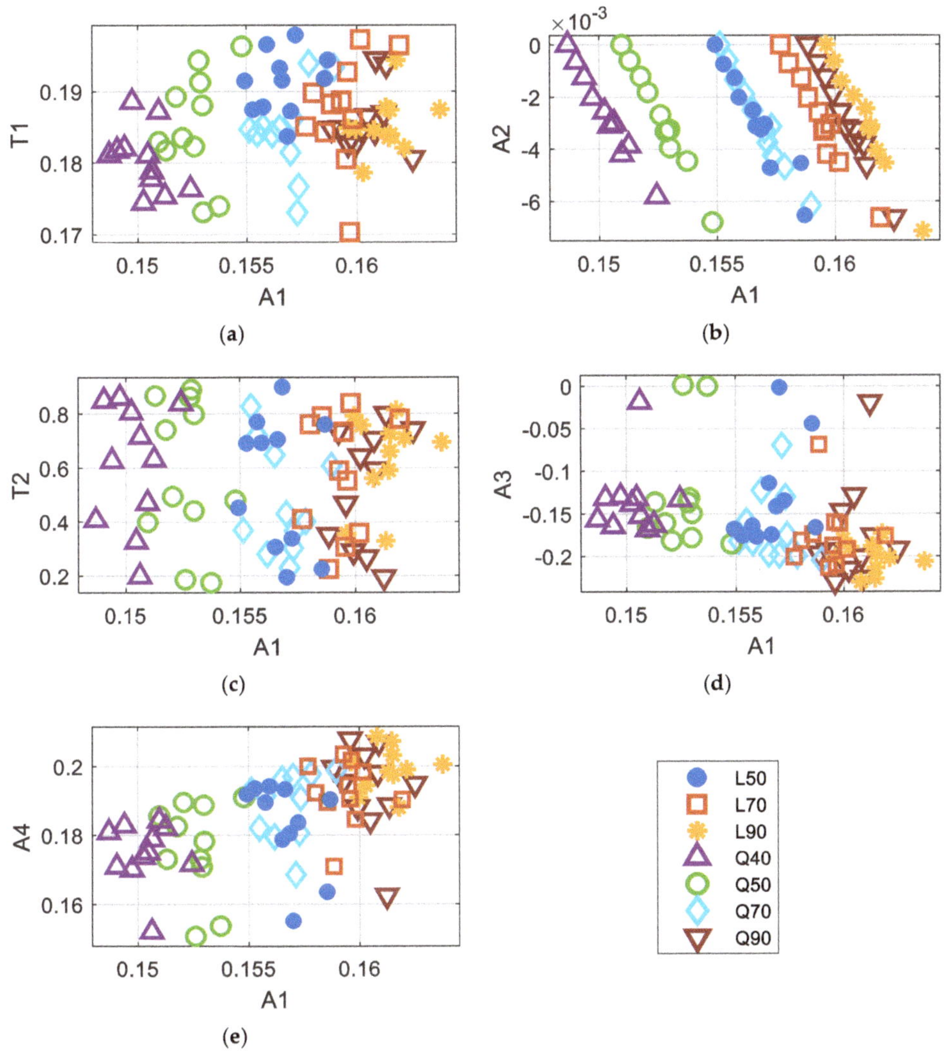

Figure 7. The distribution features over the range of A1 (Maximum Amplitude); (**a**) Relationship between A1-T1 (Maximum Time); (**b**) Relationship between A1–A2 (Maximum Tangent); (**c**) Relationship between A1-T2 (Time at Maximum Tangent); (**d**) Relationship between A1–A3 (Linear regression slope); and (**e**) Relationship between A1-A4 (Linear regression y-intercept).

Subsequently, the collected datasets were trained using LDA. The average success rate for using an LDA classifier was over 94%, with at most 10% noise. Table 3 is a confusion matrix when 77 training data were used. The table shows that the foundation was mostly classified, while 18% of the L50 case was misclassified as Q70, and vice versa. Note that LDA showed the highest accuracy when compared with other classification tools; the linear support vector machine showed 82% accuracy, ensemble provided 84% accuracies, while other methods showed over 40% errors.

Table 3. Confusion matrix for foundation property test.

Actual Properties	Assessed Properties					
	L50	**L70**	**L90**	**Q50**	**Q70**	**Q90**
L50	0.82	0.00	0.00	0.00	0.18	0.00
L70	0.00	1.00	0.00	0.00	0.00	0.00
L90	0.00	0.00	1.00	0.00	0.00	0.00
Q50	0.00	0.00	0.00	1.00	0.00	0.00
Q70	0.18	0.00	0.00	0.00	0.82	0.00
Q90	0.00	0.00	0.00	0.00	0.00	1.00

Now validation tests were conducted to verify the performance of the developed LDA model. For each case of the foundation, 100 test sets with 1~10% RMS noises added on the responses were generated. The accuracy of the classifier was plotted as a bar chart shown in Figure 8. As could be expected from the confusion matrix, Q70 showed the lowest accuracy, 83%, followed by L50. Overall accuracy was about 94.5% This result supports that by adopting LDA model, the vehicle responses could classify the change in the foundation properties with good accuracies.

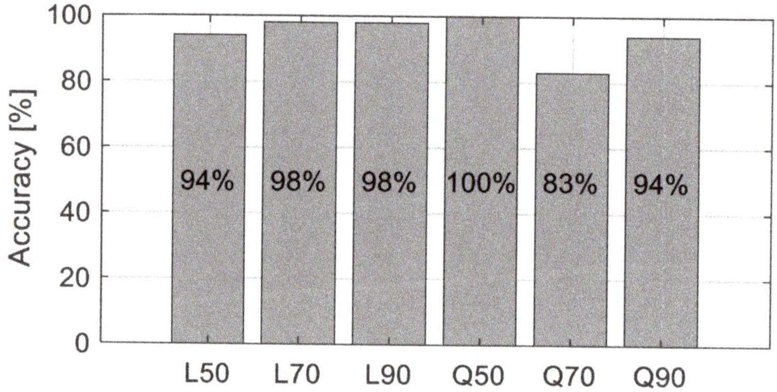

Figure 8. The success rate for identifying foundation property.

6. Conclusions

This paper presented a machine-learning-based classification of non-uniform foundation properties using vehicle responses. The dynamics response of the quarter-car model on the stochastic deformable pavement with a finite length was evaluated. A filtered white noise was used to represent the stochastic pavement roughness. The deformable subgrade was modeled by an Euler–Bernoulli beam on a Winkler-type foundation. The non-uniform characteristics were represented with varying stiffness and damping coefficients of the subgrade. Then, the vehicle–pavement–foundation interaction model was combined to yield an augmented state-space representation, which had white noise as the primary input to the system. In this study, the model could accommodate any time of foundation that was describable with a longitudinal axis, although only the impacts of linear and quadratic variations were discussed. A time-varying Lyapunov equation governing the covariance of the responses was solved to effectively obtain the response of the vehicle. From the steady-state Lyapunov solution, the solution approaches were validated. Then, various values of the subgrade's properties, along with surface roughness were compared. The parametric study showed that the stiffness produced some difference in the response profile, while the roughness produced negligible change. This fact opposed the uniform foundation case, indicating the importance of considering the non-uniform foundation.

Then, a set of simulations for measuring noise were performed and used for feature extraction for a classifier. Using an application embedded in Matlab®, linear discriminant analysis was employed to show an average accuracy of 94%. Finally, a total of 600 test sets were generated to demonstrate that the estimated foundation properties were mostly correct. Based on the outcome of this study, the contribution of the presented work and the specific conclusions are summarized as follows:

- The introduced vehicle–pavement–foundation model and nonstationary solution approach allow the investigation of the impact of nonuniform foundation characteristics on vehicle responses.
- Due to the non-stationary stochastic solution approach described, which examined the second-order statistics of the process, efficient estimation was available, where the response was determinate and unaffected by the zero-mean noises.
- The proposed approach could efficiently handle various types of vehicles, roughness, and nonlinearity of foundations.
- Based on the theoretical evaluation, a machine-learning-based classification of non-uniform foundation properties was demonstrated, which included irremovable measured noises.
- In addition to the physical realization of the presented results, future research must ensure to provide high accuracy of identification when the location of the weakened foundation is unknown, and should consider the lateral movement at the left support.
- Overall, based on the outcome of the study, the vehicle responses could be used in conjunction with machine-learning technologies for classifying the properties and types of the subgrade.

In conclusion, the presented work demonstrated the potentials of monitoring the subgrade anomalies from an inspecting vehicle that is only equipped with a set of accelerometers. Unlike the current approach of a subgrade survey that is only limited to suspicious spots, the successful realization of the presented methodology might allow a complete survey and the construction of a database for road management.

Author Contributions: R.E.K. contributed to the conceptualization, methodology, software, funding, and validation. The author has read and agreed to the published version of the manuscript.

Funding: This work was supported by the research fund of Hanyang University (HY-2019).

Conflicts of Interest: The author declares no conflict of interest.

References

1. Federal Highway Administration (US); Federal Transit Administration (US) (Eds.) *2015 Status of the Nation's Highways, Bridges, and Transit Conditions & Performance Report to Congress*; Government Printing Office: Washington, DC, USA, 2017.
2. Rada, G.R.; Perera, R.W.; Prabhakar, V.C.; Wiser, L.J. Relating ride quality and structural adequacy for pavement rehabilitation and management decisions. *Transp. Res. Rec.* **2012**, *2304*, 28–36. [CrossRef]
3. Shahin, M.Y.; Michael, I.D.; Starr, D.K. *Development of a Pavement Condition Index for Roads and Streets*; No. CERL-IR-M-232; Construction Engineering Research Lab (Army) Champaign Illinois: Champaign, IL, USA, 1978.
4. Chung, H.C.; Girardello, R.; Soeller, T.; Shinozuka, M. Automated management for pavement inspection system (AMPIS). In *Smart Structures and Materials 2003: Smart Systems and Nondestructive Evaluation for Civil Infrastructures*; International Society for Optics and Photonics: San Diego, CA, USA, 2003; Volume 5057.
5. Huang, Y.; Xu, B. Automatic inspection of pavement cracking distress. *J. Electron. Imaging* **2006**, *15*, 013017. [CrossRef]
6. Chen, Y.L.; Jahanshahi, M.R.; Manjunatha, P.; Gan, W.; Abdelbarr, M.; Masri, S.F.; Becerik-Gerber, B.; Caffrey, J.P. Inexpensive multimodal sensor fusion system for autonomous data acquisition of road surface conditions. *IEEE Sens. J.* **2016**, *16*, 7731–7743. [CrossRef]
7. Al-Suleiman, T.I.; Hamici, Z.M.; Bazlamit, S.M.; Ahmad, H.S. Assessment of the effect of alligator cracking on pavement condition using WSN-image processing. In *International Conference on Engineering, Project, and Product Management*; Springer: Cham, Switzerland, 2017.

8. Lee, Y.J.; Kim, R.E.; Suh, W.; Park, K. Probabilistic fatigue life updating for railway bridges based on local inspection and repair. *Sensors* **2017**, *17*, 936. [CrossRef] [PubMed]
9. Silva, N.; Shah, V.; Soares, J.; Rodrigues, H. Road anomalies detection system evaluation. *Sensors* **2018**, *18*, 1984. [CrossRef]
10. Sebaaly, P.; Tabatabaee, N.; Bonaquist, R.; Anderson, D. Evaluating structural damage of flexible pavements using cracking and falling weight deflectometer data. *Transp. Res. Rec.* **1989**, *1227*, 115–127.
11. Chen, D.H.; Fernando, E.; Murphy, M. Application of falling weight deflectometer data for analysis of superheavy loads. *Transp. Res. Rec.* **1996**, *1540*, 83–90. [CrossRef]
12. Vyas, V.; Singh, A.P.; Srivastava, A. Prediction of asphalt pavement condition using FWD deflection basin parameters and artificial neural networks. *Road Mater. Pavement Des.* **2020**, 1–19. [CrossRef]
13. Spencer, B.F., Jr.; Hoskere, V.; Narazaki, Y. Advances in computer vision-based civil infrastructure inspection and monitoring. *Engineering* **2019**, *5*, 199–222. [CrossRef]
14. Sun, K.; Zhang, W.; Ding, H.; Kim, R.E.; Spencer, B.F. Autonomous evaluation of ambient vibration of underground spaces induced by adjacent subway trains using high-sensitivity wireless smart sensors. *Smart Struct. Syst.* **2017**, *19*, 1–10. [CrossRef]
15. Kim, R.E.; Li, J.; Spencer, B.F.; Nagayama, T.; Mechitov, K.A. Synchronized sensing for wireless monitoring of large structures. *Smart Struct. Syst.* **2016**, *18*, 885–909. [CrossRef]
16. Cebon, D. *Handbook of vehicle-road interaction*; CRC Press: Boca Raton, FL, USA, 1999; ISBN 9789026515545.
17. Sun, L.; Luo, F. Nonstationary dynamic pavement loads generated by vehicles traveling at varying speed. *J. Transp. Eng.* **2007**, *133*, 252–263. [CrossRef]
18. Sun, L. A closed-form solution of beam on viscoelastic subgrade subjected to moving loads. *Comput. Struct.* **2002**, *80*, 1–8. [CrossRef]
19. Wedig, W.V. Dynamics of Half-Car Models under Stochastic Road Excitations. In Proceedings of the ASME 2003 International Mechanical Engineering Congress and Exposition American Society of Mechanical Engineers Digital Collection, Washington, DC, USA, 15–21 November 2003; pp. 269–275, ISBN 0-7918-3709-2.
20. Hardy, M.S.A.; Cebon, D. Response of continuous pavements to moving dynamic loads. *J. Eng. Mech.* **1993**, *119*, 1762–1780. [CrossRef]
21. Hardy, M.S.A.; Cebon, D. Importance of speed and frequency in flexible pavement response. *J. Eng. Mech.* **1994**, *120*, 463–482. [CrossRef]
22. Yongchen, L.; Li, S.; Kun, L.; Lichao, X. A method of acquiring dynamic road roughness based on vehicle-road vertical coupling. *Open Autom. Control Syst. J.* **2014**, *6*, 616–620. [CrossRef]
23. Akbarian, M.; Moeini-Ardakani, S.S.; Ulm, F.J.; Nazzal, M. Mechanistic approach to pavement–vehicle interaction and its impact on life-cycle assessment. *Transp. Res. Rec.* **2012**, *2306*, 171–179. [CrossRef]
24. Kim, R.E.; Kang, S.; Spencer, B.F.; Al-Qadi, I.L.; Ozer, H. Impact of Pavement Roughness and Deflection on Fuel Consumption Using Energy Dissipation. *J. Eng. Mech.* **2019**, *145*, 04019080. [CrossRef]
25. Franklin, J.N.; Scott, R.F. Beam equation with variable foundation coefficient. *J. Eng. Mech. Div.* **1979**, *105*, 811–827.
26. Lentini, M. Numerical solution of the beam equation with nonuniform foundation coefficient. *J. Appl. Mech.* **1979**, *46*, 901–904. [CrossRef]
27. Clastornik, J.; Eisenberger, M.; Yankelevsky, D.Z.; Adin, M.A. Beams on Variable Winkler Elastic Foundation. *J. Appl. Mech.* **1986**, *53*, 925–928. [CrossRef]
28. Lee, S.Y.; Ke, H.Y. Free vibrations of non-uniform beams resting on non-uniform elastic foundation with general elastic end restraints. *Comput. Struct.* **1990**, *34*, 421–429.
29. Kacar, A.; Tan, H.T.; Kaya, M.O. Free vibration analysis of beams on variable winkler elastic foundation by using the differential transform method. *Math. Comput. Appl.* **2011**, *16*, 773–783. [CrossRef]
30. Kumar, Y. Free vibration analysis of isotropic rectangular plates on winkler foundation using differential transform method. *Int. J. Appl. Mech. Eng.* **2013**, *18*, 589–597. [CrossRef]
31. Abdelghany, S.M.; Ewis, K.M.; Mahmoud, A.A.; Nassar, M.M. Dynamic response of non-uniform beam subjected to moving load and resting on non-linear viscoelastic foundation. *Beni-Suef Univ. J. Basic Appl. Sci.* **2015**, *4*, 192–199. [CrossRef]
32. Ansari, M.; Esmailzadeh, E.; Younesian, D. Internal-external resonance of beams on non-linear viscoelastic foundation traversed by moving load. *Nonlinear Dyn.* **2010**, *61*, 163–182. [CrossRef]

33. Ansari, M.; Esmailzadeh, E.; Younesian, D. Frequency analysis of finite beams on nonlinear Kelvin–Voight foundation under moving loads. *J. Sound Vib.* **2011**, *330*, 1455–1471. [CrossRef]
34. Oni, S.T.; Awodola, T.O. Dynamic behaviour under moving concentrated masses of simply supported rectangular plates resting on variable Winkler elastic foundation. *Lat. Am. J. Solids Struct.* **2011**, *8*, 373–392. [CrossRef]
35. Kim, R.E.; Kang, S.; Spencer, B.F.; Al-Qadi, I.L.; Ozer, H. New stochastic approach of vehicle energy dissipation on nondeformable rough pavements. *J. Eng. Mech.* **2017**, *143*, 04016118. [CrossRef]
36. Kim, R.E.; Kang, S.; Spencer, B.F.; Ozer, H.; Al-Qadi, I.L. Stochastic analysis of energy dissipation of a half-car model on nondeformable rough pavement. *J. Transp. Eng. Part B Pavements* **2017**, *143*, 04017016. [CrossRef]
37. Soong, T.T.; Grigoriu, M. *Random Vibration of Mechanical and Structural Systems*; Prentice-Hall: Englewood Cliffs, NJ, USA, 1993.
38. Wachspress, E.L. Iterative solution of the Lyapunov matrix equation. *Appl. Math. Lett.* **1988**, *1*, 87–90. [CrossRef]
39. Louhghalam, A.; Akbarian, M.; Ulm, F.J. Roughness-induced pavement–vehicle interactions: key parameters and impact on vehicle fuel consumption. *Transp. Res. Rec.* **2015**, *2525*, 62–70. [CrossRef]
40. Louhghalam, A.; Tootkaboni, M.; Ulm, F.J. Roughness-induced vehicle energy dissipation: Statistical analysis and scaling. *J. Eng. Mech.* **2015**, *141*, 04015046. [CrossRef]
41. Chatti, K.; Zaabar, I. *Estimating the Effects of Pavement Condition on Vehicle Operating Costs*; Transportation Research Board: Washington, DC, USA, 2012; Volume 720.
42. Kropáč, O.; Múčka, P. Deterioration model of longitudinal road unevenness based on its power spectral density indices. *Road Mater. Pavement Des.* **2008**, *9*, 389–420. [CrossRef]
43. Gillespie, T.D.; Sayers, M.W.; Segal, L. *Calibration of Response-Type Road Roughness Measuring Systems*; NCHRP Rep. 228; Transportation Research Board: Washington, DC, USA, 1980.
44. Xu, L. Typical Values of Young's Elastic Modulus and Poisson's Ratio for Pavement Materials. 2015. Available online: https://www.academia.edu/15702729/Typical_Values_of_Youngs_Elastic_Modulus_and_Poissons_Ratio_for_Pavement_Materials.PDF (accessed on 6 May 2018).
45. Farrar, C.R.; Worden, K. *Structural Health Monitoring: A Machine Learning Perspective*; John Wiley & Sons: Hoboken, NJ, USA, 2012.
46. Matlab. Statistics and Machine Learning Toolbox. Available online: https://mathworks.com/products/statistics.html (accessed on 10 December 2017).
47. Gorban, A.N.; Golubkov, A.; Grechuk, B.; Mirkes, E.M.; Tyukin, I.Y. Correction of AI systems by linear discriminants: Probabilistic foundations. *Inf. Sci.* **2018**, *466*, 303–322. [CrossRef]
48. Fisher, R.A. The use of multiple measurements in taxonomic problems. *Ann. Eugen.* **1936**, *7*, 179–188. [CrossRef]

sensors

MDPI

Article

Multi-Frame Star Image Denoising Algorithm Based on Deep Reinforcement Learning and Mixed Poisson–Gaussian Likelihood

Ming Xie, Zhenduo Zhang, Wenbo Zheng, Ying Li * and Kai Cao

Navigation College, Dalian Maritime University, Dalian 116026, China; mingxie@dlmu.edu.cn (M.X.); zhangzhenduo@dlmu.edu.cn (Z.Z.); zwb123456789@dlmu.edu.cn (W.Z.); caokai@dlmu.edu.cn (K.C.)

* Correspondence: yldmu@dlmu.edu.cn

Received: 26 September 2020; Accepted: 20 October 2020; Published: 22 October 2020

Abstract: Mixed Poisson–Gaussian noise exists in the star images and is difficult to be effectively suppressed via maximum likelihood estimation (MLE) method due to its complicated likelihood function. In this article, the MLE method is incorporated with a state-of-the-art machine learning algorithm in order to achieve accurate restoration results. By applying the mixed Poisson–Gaussian likelihood function as the reward function of a reinforcement learning algorithm, an agent is able to form the restored image that achieves the maximum value of the complex likelihood function through the Markov Decision Process (MDP). In order to provide the appropriate parameter settings of the denoising model, the key hyperparameters of the model and their influences on denoising results are tested through simulated experiments. The model is then compared with two existing star image denoising methods so as to verify its performance. The experiment results indicate that this algorithm based on reinforcement learning is able to suppress the mixed Poisson–Gaussian noise in the star image more accurately than the traditional MLE method, as well as the method based on the deep convolutional neural network (DCNN).

Keywords: star image; image denoising; reinforcement learning; maximum likelihood estimation; mixed Poisson–Gaussian likelihood

1. Introduction

A star image is obtained from star sensor, which is a high-accuracy attitude determination instrument. The three-axis attitude and spatial position of the star sensor can be calculated based on the reference of stars recognized in the digitized star image. Being able to accurately collect and process a star image at all times, however, is one of the challenges in the application of a star sensor due to the brightness of the sky's background and the complicated mixture of noise [1]. These conditions lead to a low Signal-to-Noise Ratio (SNR) in the star image obtained in the daytime, and thus have negative influences on the calculation of attitude and position. Similarly, it is an important basis for the star sensor to work efficiently that noise is effectively suppressed and the star targets are accurately recognized in the star image. Researchers have proposed various types of algorithms to suppress the effects of salt-and-pepper noise, strip noise, speckle noise, and defective pixels [1–5]. Nevertheless, mixed Poisson–Gaussian noise remains after these denoising methods are applied and affects the extraction of the true value of star points [6].

The traditional method suppresses the mixed Poisson–Gaussian noise simply by taking the arithmetic mean of multiple frames of star images. However, the restoration through such methods is not likely to be accurate without a large number of star images that are taken in a short period or by a stational star sensor [7]. Considering the movement in the carrier of the star sensor, such conditions could sometimes be unrealistic. Maximum likelihood estimation (MLE) is one of widely used methods that can restore a star image with limited number of frames. In the MLE method, the distribution model of the noise is constructed as a likelihood function, and the image that is most likely to produce the given observations is termed as the restored image. However, mixed Poisson–Gaussian noise is still difficult to be suppress through the MLE method due to its complicated likelihood function. This will be discussed in detail in the next section.

Towards the goal of high-accuracy star image restoration, this study integrated deep reinforcement learning techniques with the MLE method and developed an iterative denoising algorithm that can accurately suppress mixed Poisson–Gaussian noise. Generally, an agent proposes "restored images", which are evaluated through a modified likelihood function. Borrowing the idea of the MLE method, the mission of the agent is to find the image that returns the highest probability according to the modified likelihood function. Based on the gradient descent algorithm in the deep reinforcement learning method, the agent efficiently learns to propose a more appropriate "restored image" until it gets close enough to the ground truth. Thus, the aim and contribution of this study is to build a fully automated denoising algorithm that can accurately suppress the mixed Poisson–Gaussian noise in a star image.

2. Related Works

2.1. The Application of MLE in Star Image Denoising

The MLE method regards the image denoising problem as the optimization process of the likelihood function. Moreover, other regularization terms can be added to the likelihood function as the constraints on the estimator and form a denoising algorithm together with the likelihood function. Based on the MLE method and the image blur model with Gaussian noise, Katsaggelos proposed an image restoration algorithm using blur coefficients identification and expectation-maximization (EM) [8]. Llacer and Nunez applied the MLE method to restore the astronomic images obtained from the Hubble Space Telescope and proposed an iterative MLE and Bayesian algorithm [9]. Similarly, Synder discussed the convolution of astronomic image obtained from Charge Coupled Device (CCD) camera and suggested that the restoration of such astronomic image could potentially be achieved via EM method [10]. Benvenuto further explored the image denoising problem specifically for star images. They gave an approximate model of the flux of photons and verified the existence of its solution [11]. Li et al. constructed the likelihood function for multi-frame Adaptive Optics (AO) image according to the Poisson distribution model, based on which they proposed an AO image denoising method that achieved accurate results [12]. However, this algorithm is not able to restore the image with mixed Poisson–Gaussian noise, which has a more complicated distribution model. Zhang et al. approximated the mixed Poisson–Gaussian noise model using generalized Anscombe transformation approximation Fourier ptychographic (GATFP), and then solved the transformed likelihood function with the MLE method [6]. Although the GATFP method is able to optimize the complicated distribution model of mixed Poisson–Gaussian noise, it cannot avoid the error introduced during the approximation.

2.2. Reinforcement Learning and Its Application in Image Processing

Derived from behavioral psychology, reinforcement learning is a classical topic in the studies of artificial intelligence (AI) [13,14]. Various algorithms have been developed for reinforcement learning, but they share a major process in common: an agent learns through the interactions with certain environment, and gradually finds an optimal solution or approach. The Markov Decision Process (MDP) is a commonly used method of reinforcement learning [15]. In MDP, an agent takes limited choices of actions to update the state of the environment, then receives encouragement or punishment according to a reward function that evaluates the chosen action, as well as how it changed the state [16]. As it can be inferred from the learning process, MDP is usually accomplished through an iterative algorithm.

Due to the development of AI technology in the last decade, the machine learning algorithm has been applied in many subjects, including object detection [17–19], data mining [20,21], image processing [22,23], etc. Typically, the convolutional neural network (CNN) has been successfully applied in various image processing tasks, including deblurring [24–26], denoising [27–29], JPEG artifacts reduction [30–32], and super-resolution [33–40], owing to its ability to efficiently handle imagery information. Incorporated with the deep learning algorithm, traditional reinforcement learning also rapidly developed [41,42] and showed its potential of application in image processing [43,44]. Yu et al., designed a toolbox for image processing, and then developed an image restoration algorithm based on reinforcement learning using the tools within the toolbox [44]. Despite its various applications in image processing, the machine learning method is not widely utilized in star image restoration. This is partly because the star image restoration problem is different from the general image restoration problem: the goal is to restore the value of a star point as close to the true value as possible, rather than make the image visually fine. A deep convolutional neural network (DCNN) model proposed by Liu et al. is one of the limited examples of applying deep learning method in star image denoising [45]. Liu et al. trained a supervised DCNN model with simulated star image and achieved better restoration than traditional method.

2.3. Mixed Poisson–Gaussian Noise in Star Image

Gaussian noise in the star image can be introduced by many aspects including the noise of circuit components, the temperature change in image sensor, etc. The intrinsic noise introduced during the photon counting process of the photosensitive component, however, depends on the signal value, thus follows Poisson distribution. Because the measurements from all pixels are independent, for a star image with $M \times M$ pixels and N observations, the likelihood function that describe the possibility of receiving the known N observations is given as Equation (1) [46]:

$$p(z_q|b_q) = \prod_{q=1}^{N}\prod_{i=1}^{M^2}\Big(\sum_{j=1}^{+\infty}\frac{e^{-[b_q]_i}[b_q]_i^j}{j!}\frac{e^{-\frac{1}{\sigma^2}([z_q]_i-j)^2}}{\sqrt{2\pi\sigma^2}}\Big) \tag{1}$$

where z_q is the noise image, b_q is the ground truth, σ is the variance of Gaussian distribution, subscript i is an index of the number of pixels, q is an index of the number of observations, and j is an index of the possible observation of the gray value. As discussed above, MLE method is intended to find a restore image b'_q that returns the highest value of the likelihood function.

Taking the negative logarithm of the likelihood function makes it easier to solve this problem. Thus, the optimized likelihood function is given as:

$$\min f(z_q|b'_q) = -\log\Big(\prod_{q=1}^{N}(p(z_q|b'_q)\Big) = -\sum_{q=1}^{N}\sum_{i=1}^{M^2}\log\Big(\sum_{j=1}^{+\infty}\frac{e^{-[b'_q]_i}[b'_q]_i^j}{j!}\frac{e^{-\frac{1}{\sigma^2}([z_q]_i-j)^2}}{\sqrt{2\pi\sigma^2}}\Big) \tag{2}$$

Nevertheless, it is still extremely difficult to directly solve this likelihood function. Marnissi et al., further optimized the likelihood function using generalized Anscombe transformation (GAT) approximation, and form the likelihood function as Equation (3) [47]:

$$p(z_q|b_q) = \prod_{i=1}^{M^2}\frac{1}{\sqrt{2\pi}}\exp\Big(-\frac{1}{2}\Big(z_q - 2\sqrt{[b_q]_i + \frac{3}{8} + \sigma^2}\Big)^2\Big) \tag{3}$$

As indicated in Equation (3), GAT method simplifies the mixed Poisson–Gaussian likelihood as a Gaussian-like distribution. Although it is able to provide an approximate solution to the likelihood function, the tails of variance stabilized coefficients distribution are still empirically longer than normality [48] (which means the variance of the transformed noise still depends on signal intensity). In this study, the mixed Poisson–Gaussian likelihood function is directly introduced in the reinforcement learning algorithm without approximation. Therefore, it is expected that the likelihood function given by Equation (2) can be solved more accurately, and the mixed Poisson–Gaussian noise in the star image can be suppressed more effectively.

3. Methodology

3.1. Dataset

In this study, simulated star images are produced as the dataset for training and testing the model. Simulated star image has been successfully utilized to train machine learning model owing to its ability of generating a variety of star images under different conditions at very low cost [45]. Xu et al., studied the stellar radiation model and proposed a star image simulation method based on calibration coefficient, which relates to stellar magnitude m_v and color temperature T [49]. Generally, the total energy of photoelectrons generated by an observed star can be calculated by Planck's black body radiation law as:

$$E_s(\lambda, T) = \frac{2\pi hc^2}{\lambda^5[\exp(hc/\lambda k_B T) - 1]} \tag{4}$$

where λ is wavelength, k_B is Boltzmann constant, h is Planck constant, E_s is the stellar radiation energy per unit area, per unit time, and per incremental wavelength (W/m^3).

The energy of star point is then dispersed into several pixels based on point spread function (PSF) of the optical system. After that, the star points are integrated with the background simulated based on MODTRAN software to generate the simulated star image. Finally, mixed Poisson–Gaussian noise is added to the image. This simulation method is adopted in this study to produce the training dataset. More specifically, three levels of noise images are produced in the dataset as mild noise, moderate noise, and severe noise, in which the variances of Gaussian noise are 5, 10, and 20, respectively. The noise in each image is randomly generated and independent from each other in order to avoid having the same noisy instantiation among the different images in the dataset. The simulated image of ground truth and three sets of star images with different amount of noise is shown as Figure 1.

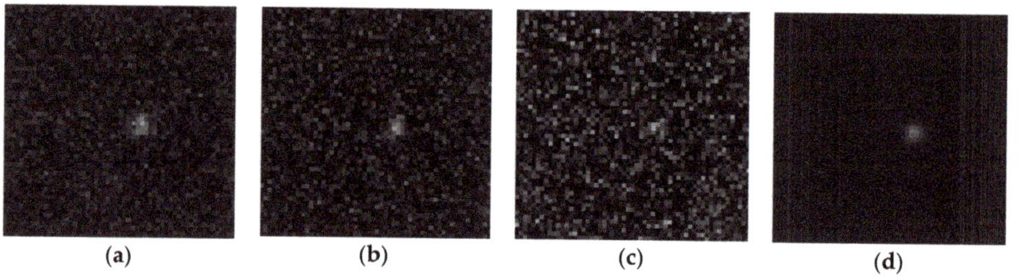

Figure 1. Simulated star images with: (**a**) mild noise, (**b**) moderate noise, (**c**) severe noise, (**d**) ground truth.

Two thousand different simulated star images are produced as the images of ground truth. For each star image of the ground truth, 10 noise images are produced for each type of noise. Thus, there are 6000 sets of noise images that contain three levels of noise and are produced based on 2000 different images of ground truth. Each set includes 10 frames of noise image, thus the training and testing for certain image with each ground truth image that contains each type of noise is carried out based on 10 frames of noise images. The image sets are randomly shuffled for model training. Seventy percent of the images are used for training, 10% of the images are used for validation, and 20% of the images are used for testing. The star images with different types of noise are distributed in the training, validation, and testing dataset as evenly as possible. All of the tests on hyperparameters are carried out separately with the testing images of these three types of noise.

3.2. Reinforcement Learning Algorithm

Different from existing star image denoising methods (such as GAT) that try to approximately solve the likelihood function given by Equation (2), we intend to solve the denoising problem through MDP, which include an environment part (state), a decision-making part (agent), a rewarding part, and a stopping part (Figure 2). The decision-making part receives the noise image and treats it as the initial state, makes a sequence of actions on the image and produces a processed image. The rewarding part mainly consists of a reward function that returns a higher value when the processed image is closer to the ground truth. In the case of the star image denoising in this study, the reward function is directly derived from the likelihood function that is given by Equation (2). The stopping part evaluates the processed image and determines whether it is close enough to the ground truth. If the processed image does not satisfy the stopping part, it will be sent back to the decision-making part as an updated state. Otherwise, the stopping part will end the iterations and return the processed image as the final restored image. The detail settings of these parts are discussed as follows.

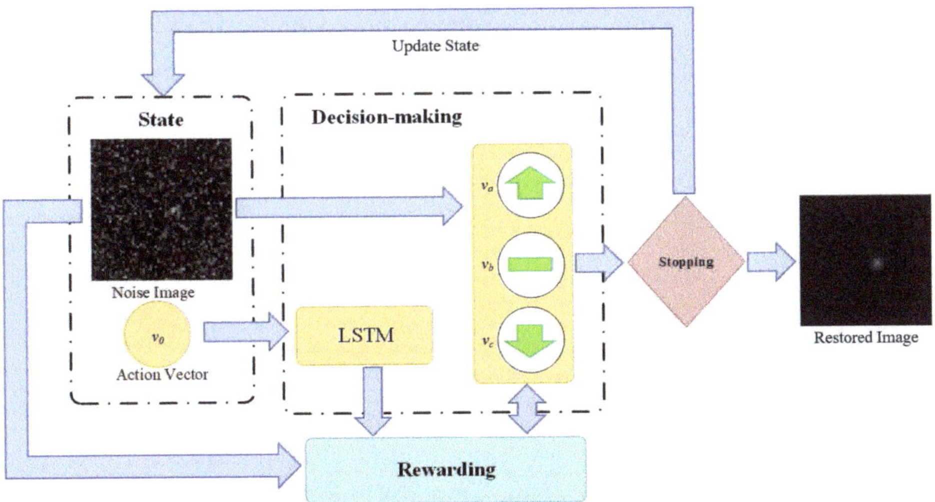

Figure 2. Flowchart of the star image denoising algorithm based on reinforcement learning.

3.2.1. State and Initial State

The state is referred as a set of input information for the decision-making part. In this study, the state consists of two sections: (1) the current input image, on which the selected actions will be directly applied; (2) the historical action vector, which is the action selected in the previous iteration. In the iterative image denoising model proposed by Yu et al., a past historical action vector is also added as a reference to the action selection at current iteration [44]. They suggest that the restoration results are improved compared with using current input image only.

The input image of the initial state is set as the arithmetic mean of the multi-frame star images. Because the average of multi-frame star images is usually not too far away from the ground truth (as discussed in the introduction, this is the traditional way to restore star image), using it as the initial input image reduces the numbers of iterations (NOI), and also avoids the local optimum. The action vector of the initial state is set as a zero vector, since there is no action taken before the initial state.

3.2.2. Decision-Making Part

The decision-making part receives the input information and performs predefined choices of actions on the image. There are two modules in the decision-making part corresponding to the two sections of input state. The first module processes the input image. The star images are 8-bit or 16-bit grayscale images, of which the gray value is determined by the number of photons counted by the star sensor. Therefore, the decisions for any given pixels can be simply defined as: (1) increasing gray value by w; (2) decreasing gray value by w; (3) stay the same. The parameter w is the searching radius, which is (similar as the learning rate in deep neural network) a hyperparameter that can be adjusted based on the property of image. For the 8-bit grayscale images with mild noise, a searching radius can be directly set as 1, since the initial input image is usually not too far away from the ground truth. For the 16-bit gray-scale images or images with severe noise, searching radius could be set as a variable that is larger in the first few iterations, and then gradually decreases to 1 as the iterations go on. Increasing the initial searching radius (ISR) can reduce NOI when the amount of noise is relatively large, but it has to decrease to 1 in order to produce accurate restored image.

The second module analyzes the historical action vectors based on long short-term memory (LSTM) [50]. LSTM is a special type of recurrent neural network (RNN), of which the output of current state partly depends on the previous input. Compared with regular RNN, LSTM includes an additional hidden state that decides whether the previous input should be "remembered" or "forgotten" based on a gating function. By abandoning unnecessary inputs, LSTM overcomes the gradient vanishing/explosion problem in regular RNN, and achieves better performance in processing long sequence data [51]. LSTM is introduced in the model because the iterative actions performed by the decision-making part form a sequence dataset, which can be effectively processed by LSTM. By storing historical images and corresponding actions, LSTM provides additional information and references for current action selection, and enable the model to learn from previous actions.

3.2.3. Rewarding Part

Theoretically, any functions that fulfill the requirement that returns a higher value as it is getting closer to the ground truth could potentially be a reward function. However, the choice of reward function significantly influences the performance of the restoration.

In the traditional reinforcement learning algorithm for image denoising, reward function is usually derived from Peak Signal-to-Noise Ratio (PSNR) or Structural Similarity (SSIM) [44]. In this study, the likelihood function of mixed Poisson–Gaussian noise, which was approximately solved in the previous studies [5,47,48], is introduced as the reward function without approximation. Specifically, the reward function is showing as follows:

$$\max f(z_q|b'_q) = \log(\prod_{q=1}^{N}(p(z_q|b'_q)) = \sum_{q=1}^{N}\sum_{i=1}^{M^2}\log(\sum_{j=1}^{+\infty}\frac{e^{-[b'_q]_i}[b'_q]_i^j}{j!}\frac{e^{-\frac{1}{\sigma^2}([z_q]_i-j)^2}}{\sqrt{2\pi\sigma^2}}) \tag{5}$$

There are two advantages of applying this likelihood function as reward function: (1) it returns highest value at the ground truth, which fulfills the requirement of reward function for reinforcement learning; (2) by directly applying the likelihood function as the reward function rather than making an approximate solution (like GAT method), it is expected to achieve more accurate restoration on the star image.

In practice, the variance of the Gaussian noise is unknown, but is required to calculate the rewards. This term needs to be calculated from the variance of the marginal pixels that are far away from stars. Because the signal for the background is expected to be close to 0 and Poisson noise is related to signal intensity, ideally the Poisson noise in the marginal pixels are ignorable, and the variance in the marginal pixels are mostly contributed by that of the Gaussian noise.

It should also be noted that there is an infinite term in the function. This term refers as the probability of the showing certain gray value at a given pixel. Therefore, it usually returns high value when the gray value is close to the ground truth, and then decreases to almost 0 when the difference is too large (it is less likely that the observed value has too much difference from the ground truth). Therefore, in our algorithm, this term is not taken into account when it is smaller than 10^{-8}—the probability less than that has little influence on the reward function and is ignorable. Moreover, in practice, the differences between the possibilities calculated from Equation (5) could be very small (especially when the variance of the noise is relatively large and the curve of the likelihood function is "flat"), which means the differences between the rewards of actions are not obvious. As a result, the decision-making part may fail to determine the appropriate action due to the vague rewarding. In order to solve this problem, the values returned from the likelihood function may need to be normalized to a larger scale, so as to magnify the differences between the rewards of different states and actions.

3.2.4. Stopping Part

An automatic stopping part is designed in the algorithm in order to determine whether the processed image is close enough to the ground truth, and if yes, automatically stop the iterations. According to the decision-making part and reward function described in the previous sections, it can be easily inferred that when the processed image reaches the ground truth, the decision-making part would choose to stay at the same gray value, because either increasing or decreasing the gray value would return negative rewards. Thus, stopping index (SI), defined as the numbers of times that the decision-making part has continuously chosen to stay at the same gray value, is designed as the main algorithm of the stopping part. SI is also a hyperparameter that should be adjusted based on property of the star images. For the reason that the decision-making part could possibly choose random actions, increasing SI in some extent can improve the accuracy of the restoration results, but at the same time, increase number of iterations (NOI). For the same reason, the maximum number of iterations (MNI) may also need to be defined, so as to avoid infinite loop. The detail test on SI and MNI will be discussed in the next section.

3.3. *Implementation Details*

The algorithm in this study is built using Tensorflow backend [52] and in Python 3.6 environment. Deep Q-learning [41] is adopted in this study for the training process. The ε-greedy is set relatively high as 0.95, because the rewarding function is continuous and has only single peak, which means taking the action with higher rewarding is a better choice than taking a random action for the most of the cases. Adam [53] optimizer is applied in the model and the batch size is set at 32. The hyperbolic tangent function is applied as the gating function of LSTM.

In order to find the favorable choice of hyperparameters, two sets of experiments were conducted to test the hyperparameters introduced above. For the tests on SI, the searching radius was set as a constant of 1, and MNI was set as a constant of 500. Six experiments were conducted for SIs, varying from 2 to 7 with an increment of 1. For the tests on the ISR, the SI is set as a constant of 3, and the MNI was set as a constant of 500. Four experiments were conducted for the test on ISR. In one of the experiments, the searching radius is set as a constant of 1. While in the other three experiments, the ISR is set at 5, 10, and 20, respectively, and then decreased at a factor of 0.5 (rounding up) every 10 iterations. The working flow and specific implementations for the algorithm proposed in this study is summarized in Algorithm 1.

Algorithm 1. Steps for the proposed denoising algorithm

Step 1: Create initial input image as the arithmetic mean of N frames of noise image; define initial action vector v_0 as a zero vector; initialize stopping counter c_s = 0; setup hyperparameter: *SI, MNI, ISR*	* Initialize operation
Step 2: Calculate the variance of Gaussian noise using the boundary pixels	
Step 3: Define action table ($a_1 = + w$, $a_2 = - w$, $a_3 =$ *no action*, initialize w equal to *ISR*); initialize action-value function Q with random weight θ	* Setup deep Q-learning
Step 4: **for** each of the $M \times M$ pixels: initialize sequence of state $\{b_i\}$	* Start iteration
Iterate through $i = 1, 2, \ldots, MNI$:	
Process v_{i-1} through LSTM and obtain v'_{i-1}	
With probability ε select $v_i = \text{argmax } Q(b_{i-1}; v'_{i-1}; \theta)$	
Otherwise select a random action v_i from action table	
Execute action v_i, update environment $b_i = b_{i-1} + v_i$, and observed reward r_i	
Update action-value function Q with the observed $(b_i; v_i; r_i)$	
After every 10 iterations $w \leftarrow$ rounding up $(w/2)$	
if a_3 is chosen in this loop **then**: $c_s \leftarrow c_s + 1$	
else: $c_s \leftarrow 0$	
end if	
if $c_s \geq SI$ **or** $I > MNI$ **then**: output the state b_i as the restored result, end the iteration	* Automatic stopping
end if	
end for	

4. Results

Hyperparameters were tested and calibrated so as to provide appropriate settings to the model. SI, MNI and ISR were tested for simulated star images of different amount of noise. Accuracy and running time are the two aspects tested in the experiments. The running time is evaluated by NOI—obviously, more iterations are more time-consuming. Concerning the goal of star image restoration, the accuracy was evaluated by mean squared error (MSE), PSNR, and SSIM, which are defined in Equations (6)–(8) as follow:

$$MSE = \frac{1}{M^2}\sum_{i=1}^{M^2}([b'_q]_i - [b_q]_i)^2 \tag{6}$$

$$PSNR = 10\log_{10}(\frac{MAX^2}{MSE}) \tag{7}$$

$$SSIM = \frac{(2\mu_b\mu_{b'} + (0.01MAX)^2)(2\sigma_{bb'} + (0.03MAX)^2)}{(\mu_b{}^2 + \mu_{b'}{}^2 + (0.01MAX)^2)(\sigma_b{}^2 + \sigma_{b'}{}^2 + (0.03MAX)^2)} \tag{8}$$

where MAX is the possible maximum gray value (which is 255 for 8-bit and 65,535 for 16-bit grayscale image), b represents the ground truth and b' represents the restore image (also in subscript), μ is the average and σ is the variance, $\sigma_{bb'}$ is the covariance of restore image and ground truth. These three evaluation indicators are calculated for each set of noise images, and then averaged with other sets of noise images with the same type of noise. As described in Section 3.1, there are 1200 sets of noise images with three different levels of noise utilized as the testing data. Thus, the three evaluation indicators are averaged among 400 tests (which is the value range of q in Equation (6)) in this study.

The results of experiments on SI and MNI are shown in Figure 3. It turns out that the three evaluation indicators share the similar trend with the change of SI. Specifically, the accuracy of the model improves significantly in a limited range with the increase of SI regardless of the amount of noise in the star images. That means increasing SI in a certain range can improve the quality of restoration results. However, as shown in the same figure, the increase of the SI also leads to more NOI (longer running time). When the SI exceeds certain threshold, the model ends up in infinite loop and has to be stopped by MNI. In that vein, the appropriate SI setting should be large enough to achieve accurate restoration, but small enough to avoid unnecessary iterations.

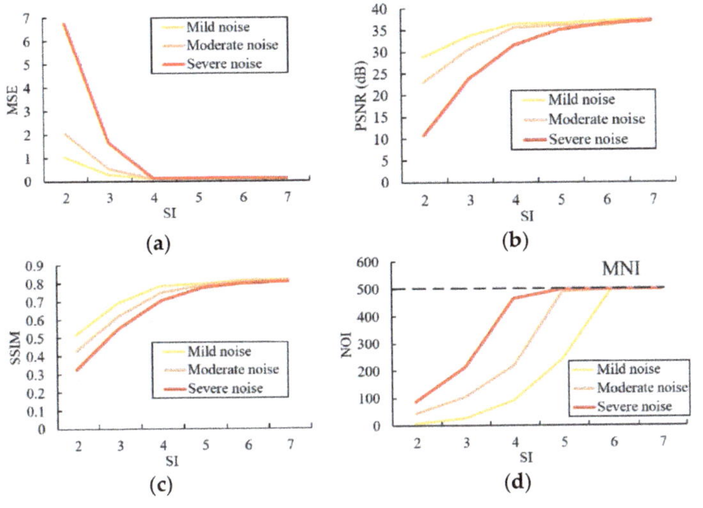

Figure 3. Test results of (**a**) mean square error (MSE); (**b**) peak signal-to-noise ratio (PSNR); (**c**) structure similarity (SSIM); (**d**) number of iterations (NOI) at different settings of stopping index (SI)) for three levels of noise.

The results of experiments on the ISR are shown in Figure 4. It turns out that ISR does not have significant influence on the accuracy of restoration results, but affects the running time. Moreover, its influences on running time vary for star images with different amounts of noise: for the star images with mild noise, increasing the ISR does not reduce running time; for the star images with moderate and severe noise, however, the NOI significantly reduced with a larger ISR.

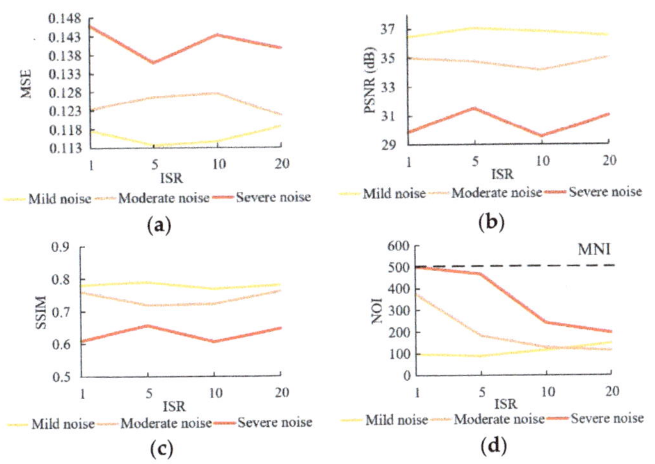

Figure 4. Test results of (**a**) MSE; (**b**) PSNR; (**c**) SSIM; (**d**) NOI at different settings of initial searching radius (ISR)) for three levels of noise.

The performance of the star image denoising method proposed in this study is compared with GATFP method proposed by Zhang et al. [6] and DCNN algorithm proposed by Liu et al. [45] on the same testing dataset. For the GATFP method, the denoising process is conducted based on the optimized likelihood function given in Equation (3). The implementation of the DCNN model is a setup based on the structure described in Reference [45], which includes 17 convolutional layers and 256 neurons in each layer. Rectified linear unit (ReLU) [54] is applied as the activation function between the layers. The performance of these denoising methods is also evaluated with the three evaluation indicators described above. The arithmetic means of the evaluation indicators for all the testing images in each category of noise are compared in Table 1 for all the three denoising methods. Typical examples of restored images are also shown in Figure 5.

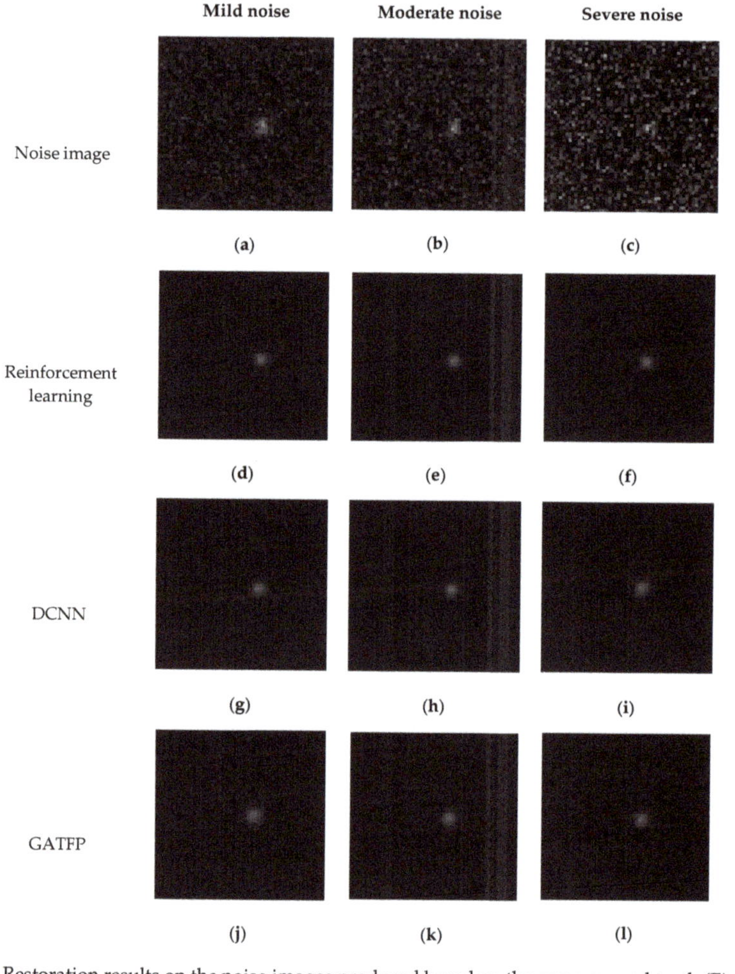

Figure 5. Restoration results on the noise images produced based on the same ground truth (Figure 1): (**a**) noise image with mild noise, (**b**) noise image with moderate noise, (**c**) noise image with severe noise, (**d**) denoising result using reinforcement learning on mild noise, (**e**) denoising result using reinforcement learning on moderate noise, (**f**) denoising result using reinforcement learning on severe noise, (**g**) denoising result using DCNN [45] on mild noise, (**h**) denoising result using DCNN [45] on moderate noise, (**i**) denoising result using DCNN [45] on severe noise, (**j**) denoising result using GATFP [6] on mild noise, (**k**) denoising result using GATFP [6] on moderate noise, (**l**) denoising result using GATFP [6] on severe noise.

Table 1. Arithmetic means of evaluation indicators for all the testing images with three levels of noise using three different denoising methods.

	Mild Noise			Moderate Noise			Severe Noise		
	MSE	**PSNR(dB)**	**SSIM**	**MSE**	**PSNR(dB)**	**SSIM**	**MSE**	**PSNR(dB)**	**SSIM**
Reinforcement learning	0.1147	37.57	0.8215	0.1149	37.10	0.8150	0.1440	31.58	0.7060
DCNN [45]	0.1238	35.43	0.7776	0.1260	35.06	0.7602	0.1587	30.81	0.6816
GATFP [6]	0.1402	31.18	0.7153	0.1638	27.94	0.6384	0.1925	23.05	0.5483

5. Discussion

5.1. Hyperparameter Settings

5.1.1. SI and MNI

As indicated in the experiment results, SI influences both accuracy and running time. When the SI is set at 2 or 3, the accuracy is relatively low because the decision-making part randomly chooses to stay at the same value. But this possibility quickly decreases as the SI increases to 4. That means the accuracy of the denoising model can be improved by increasing SI in a limited extent: as indicated in Figure 3, the increase of SI does not improve the model accuracy when it is set beyond 5.

NOI also increases with SI. In fact, when SI is larger than 6, the model easily gets into infinite loop and has to be stopped by MNI. This is because in order to stop the iterations without MNI, the decision-making part has to avoid randomly choosing to increase or decrease the gray value for number of times equals to SI after reaching the appropriate restored value—that possibility also decreases with the increase of SI.

It is also notable that the SI has little relation with the amount of noise or searching radius (with adequate iterations). Because the amount of noise decreases with the iterations, and SI does not start to work until the decision-making part reaches to the appropriate restored value. In that vein, 4 or 5 seems to be the appropriate value for SI regardless of the amount of noise in the star image. The MNI setting, however, depends on the amount of noise in the image, since NOI increases with the amount of noise. A short experiment under low SI (for example, 3 or 4) may be necessary to figure out approximately how many iterations the decision-making part needs to form the restored image before formally running the model in order to find the appropriate setting of MNI.

5.1.2. Initial Searching Radius

As indicated in the experiment results, ISR does not affect the accuracy significantly, since it will decrease to 1 in the end. But it affects the NOI. For the images with mild noise, an ISR larger than 1 is unnecessary, and sometimes even causes more iterations, especially when ISR is larger than the noise. For the images with moderate and severe noise, a larger ISR reduces NOI significantly, and therefore, restores the images faster. In practice, however, the quality of observed star image is unknow. In that vein, searching radius is a hyperparameter that needs to be calibrated with actual input star image.

According to discussion of the experiments above, the recommended setting of the hyperparameter for the noise images with different amount of noise using the hyperparameter settings shown in Table 2. This setting is also applied to the training of the reinforcement learning model when comparing its performance with other two methods.

Table 2. Hyperparameter settings for the star images of three different amount of noise.

	Mild Noise	Moderate Noise	Severe Noise
Stopping index (SI)	4	4	4
Initial searching radius (ISR)	1	5	10
Maximum number of iterations (MNI)	300	500	500

5.2. Comparisons with Existing Methods

The algorithm proposed in this article is compared with GATFP [6] and DCNN [45] methods. The performances of the two methods on the dataset used in study generally confirm with those presented in the previous studies. As indicated in Table 1 and Figure 5, these denoising methods are able to suppress the noise in a star image to some extent. The three denoising methods have better performance on the suppression of mild noise than that of moderate noise restoration, which is better than that of extreme noise. This means the images with less noise are naturally easier to be accurately denoised. Although it can hardly be identified from the restored image (Figure 5) by human eyes, the three methods have different performances in term of retrieving the true value of star target. As shown in Table 1, all the three of the evaluation indicators shows that the reinforcement learning method proposed in this study provides more accurate restoration results than that of the DCNN method, which is more accurate than that of GATFP method. That means the algorithm proposed in this article is able to achieve more accurate restoration than the existing star image denoising methods.

The fact that the two machine learning methods achieve better restoration results, especially when the star images are more distorted, is possibly because they overcome the error introduced by the approximation in GAT method. Moreover, the method proposed in this article can perform more accurate restorations than that based on DCNN. It should be noted that the advantage in accuracy of the restoration results is probably caused by the reward function applied in the reinforcement learning algorithm, which is directly linked with the distribution pattern of the noise in the star image. Although the reinforcement learning algorithm proposed in this study is designed more specifically for star image with mixed Poisson–Gaussian noise, it certainly has the potential to be applied to suppress other type of image noise, as long as the corresponding reward function is defined.

In comparing the two machine learning models, the DCNN model [45] includes a series of convolutional layers for image feature extraction and a deep neural network (DNN) for denoising calculation, while the reinforcement learning model presented in this study is an iterative MDP model that is integrated with DNN for action-value regression. The different structures of the two models lead to different training processes. The weights and bias of the neurons are trained in DCNN model, while the optimal denoising process is trained through iterative trials in reinforcement learning model. As a result, the reinforcement learning model is much simpler than the DCNN model in terms of computational complexity. As shown in Table 3, the parameters that need to be trained in the reinforcement learning model is less than one half of those in DCNN model, and the number of computations in training the reinforcement learning model is only about one fourth of that in training DCNN model. This results generally conforms with a comparison between reinforcement learning model and DCNN model by Yu et al. [44].

Table 3. Comparison of the computational complexity between the reinforcement learning model proposed in this study and DCNN model [45].

	Reinforcement Learning	DCNN [45]
Parameters ($\times 10^5$)	1.10	2.47
Calculations ($\times 10^8$)	2.26	9.02

Moreover, large numbers of noise images for different ground truth are necessary to train DCNN model, while the reinforcement learning model requires much less training dataset: theoretically, the agent can finally propose the restored image even with only one set of noise images with adequate numbers of trials. Thus, the star image denoising model proposed in this study depends on less training dataset, and may be more practical to extract the true value of star target, especially when the observations are limited.

6. Conclusions and Future Studies

A novel approach for star image denoising based on reinforcement learning is presented in this study. Compared with the existing methods based on DCNN, this model learns to dynamically search for the appropriate restored image based on reward function derived from mixed Poisson–Gaussian likelihood function. Thus, the mechanism behind our algorithm is more intuitive, and leads to more accurate restoration, than those that are based on DCNN. Compared with MLE methods that try to approximately solve the likelihood function (such as GAT), the same likelihood function is applied as the reward function without approximation. In that way, the likelihood function is solved by the computer through the MDP more accurately.

The limitation of this algorithm is that we didn't add any kind of convolution layer to the input image. As a result, when the star image is rather huge, the restoration process becomes time-consuming (the accuracy is not affected though). The reason we did not apply CNN in the algorithm is that the reward function derived from the likelihood function only works for the gray value of the star image, but not the image feature as captured by the CNN. An appropriate way to integrate CNN in the algorithm, and thus further improve its efficiency, could be a topic of our study in the future.

It is also a potential topic to extend the application of the reinforcement learning model proposed in this study into the denoising practice of broader types of images, which could have different numbers of color depths. Processing the noise image depth by depth may not be a favorable approach, since it disregards the correlations between the color depths. This problem may also be solved by a properly designed CNN that can extract the features of a noise image.

Author Contributions: Conceptualization, M.X., Z.Z. and Y.L.; methodology, M.X.; software, M.X. and K.C.; validation, M.X., W.Z. and K.C.; formal analysis, M.X.; investigation, M.X.; data curation, M.X. and W.Z.; writing—original draft preparation, M.X. and W.Z.; writing—review and editing, M.X., Z.Z., W.Z., Y.L., and K.C.; visualization, M.X.; supervision, Y.L. and Z.Z.; project administration, Y.L.; funding acquisition, Y.L. and M.X. All authors have read and agreed to the published version of the manuscript.

Funding: This research was funded in part by China National Key R&D Program grant number 2018YFB1600400 and 2020YFE0201500, and in part by China Postdoctoral Science Foundation, grant number 2020M670730.

Acknowledgments: The authors would also like to thank Qinglai Yu, Zhanjun Ma from Dalian Maritime University for their help in preparing the training dataset. The authors are grateful to the anonymous reviewers for their constructive comments on this work.

Conflicts of Interest: The authors declare no conflict of interest.

References

1. Wang, W.; Wei, X.; Lin, J.; Wang, G. Noise suppression algorithm of short-wave infrared star image for daytime star sensor. *Infrared Phys. Technol.* **2017**, *85*, 382–394. [CrossRef]
2. Jin, B.; Park, N.; George, K.M.; Choi, M.; Yeary, M.B. Modeling and analysis of soft-test/repair for CCD-based digital X-ray systems. *IEEE Trans. Instrum. Meas.* **2003**, *52*, 1713–1721. [CrossRef]
3. Hwang, H.; Haddad, R. Adaptive median filters: New algorithms and results. *IEEE Trans. Image Process.* **1995**, *4*, 499–502. [CrossRef]
4. Yugander, P.; Tejaswini, C.H.; Meenakshi, J.; Samapath kumar, K.; Suresh Varma, B.V.N.; Jagannath, M. MR image enhancement using adaptive weighted mean filtering and homomorphic filtering. In Proceedings of the International Conference on Computational Intelligence and Data Science, Gurgaon, India, 20–23 February 2019; Procedia Computer Science. Elsevier: Amsterdam, The Netherlands, 2020. [CrossRef]
5. Zhou, Z.; Lam, E.Y.; Lee, C. Nonlocal means filtering based speckle removal utilizing the maximum a posteriori estimation and the total variation image prior. *IEEE Access* **2019**, *7*, 99231–99243. [CrossRef]
6. Zhang, Y.; Song, P.; Dai, Q. Fourier ptychographic microscopy using a generalized Anscombe transform approximation of the mixed Poisson-Gaussian likelihood. *Opt. Express* **2017**, *25*, 168–179. [CrossRef]
7. Bonavito, N.L.; Dorband, J.E.; Busse, T. Maximum entropy restoration of blurred and oversaturated Hubble space telescope imagery. *Appl. Opt.* **1993**, *32*, 5768–5774. [CrossRef]
8. Katsaggelos, A.K.; Lay, K.T. Maximum likelihood blur identification and image restoration using the EM algorithm. *IEEE Trans. Signal Process.* **1991**, *39*, 729–733. [CrossRef]

9. Llacer, J.; Nuiez, J. Iterative maximum-likelihood and Bayesian algorithms for image reconstruction in astronomy. In *Restoration of Hubble Space Telescope Images*; White, R.L., Allen, R.J., Eds.; Space Telescope Science Institute: Baltimore, MD, USA, 1990; pp. 62–69.
10. Snyder, D.L.; Hammoud, A.M.; White, R.L. Image recovery from data acquired with a charge-coupled-device camera. *J. Opt. Soc. Am. A* **1993**, *10*, 1014–1023. [CrossRef]
11. Benvenuto, F.; Camera, A.L.; Theys, C.; Ferrari, A.; Lanteri, H.; Bertero, M. The study of an iterative method for the reconstruction of images corrupted by Poisson and Gaussian noise. *Inverse Probl.* **2008**, *24*, 1–20. [CrossRef]
12. Li, D.; Sun, C.; Yang, J.; Liu, H.; Peng, J.; Zhang, L. Robust Multi-Frame Adaptive Optics Image Restoration Algorithm Using Maximum Likelihood Estimation with Poisson Statistics. *Sensors* **2017**, *17*, 785. [CrossRef]
13. Barto, A.G. Reinforcement Learning. In *Neural Systems for Control*; Omidvar, O., Elliott, D.L., Eds.; Academic Press: Cambridge, MA, USA, 1997; pp. 7–30.
14. Sutton, R.S.; Barto, A.G. *Reinforcement Learning: An. Introduction*; MIT Press: Cambridge, MA, USA, 1998.
15. Bellman, R.E. A Markov Decision Process. *J. Math. Fluid Mech.* **1957**, *6*, 679–684.
16. Puterman, M.L. *Markov Decision Processes: Discrete Stochastic Programing*; John Wiley & Sons, Inc.: Hoboken, NJ, USA, 2014.
17. Redmon, J.; Divvala, S.; Girshick, R.; Farhadi, A. You only look once: Unified, real-time object detection. In Proceedings of the IEEE Conference on Computer Vision and Pattern Recognition, Las Vegas, NV, USA, 27–30 June 2016; pp. 779–788.
18. Liu, W.; Anguelov, D.; Erhan, D.; Szegedy, C.; Reed, S.; Fu, C.-Y.; Berg, A.C. SSD: Single shot multibox detector. In Proceedings of the European Conference on Computer Vision, Amsterdam, The Netherlands, 8–16 October 2016; Springer: Cham, Switzerland, 2016; pp. 21–37.
19. Ren, S.; He, K.; Girshick, R.; Sun, J. Faster-RCNN: Towards real-time object detection with region proposal networks. In Proceedings of the Neural Information Processing Systems, Montreal, QC, Canada, 7–12 December 2015; pp. 265–283.
20. Zhang, P.; Cheng, H.; Zou, B.; Dai, P.; Ye, C. Load data mining based on deep learning method. In Proceedings of the 3rd International Conference on Computer Science and Application Engineering, Sanya, China, 22–24 October 2019.
21. Nguyen, G.; Dlugolinsky, S.; Bobak, M.; Tran, V.; Garcia, A.L.; Heredia, I.; Malik, P.; Hluchy, L. Machine learning and deep learning for frameworks and libraries for large-scale data mining: A survey. *Artif. Intell. Rev.* **2019**, *52*, 77–124. [CrossRef]
22. Zhang, K.; Zou, W.; Zhang, L. FFDNet: Toward a Fast and Flexible Solution for CNN based Image Denoising. *IEEE Trans. Image Process.* **2018**, *27*, 4608–4622. [CrossRef]
23. Jiao, L.; Zhao, J. A survey on the new generation of deep learning in image processing. *IEEE Access* **2019**, *7*, 172231–172263. [CrossRef]
24. Nah, J.; Kim, T.H.; Lee, K.M. Deep multi-scale convolutional neural network for dynamic scene deblurring. In Proceedings of the IEEE Conference on Computer Vision and Pattern Recognition, Honolulu, HI, USA, 21–26 July 2017; pp. 257–265.
25. Sun, J.; Cao, W.; Xu, Z.; Ponce, J. Learning a convolutional neural network for non-uniform motion blur removal. In Proceedings of the IEEE Conference on Computer Vision and Pattern Recognition, Boston, MA, USA, 7–12 June 2015; pp. 769–777.
26. Xu, L.; Tao, X.; Jia, J. Inverse kernels for fast spatial deconvolution. In Proceedings of the European Conference on Computer Vision, Zurich, Switzerland, 5–12 September 2014; pp. 33–48.
27. Chen, Y.; Yu, W.; Pock, T. On learning optimized reaction diffusion processes for effective image restoration. In Proceedings of the IEEE Conference on Computer Vision and Pattern Recognition, Boston, MA, USA, 7–12 June 2015; pp. 5261–5269.
28. Lefkimmiatis, S. Non-local color image denoising with convolutional neural networks. In Proceedings of the IEEE Conference on Computer Vision and Pattern Recognition, Honolulu, HI, USA, 21–26 July 2017; pp. 3587–3596.
29. Zhang, K.; Zou, W.; Chen, Y.; Meng, D.; Zhang, L. Beyond a Gaussian denoiser: Residual learning of Deep CNN for image denoising. *IEEE Trans. Image Process.* **2017**, *26*, 3142–3155. [CrossRef]

30. Dong, C.; Deng, Y.; Loy, C.C.; Tang, X. Compression artifacts reduction by a deep convolutional network. In Proceedings of the IEEE Conference on Computer Vision and Pattern Recognition, Boston, MA, USA, 7–12 June 2015; pp. 576–584.
31. Guo, J.; Chao, H. Building dual-domain representations for compression artifacts reduction. In Proceedings of the European Conference on Computer Vision, Amsterdam, The Netherlands, 8–16 October 2016; pp. 629–644.
32. Wang, Z.; Liu, D.; Chang, S.; Ling, Q.; Yang, Y.; Huang, T.S. D3: Deep dual-domain based fast restoration of JPEG-compressed images. In Proceedings of the IEEE Conference on Computer Vision and Pattern Recognition, Las Vegas, NV, USA, 27–30 June 2016; pp. 2764–2772.
33. Dong, C.; Loy, C.C.; He, K.; Tang, X. Image super-resolution using deep convolutional networks. *IEEE Trans. Pattern Anal. Mach. Intell.* **2017**, *38*, 295–307. [CrossRef]
34. Hui, T.W.; Loy, C.C.; Tang, X. Depth map super-resolution by deep multi-scale guidance. In Proceedings of the European Conference on Computer Vision, Amsterdam, The Netherlands, 8–16 October 2016; pp. 3539–3569.
35. Kim, J.; Lee, J.K.; Lee, K.M. Accurate image super-resolution using very deep convolutional networks. In Proceedings of the IEEE Conference on Computer Vision and Pattern Recognition, Las Vegas, NV, USA, 27–30 June 2016; pp. 1646–1654.
36. Kim, J.; Lee, J.K.; Lee, K.M. Deeply-recursive convolutional network for image super-resolution. In Proceedings of the IEEE Conference on Computer Vision and Pattern Recognition, Las Vegas, NV, USA, 27–30 June 2016; pp. 1637–1645.
37. Haris, M.; Shakhnarovich, G.; Ukita, N. Deep back-projection networks for super-resolution. In Proceedings of the IEEE Conference on Computer Vision and Pattern Recognition, Salt Lake City, UT, USA, 18–23 June 2018; pp. 1664–1673.
38. Tai, Y.; Yang, J.; Liu, X. Image super-resolution via deep recursive residual network. In Proceedings of the IEEE Conference on Computer Vision and Pattern Recognition, Honolulu, HI, USA, 21–26 July 2017; pp. 2790–2798.
39. Tai, Y.; Yang, J.; Liu, X.; Xu, C. Memnet: A persistent memory network for image restoration. In Proceedings of the IEEE International Conference on Computer Vision, Venice, Italy, 22–29 October 2017; pp. 4549–4557.
40. Wang, X.; Yu, K.; Dong, C.; Loy, C.C. Recovering realistic texture in image super-resolution by deep spatial feature transform. In Proceedings of the IEEE Conference on Computer Vision and Pattern Recognition, Salt Lake City, UT, USA, 18–23 June 2018; pp. 606–615.
41. Mnih, V.; Kavukcuoglu, K.; Silver, D.; Rusu, A.A.; Veness, J.; Bellemare, M.G.; Graves, A.; Riedmiller, M.; Fidjeland, A.K.; Ostrovski, G. Human-level control through deep reinforcement learning. *Nature* **2015**, *518*, 529–533. [CrossRef]
42. Silver, D.; Huang, A.; Maddison, C.J.; Guez, A.; Sifre, L.; Van Den Driessche, G.; Schrittwieser, J.; Antonoglou, I.; Panneershelvam, V.; Lanctot, M. Mastering the game of go with deep neural networks and tree search. *Nature* **2016**, *529*, 484–489. [CrossRef]
43. Peng, Y.; Zhang, J.; Ye, Z. Deep reinforcement learning for image hashing. *IEEE Trans. Multimed.* **2020**, *22*, 99–111. [CrossRef]
44. Yu, K.; Dong, C.; Lin, L.; Loy, C.C. Crafting a toolchain for image restoration by deep reinforcement learning. In Proceedings of the IEEE Conference on Computer Vision and Pattern Recognition, Salt Lake City, UT, USA, 18–23 June 2018; pp. 2443–2452.
45. Liu, Y.; Zhao, C.; Xu, Q. Neural network-based noise suppression algorithm for star images captured during daylight hours. *Acta Opt. Sin.* **2019**, *39*, 1–8.
46. Chouzenoux, E.; Jezierska, A.; Pesquet, J.C.; Talbot, H. A convex approach for image restoration with exact Poisson–Gaussian likelihood. *SIAM J. Imaging Sci.* **2015**, *8*, 2662–2682. [CrossRef]
47. Marnissi, Y.; Zheng, Y.; Pesquet, J.C. Fast variational Bayesian signal recovery in the presence of Poisson-Gaussian noise. In Proceedings of the IEEE International Conference on Acoustics, Speech and Signal Processing, Shanghai, China, 20–25 March 2016; pp. 3964–3968.
48. Zhang, J.; Hirakawa, K.; Jin, X. Quantile analysis of image sensor noise distribution. In Proceedings of the IEEE International Conference on Acoustics, Speech and Signal Processing, Brisbane, QLD, Australia, 19–24 April 2015; pp. 1598–1602.
49. Xu, Q.; Zhao, C.; Li, X. Stellar radiation modeling and image simulation for airborne daytime star sensor. In Proceedings of the IEEE International Conference on Acoustics, Speech and Signal Processing, Shanghai, China, 20–25 March 2016; pp. 630–635.

50. Hochreiter, S.; Schmidhuber, J. Long short-term memory. *Neural Comput.* **1997**, *9*, 1735–1780. [CrossRef]
51. Sundermeyer, M.; Schluter, R.; Ney, H. LSTM neural networks for language modeling. In Proceedings of the InterSpeech, Portland, OR, USA, 9–13 September 2012.
52. Abadi, M.; Barham, P.; Chen, J.; Chen, Z.; Davis, A.; Dean, J.; Ghemawat, S.; Irving, G.; Isard, M.; Kudlur, M.; et al. TensorFlow: A system for large-scale machine learning. In Proceedings of the 12th USENIX Symposium on Operating System Design and Implementation (OSDI), Savannah, GA, USA, 2–4 November 2016; pp. 265–283.
53. Kingma, D.; Ba, J. Adam: A method for stochastic optimization. In Proceedings of the IEEE International Conference on Learning Representation, San Diego, CA, USA, 7–9 May 2015.
54. Nair, V.; Hinton, G.E. Rectified linear units improve restricted Boltzmann machines. In Proceedings of the International Conference on Machine Learning, Haifa, Israel, 21–24 June 2010; pp. 807–814.

sensors

MDPI

Review

Review of Capacitive Touchscreen Technologies: Overview, Research Trends, and Machine Learning Approaches

Hyoungsik Nam *, Ki-Hyuk Seol, Junhee Lee, Hyeonseong Cho and Sang Won Jung

Department of Information Display, Kyung Hee University, Seoul 02447, Korea; seol8118@khu.ac.kr (K.-H.S.); jklopp1007@khu.ac.kr (J.L.); chohs0713@khu.ac.kr (H.C.); mrswjung@khu.ac.kr (S.W.J.)
* Correspondence: hyoungsiknam@khu.ac.kr; Tel.: +82-2-961-0925

Abstract: Touchscreens have been studied and developed for a long time to provide user-friendly and intuitive interfaces on displays. This paper describes the touchscreen technologies in four categories of resistive, capacitive, acoustic wave, and optical methods. Then, it addresses the main studies of SNR improvement and stylus support on the capacitive touchscreens that have been widely adopted in most consumer electronics such as smartphones, tablet PCs, and notebook PCs. In addition, the machine learning approaches for capacitive touchscreens are explained in four applications of user identification/authentication, gesture detection, accuracy improvement, and input discrimination.

Keywords: touchscreen; capacitive; display; SNR; stylus; machine learning

Citation: Nam, H.; Seol, K.-H.; Lee, J.; Cho, H.; Jung, S.W. Review of Capacitive Touchscreen Technologies: Overview, Research Trends, and Machine Learning Approaches. *Sensors* **2021**, *21*, 4776. https://doi.org/10.3390/s21144776

Academic Editor: Paweł Pławiak

Received: 4 June 2021
Accepted: 12 July 2021
Published: 13 July 2021

1. Introduction

Human beings collect a lot of information through their eyes, and many displays around us play a key role to transfer this visual information. Displays have evolved dramatically from cathode-ray tube (CRT) [1–4] via plasma display panel (PDP) [5–10] and liquid crystal display (LCD) [11–15] to cutting-edge organic light-emitting diode (OLED) [16–22] and micro-LED technologies [23–28]. This evolution has led to larger screen-size, slimmer design, lower weight, higher resolution, faster frame rate, brighter luminance, wider color gamut, longer life time, and lower power consumption in the large-size display applications such as monitors, televisions (TVs), and digital signage [29–39]. The resolutions of off-the-shelf displays have increased up to 8K (7680 × 4320) along with the high frame rate of 120 Hz and the larger screen sizes than 55-inch have taken more than 30% of overall TV set sales [40,41]. Even rollable OLED TVs were demonstrated in the consumer electronics show 2018 (CES2018) [42]. On the other side of the small-size display applications, higher density of pixels, narrower bezel, flexibility, bendability, rollability, and low power consumption have been achieved along with enhanced picture quality [43–48]. The latest smartphones contain the bezel-less screens of larger pixel densities than 450 pixel per inch (ppi) and smartphones with foldable displays are being sold on the market [49]. Recently, as augmented reality and virtual reality (AR/VR) attract substantial interest, the demand for high-performance near-eye displays is increasing further [50–57]. Consequently, the very high resolution OLED on silicon (OLEDoS) displays up to 4410 ppi have been reported [58–63].

On top of the role of a visual information provider, displays have supported the interaction with users by means of various user interfaces. Users can adjust the visual information on the screen by themselves. The very old but still popular representative user interfaces are mouse and keyboard [64–66]. There have also existed pen tablets for more elaborate works such as drawing and writing [67–70]. Because these devices work on the different planes separated from displays, additional markers such as cursors and pointers are needed. On the other hand, more intuitive input interfaces called touchscreens have been studied to directly interact with displays by touching displays [71–74]. Touchscreen technologies can be categorized into finger-touch and stylus-touch methods. While finger-touch methods include resistive, capacitive, acoustic wave, and optical approaches [75–101], stylus-touch ones cover up to electromagnetic resonance (EMR) schemes including finger-touch methodologies [102–108]. Recently, as wearable devices such as smartwatches and smartbands are becoming more popular, small-size displays are becoming further widespread with touch sensing functionality. However, because this very small-area screen cannot support multiple finger-touches and the whole area is covered even by a single finger, a variety of separate input modalities in the outside of the screen have been studied by using infrared (IR) line sensors, microphones, gaze trackers, IR proximity sensors, electric field sensors, deformation sensors, magnetic field sensors, and mechanical interfaces [109–123]. In addition, some approaches have coped with the limitation of the single touch by differentiating palm and finger or identifying pad, nail, tip, and knuckle of a finger [124,125]. Especially, because AR/VR displays are placed near to eyes, it is impossible to touch the screen directly. Therefore, other input tools using various sensors such as leap motion sensors, electromyograph sensors, inertial measurement units, eye-trackers, IR facial gesture sensors, cameras, and axis-tilt sensors, have been employed [126–134].

There have been also efforts to integrate machine learning (ML) approaches into touchscreen technologies. These ML networks are employed to add extra input tools, to improve the touch-sensing performance, to support the user identification/authentication, to discriminate finger-touches from others, and to capture the gestures [135–164].

There have been brief reviews of touchscreen technologies [76,96]. Walker [165] published many overview papers about a variety of touchscreen technologies from resistive to optical and electromagnetic resonance (EMR) stylus schemes. Those papers explained their histories, principles of operation, pros and cons, and applications. However, the technological details have not been handled such as algorithms, driving circuits, and ML approaches. Kwon et al. [166] reviewed capacitive touchscreen technologies including sensors, driving circuits, sensing methods, and stylus schemes in more detail. However, ML approaches were not introduced. Bello et al. [164] summarized ML approaches to improve security on touchscreen devices without addressing the touchscreen technologies. A variety of ML applications only for the security issues were addressed. This paper provides a unified and broader view of the touchscreen technologies with the detailed explanation and ML approaches in various scenarios.

The contributions of this paper are as follows:

- Providing the most comprehensive review about the touchscreen technologies. In particular, this describes various studies on sensing methods and ML approaches.
- Supplementing capacitive touchscreen techniques of the previous review paper [166] by focusing on research topics and results.
- Including various ML methods for user identification/authentication, gesture detection, accuracy improvement, and input discrimination.
- Proposing future directions for researches on touchscreen technologies integrated with ML networks.

This paper is organized as follows. Section 2 addresses the overview of the touchscreen technologies, and then Section 3 describes various studies on capacitive touchscreen applications that are integrated in most smartphone and notebook displays. Section 4 shows the ML approaches working with existing capacitive touchscreen technologies. Section 5 concludes this paper with some suggestions of the future directions.

2. Overview of Touchscreen Technologies

In this section, touchscreen technologies for finger as well as stylus have been simply addressed in terms of principles of operation, advantages, and drawbacks. We categorize the touchscreen technologies into four categories of resistive, capacitive, acoustic wave, and optical, and address further various techniques in each category as shown in Figure 1. Table 1 compares their specifications.

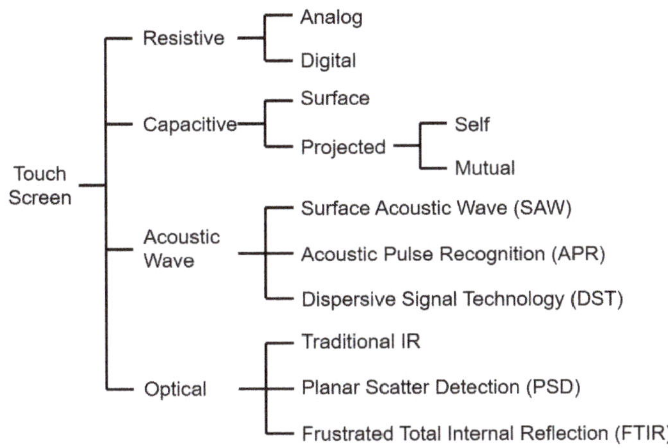

Figure 1. Categories of touchscreen technologies. This figure excludes the touchscreen techniques embedded in pixel areas in a display panel.

Table 1. Comparisons of touchscreen technologies.

Category	Resistive	Capacitive	Acoustic	Optical
# Layers	2	2	1	0 (Traditional) 1 (PSD, FTIR)
Touch Diversity	High	Low	Average	High
Image Clarity (Transmittance)	75–85%	85–90%	92–%	88–%
Multi-Touch	No (Analog) Yes (Digital)	No (Surface, Self) Yes (Mutual)	No	No (Traditional) Yes (PSD, FTIR)
Durability	Poor	Good	Best	Best
Computational Power	Low	Average	High	Low (Traditional) High (PSD, FTIR)
Dimension	Small	Small (ITO) Large (Metal Mesh)	Large	Large
Touch/Tap Force	Strong	Light	Average	Light
Resistance to Contaminants	Best	Poor	Good	Good
Holding Function	Yes	Yes	Yes (SAW) No (APR, DST)	Yes
Mounting Dependency	No	No	No (SAW) Yes (APR, DST)	No

2.1. Resistive Touchscreen

An analog resistive scheme is the oldest touchscreen technology [165]. It extracts touch coordinates by sampling the voltage at the touched area. The voltage is proportional to the location of the screen due to the voltage division based on the ratio of resistances from the current position to two opposite sides [78]. The most popular resistive touchscreen panels are fabricated by 4-wire and 5-wire architectures [79]. Both methods estimate x-axis and y-axis coordinates of a touch position sequentially. Normally, two separate layers are coated by the conductive films only at one side, and one layer should be composed of a flexible material. When the touch force is applied, the flexible layer is pressed to contact the other layer and to obtain the voltage at the contacted area. Four-wire structures use both layers to generate the voltage slopes as well as to sense the voltage as illustrated in Figure 2a. For example, after the flexible layer (Layer #1) generates the voltage slope at an x-axis and the other (Layer #2) senses the voltage, Layer #2 generates the voltage slope at an y-axis and Layer #1 senses the voltage. Five-wire ones apply voltages only to one specific layer (Layer #2) and use the other layer (Layer #1) only to sense the voltage as depicted in Figure 2b. Therefore, it is known that 5-wire schemes usually have a longer life time.

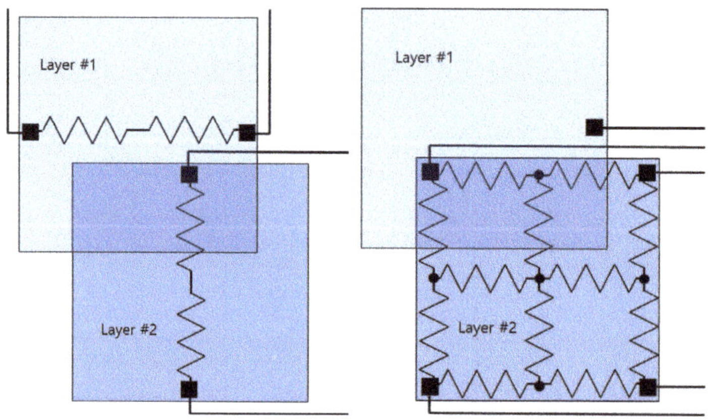

Figure 2. Two most popular resistive touchscreen architectures. (**a**) Four-wire. (**b**) Five-wire.

The advantages of the resistive touchscreen technology are to be able to work with anything, to be fabricated at the lowest cost, to be insensitive to any contaminants, and to consume low power. However, it has drawbacks of the only single touch support, the poor durability due to scratches, poking, and sharp objects, the poor optical clarity, and the relatively high touch force requirement [80,165].

On the other hand, there have been efforts to support multi-touch capability. Some researchers were trying to add the multi-touch functionality to a conventional structure by sensing the current consumption at voltage sources [167–169]. Whereas, other researchers divide the conductive films into multiple lines and columns that give rise to many separate overlapped areas [170–172], where each area can detect touches separately. This scheme is named as the digital resistive touchscreen [165]. Since the resistive touchscreen methods fall short of the capacitive schemes, the resistive touchscreen panels are being applied to the limited areas such as toys, office electronics, and card payment machines.

2.2. Capacitive Touchscreen

Capacitive touchscreens sense the change of the capacitance caused by the finger to estimate the touch position. While resistive schemes need the pressing force to make the actual contact between two conductive layers, capacitive methods can obtain the capacitance change just by the light touch on the screen. Consequently, it enables the smooth and fast scrolling, high durability, and excellent optical performance. In addition, any materials can be adopted for layers, for example, glasses and plastics, while resistive technologies require one flexible layer at least. Because the parasitic capacitance added by fingers is very small, large-size capacitive touchscreen panels are very difficult to implement and contaminants such as water and dusts can be also recognized as touches. Recently, the large size capacitive touchscreens have been reported based on the metal mesh structure [108,173]. It can support only capacitive input tools including fingers to make parasitic capacitors with electrodes of the touchscreen panel.

The capacitive scheme is divided into surface-capacitive [81,83] and projected-capacitive methods [82,84]. Surface-capacitive touchscreens consist of one conductive layer of which four corners are connected to four perfectly synchronized alternative current (AC) voltage signals as described in Figure 3. While any difference does not occur without touches at these voltage sources, the finger touching the screen brings out the current difference in four voltage sources. As the voltage source is located nearer to the touch point, the current variation becomes larger due to the smaller resistive load. As a result, the touch locations are extracted from the ratio of the currents over four voltage sources. Even though it cannot deal with multiple touches at the same time, its high durability enables the integration in automated teller machines (ATMs).

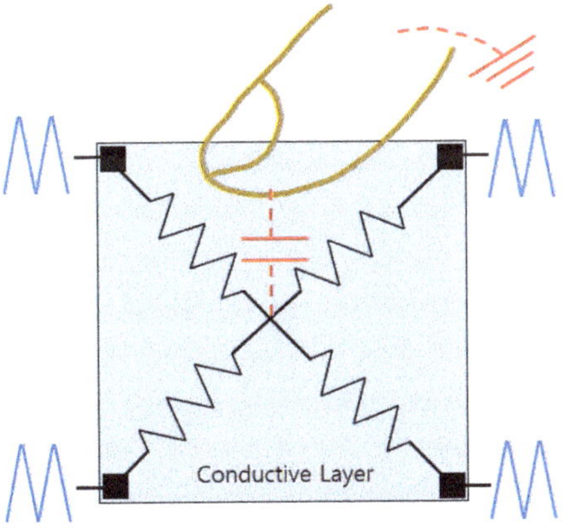

Figure 3. Surface-capacitive touchscreen. The touch location can be estimated from the current variation at four corner AC voltage sources caused by the finger touch.

The projected-capacitive methods can be further divided into self-capacitance and mutual capacitance architectures. Especially, the mutual capacitance has been the mainstream technology used in most consumer electronics such as smartphones, tablet PCs, and notebook PCs since the appearance of iPhones in 2007, because it can support multi-touch functions along with high durability and good optical clarity.

In general, the projected-capacitive touchscreen panels use two patterned conductive layers that are separated and crossed to each other in the shape of a matrix. Horizontal and vertical patterns correspond to the position information of the touch event. While the self-capacitance senses the capacitance between layers and ground as shown in Figure 4a, the mutual capacitance measures the capacitance at the overlapped areas of horizontal and vertical patterns as presented in Figure 4b. Consequently, the finger touch increases the self-capacitance due to the additional parasitic capacitor in parallel and decreases the mutual capacitance due to the electric field loss by the finger placed between two electrodes.

The self-capacitance estimates x-axis and y-axis coordinates sequentially by measuring the capacitance of vertical and horizontal electrodes over the ground, respectively. Consequently, the multiple touches may cause ghost touches. For example, when there are two touches at locations of (x1, y1) and (x2, y2), the self-capacitance can figure out that there are touches at x1, x2, y1, and y2, separately, and then it provides two correct locations of (x1, y1) and (x2, y2) along with two additional ghost locations of (x1, y2) and (x2, y1) by four possible combinations of two x-axis data and two y-axis data. Thus, the self-capacitance has difficulty to support multi-touch functionality. To cope with this ghost touch issue, some panel makers use separate self-capacitance cells directly connected to the touchscreen controller that senses each capacitance variation, respectively, [174]. This approach has been implemented in the off-the-shelf smartphones.

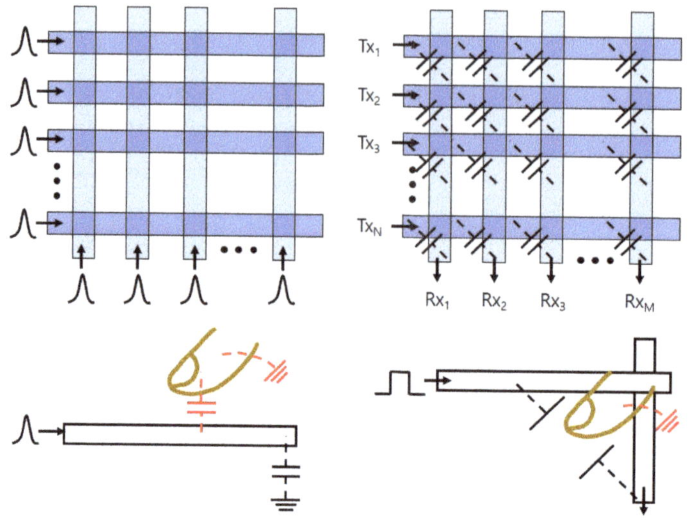

Figure 4. Projected-capacitive methods. (**a**) Self-capacitance. (**b**) Mutual capacitance.

On the other hand, because the mutual capacitance measures the overlap capacitance separately between vertical and horizontal conductive patterns, it can support multi-touch functions without any limits on the number of fingers. Therefore, it has become the widely used touchscreen technology today. The excitation pulses are applied to horizontal patterns and the transferred charges are measured through charge amplifiers at the ends of the vertical patterns. Since the amount of transferred charges is proportional to the mutual capacitance, the variation of capacitance can be detected. Section 3 will address the mutual capacitance approaches in more details.

While additional touchscreen panels on the displays require further electronics, the embedded touchscreen solutions that are called an in-cell touch can merge panel and touchscreen electronics into a single driver integrated circuit. Therefore, various in-cell approaches have been developed including self-capacitance cells and capacitive sensors embedded in pixel areas [174–179].

2.3. Acoustic Wave Touchscreen

The acoustic wave scheme is composed of a wave guide, sound wave sources, and receivers. The well-known technology is a surface acoustic wave (SAW) touchscreen as depicted in Figure 5 [85–87,91]. The SAW contains two pairs of ultrasonic transmitters and receivers to calculate x-axis and y-axis coordinates of touch locations, respectively. The reflectors in the bezel area generate multiple horizontal and vertical acoustic wave paths that have different arrival times at receivers. When a finger is placed in a certain path, the signals attenuated by that touch arrive at the receiver with corresponding delays that are converted into the position coordinates. Because the SAW needs only one wave guide layer, it has the most excellent optical performance. In addition, large size touchscreen and high durability are achievable. However, its sequential estimation of x-axis and y-axis coordinates gives rise to the same ghost touches as the self-capacitance method. It can also detect some input tools of soft materials to absorb waves and the sensing performance is sensitive to contaminants on the screen.

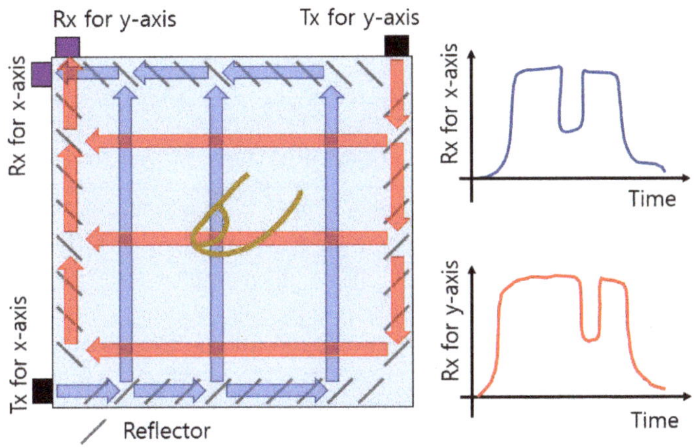

Figure 5. Surface acoustic wave (SAW) touchscreen. The ultrasonic waves move through multiple horizontal and vertical paths. The finger touch attenuates the received signal strength in the contacted wave paths and their delay information is converted to the touch locations.

The another one is a bending wave scheme, where the sound wave caused by tapping on the screen is used as the sound source as well as the touch signal [88,89,89,90]. There are two methods of acoustic pulse recognition (APR) and dispersive signal technology (DST) [180]. The APR senses the bending waves by multiple piezoelectric transducers and processes them with the data stored in the memory to extract the touch positions. Therefore, the APR needs a prior process to sample and store the large amount of bending wave data at enough number of positions over the screen. However, because the bending wave characteristics are not deterministic, the resultant coordinates have some variance, leading to errors on the location estimation. Furthermore, the enough bending wave strength is required for the sensors to detect. The bending wave characteristics are dependent of the mounting structure and material. Since too large an amount of data is necessary for multi-touch cases, it supports only a single-touch input.

To cope with the requirement of the prior process to store the bending wave data in the APR, the DST extracts touch locations directly only from the measured bending wave data. Because the signal delay is affected by its frequency, the measured time and frequency information is used to reconstruct the bending wave pattern on the screen, which is converted to the touch coordinates. However, it also has several drawbacks such as only single touch support, high tapping strength, measurement variance, mounting dependency, as well as high computational power. In addition, both APR and DST cannot support the holding function because only the tapping action generates the sound waves.

2.4. Optical Touchscreen

The optical touchscreens are developed based on the invisible infrared (IR). The traditional IR-based touchscreen places transmitters at two sides and receivers at their opposite sides without any additional layers. Because the touches block the light path over the screen between a pair of transmitter and receiver, x-axis and y-axis coordinates can be obtained by finding the receivers' positions that do not receive IR. While large-size displays and excellent optical clarity can be supported, the bezel needs some height over the screen for IR transmitters and receivers and the multiple touches cause the ghost touch issue.

The other IR-based schemes such as planar scatter detection (PSD) [100,101] and frustrated total internal reflection (FTIR) [75,95,98,99] are similar to the acoustic wave approaches except for the use of IR instead of the sound wave. In the PSD, while the transmitters send the IR lights through the wave guide at the total internal reflection (TIR) condition, receivers sense them. When any touches are applied on the wave guide plate, it breaks the TIR condition out, therefore, the scattered and remaining TIR lights arrive at multiple receivers as described in Figure 6a, leading to the extraction of the touch location by the complex analysis. The PSD can support multi-touch and high image clarity, but the larger-size touchscreens require higher computational power to extract the touch location. The FTIR also makes use of the TIR condition, but the touch location is attained from the lights escaped toward the opposite plane to the touched one as depicted in Figure 6b. Those lights are captured by the external camera or vision sensors and the resultant images provide the information of touch locations. There also exist the embedded LCD solutions, where IR transmitters are allocated in the backlight and the vision sensors are placed in the pixel areas.

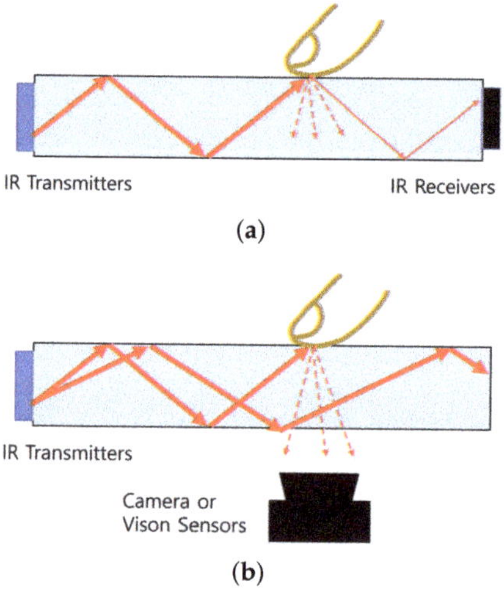

Figure 6. TIR-based IR touchscreen technologies. (**a**) PSD. (**b**) FTIR.

3. Main Research Trends in Mutual Capacitance Capacitive Touchscreen Technologies

As explained in the previous section, there have exist various touchscreen technologies by means of resistance, capacitance, sound wave, and IR. Among them, the capacitive touchscreen has become a mainstream scheme, especially, the mutual capacitance touchscreen is the most widely used technology on many consumer electronics such as smartphones, notebook PCs, tablet PCs, and smartwatches, because of its multi-touch support, slim form factor, high optical quality, excellent durability, smooth scrolling, and so on. Particularly, this section addresses the mutual capacitance capacitive touchscreens in more details. Unlike the self-capacitance method where the parasitic capacitor of a finger touch is connected to the self-capacitor in parallel, the mutual capacitance scheme experiences the capacitance reduced by electric field leakages into a finger. As a result, the touch location can be found out by searching the position which mutual capacitance is reduced.

As shown in Figure 7, a conventional mutual capacitance capacitive touchscreen panel is composed of excitation (EX) electrodes and sensing (SE) electrodes, which give rise to the mutual capacitor array at their intersection areas [181,182]. Excitation drivers generate EX pulses sequentially in the way of line-by-line that arrive at charge amplifiers attached to SE lines through mutual capacitors. The non-inverting input terminals of these charge amplifiers are connected to the reference voltage (V_{REF}) and the charge transferred through a mutual capacitor (C_m) is converted through a feedback capacitor (C_f) into analog voltages (V_{OUT}) that are proportional to the mutual capacitance as presented in Equation (1). V_{EX} is the amplitude of the EX pulse. When a user touches on the screen with a finger, the reduction on the mutual capacitance is sensed as the different output voltage of the charge amplifier from the voltage level obtained without any touches as illustrated in Figure 8. To improve the precision of the touch detection, the transferred charge is accumulated at the charge amplifiers over multiple EX pulses. In addition, a multiplexer (MUX) allows one analog-to-digital converter (ADC) to sample the output voltages of charge amplifiers in all SE lines sequentially. Finally, a host processor handles the digital data to determine the touch locations and it also controls excitation drivers.

$$V_{OUT} = V_{REF} - \frac{C_m}{C_f} V_{EX}. \tag{1}$$

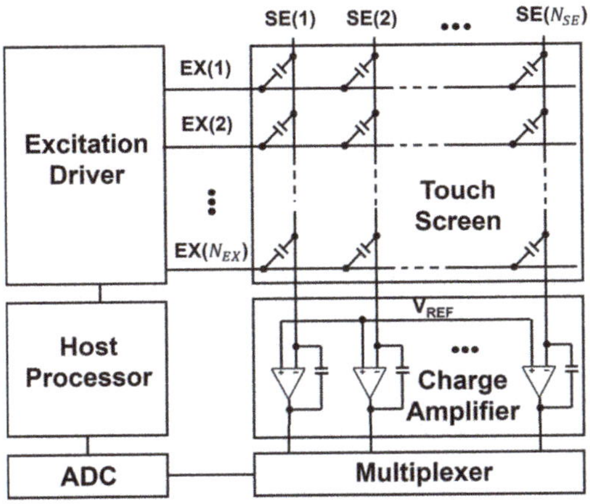

Figure 7. Block diagram of a conventional mutual capacitance capacitive touchscreen system. N_{EX} and N_{SE} are the numbers of excitation and sensing electrodes, respectively. The touch locations are estimated from the mutual capacitor array formed at the intersection areas of EX and SE lines.

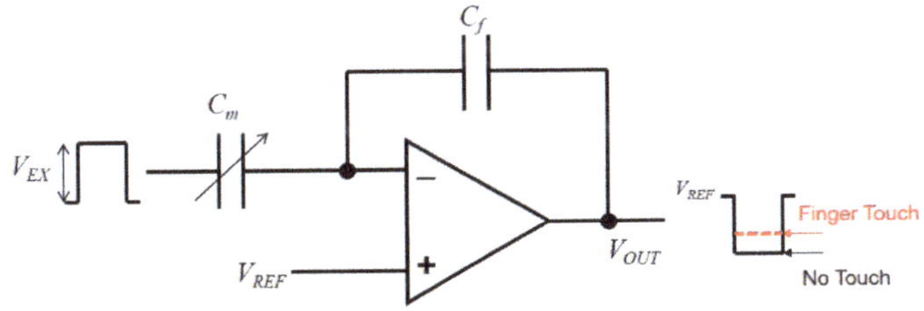

Figure 8. Charge amplifier circuit. The change of C_m causes different output voltage levels (V_{OUT}).

The mainstream studies in mutual capacitance schemes are (a) improving signal-to-noise ratio (SNR) to achieve higher accuracy as well as robustness over the noises and (b) utilizing additional input tools such as styli besides fingers. In addition, it is another research trend to integrate the pressure-sensing capability. However, the most approaches support this pressure sensing function through additional sensors [183,184], the separation distance changes [185,186], or the internal circuit of the stylus [108,187]. Because additional sensors and separation distance changes are out of this review's scope, the stylus technologies are addressed along with their pressure sensing schemes.

3.1. SNR Improvement

For the SNR improvement, various noises such as display noises, charger noises, and hum noises need to be addressed. Usually, the most sensing circuits employ the voltage accumulation at the output of the charge amplifier as shown in Figure 9, to suppress the noise power over the main signal, based on the assumption that the noise is independent and identically distributed [188]. Because the noise power and the signal power are proportional to the number of pulses and its squared value, respectively, the SNR improvement is achieved.

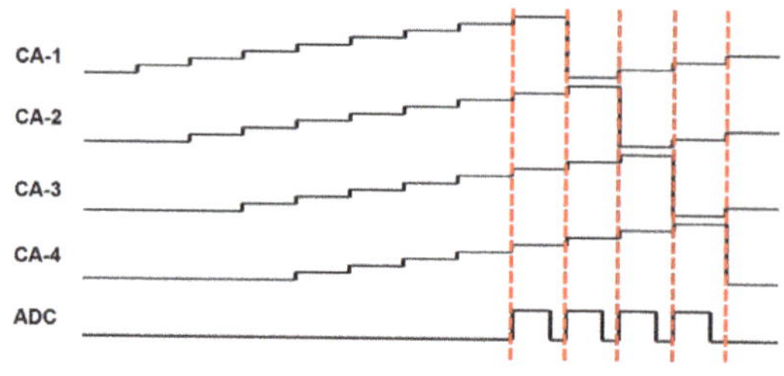

Figure 9. The SNR improvement is achieved by accumulating output pulses of charge amplifiers. CA-1 to CA-4 are the accumulated voltages for the outputs of four charge amplifiers. The outputs accumulated over several pulses are sampled by the ADC.

Yang et al. [189] employed the differential-ended charge amplifier with two out-of-phase excitation pulses (EX, EXb) as depicted in Figure 10. With the single-ended amplifier, the output dynamic range (DR) is limited by the default mutual capacitance without touches. However, the proposed differential structure reflected only the difference of the mutual capacitance (C_m) over the adjacent line's capacitor on the output voltages. In addition, the differential-ended amplifier gave rise to the doubled DR by non-inverting and inverting outputs (V_{outp}, V_{outn}). Therefore, the increased signal power led to the improved SNR performance.

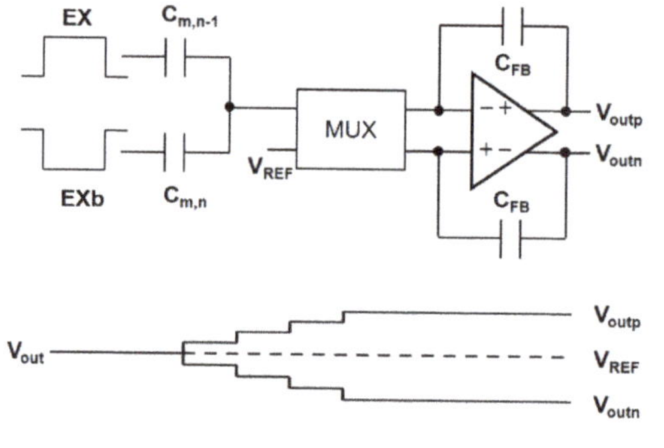

Figure 10. The dynamic range of the output voltage is enhanced by using differential-ended charge amplifier and out-of-phase excitation pulses.

Kim et al. [190] proposed the common-mode noise cancellation by subtracting the signals of the adjacent EX lines. Since the parasitic capacitance between neighboring EX lines of a touchscreen panel and display panel are almost equal, the injected noises from the display to the touchscreen would be similar, therefore, the differential sensing method over EX pulses of two neighboring lines could eliminate the common-mode display noise. Yang et al. [191] added a charge-interpolation filtered-delta-integration (CI-FDI) scheme to cancel the charger noise. The large noise is detected, and then the noise is prevented by the charge-interpolation.

As the other method to reduce the display noise components, it was proposed that the sensing operation was conducted only during the vertical blank interval as presented in Figure 11a. However, the sensing operation over the whole touchscreen area should be finished within very short period of time at the end of a frame time. Since it could not support smooth scrolling motions, the time division multiple sensing (TDMS) was introduced to spread the touch sensing functions evenly over a frame time [174,192], where the divided vertical blank parts were added in the middle of the active interval by stopping the scanning operations as illustrated in Figure 11b.

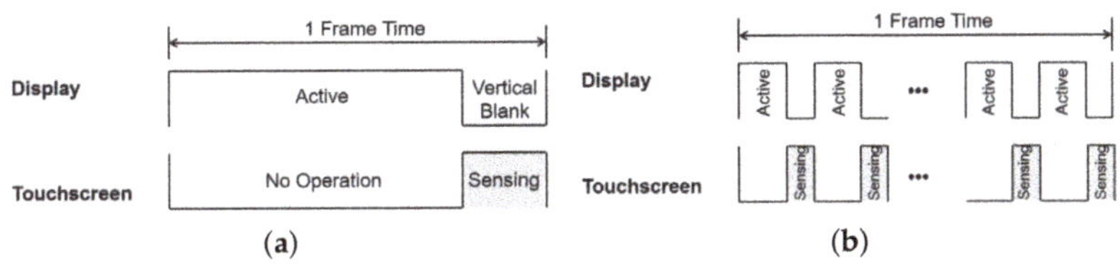

Figure 11. Touch sensing methods during the vertical blank periods. (**a**) Sensing during one vertical blank period. (**b**) TDMS sensing during divided vertical blank periods in the middle of a frame.

Miura et al. [193] adopted a two-step dual-mode scheme that performed self- as well as mutual-capacitance measurements. After the self-capacitance measurement found the touched areas, the mutual-capacitance measurement provided the fine touch location over the touched areas. Therefore, it achieved the high resolution of 80 × 80 and the high scan rate of 322 Hz.

An output accumulation sensing method can improve the SNR by applying multiple EX pulses per one touch position estimation, but lowers the scan rate inevitably. To improve SNR as well as scan rate simultaneously, Shin et al. [194] implemented a code-division multiple-sensing (CDMS) method. While multiple EX pulses per one EX line are used for sensing one touch position like the output accumulation sensing method, multiple EX electrodes are excited with orthogonal patterns simultaneously as illustrated in Figure 12. Then, the touch information over multiple positions was obtained at the same time through the decoding process. As a result, the CDMS method achieves a much higher scan rate without the SNR degradation [195,196].

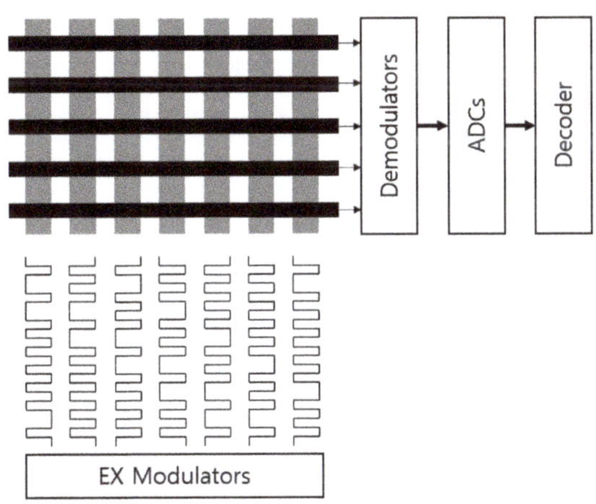

Figure 12. Code division multiple sensing. CDMS enables the multiple capacitance sensing at the same time by transmitting orthogonal codes through multiple EX lines. The charge amplifiers' outputs are demodulated and converted into digital data that are decoded as multiple simultaneous touch locations.

Park et al. [197] used the delta-sigma modulator in the ADC to move the low frequency noise to the high frequency region as shown in Figure 13. By applying a low-pass filter to remove the high frequency noise components, the noise power within the touch signal band was substantially reduced, leading to the improved SNR performance.

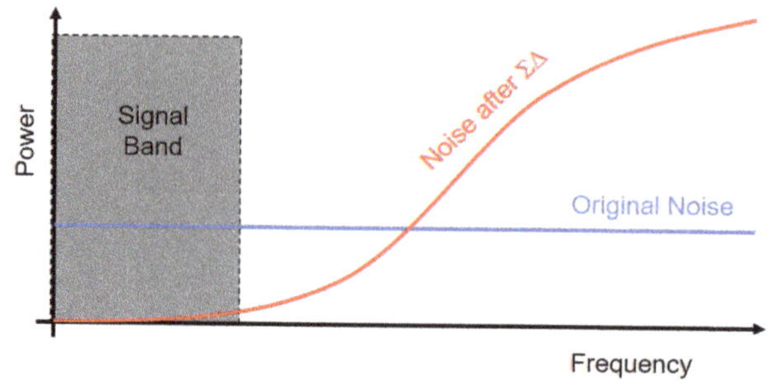

Figure 13. The sigma delta modulator shapes the noise power by moving the low frequency noises to the high frequency region. Therefore, the SNR at the signal band is improved by means of low-pass filtering.

An et al. [108] introduced the multiple-frequency driving scheme. A fast Fourier transform (FFT) was applied to find the touch locations because EX pulses had different frequencies for each EX line, that is, the number of EX frequencies is equal to the number of EX lines. Furthermore, a spectrum of external noises was acquired, and then the frequencies of EX signals were located in the low noise region for the high SNR. However, in order to implement the FFT functionality, EX signals must be driven at a very high frequency for the large number of EX lines, which leads to the increased power consumption and hardware complexity. In addition, An et al. [198] integrated the amplitude-modulation to the multiple-frequency driving scheme to reduce the charge-overflow. The excitation pulses were amplitude-modulated to reduce their amplitudes with the same periodicity. It achieved 33.9 dB charge-overflow reduction, leading to the high SNR performance.

The above SNR improving technologies are summarized in Table 2 including SNR, scan rate, touchscreen resolution, year, and reference.

Table 2. Specification comparison of capacitive touchscreen technologies.

Approach	SNR (dB)	Scan Rate (Hz)	Resolution	Year	Reference
Accumulation	24	65	20 × 16	2010	[188]
High DR	12.6	140	53 × 29	2011	[189]
Noise Cancel	35	120	30 SE	2012	[190]
CI-FDI	39	27	43 × 24	2013	[191]
CDMS	55 72	240	30 × 24 32 × 10	2013 2016	[194] [196]
Dual-mode	41	322	80 × 80	2015	[193]
TDMS	52	120	80 × 45	2015	[174]
Noise Shaping	40 67	6300 50	8 × 12	2014	[197]
Multiple Frequency	61 61.6	3900 2930	64 × 104	2017 2020	[108] [198]

3.2. *Stylus Support*

Handwriting and drawing applications on touchscreens require more elaborate input tools than a finger. The representative input tool is a stylus that has the shape of a pen. The simplest stylus implementation called a passive stylus is based on the conductive tip that imitates a finger touch as depicted in Figure 14a [103]. However, since the contact area must be large enough to be comparable to that of a finger, it is not adequate to elaborate works.

In addition, it cannot support smooth scrolling and high durability due to its rubber-type tip. Additionally, the passive stylus is not distinguished from a finger. Lin et al. [199] showed that the pressure level could be sensed in the passive stylus due to the contact area change of the deformable head proportional to the pressing force.

Active stylus schemes have been introduced to support elaborate works even with small-sized tips as illustrated in Figure 14b. The early active stylus [105] senses EX pulses from the touchscreen and transmits its inverted and boosted pulses back to the touchscreen via a tip. As these inverted pulses reduce the amount of charges transferred to the charge amplifiers, the resultant outputs of charge amplifiers become equivalent to the voltage levels caused by the reduced mutual capacitance. The process of amplifying the sensed EX signal allows for the use of much smaller radius tips than the passive stylus.

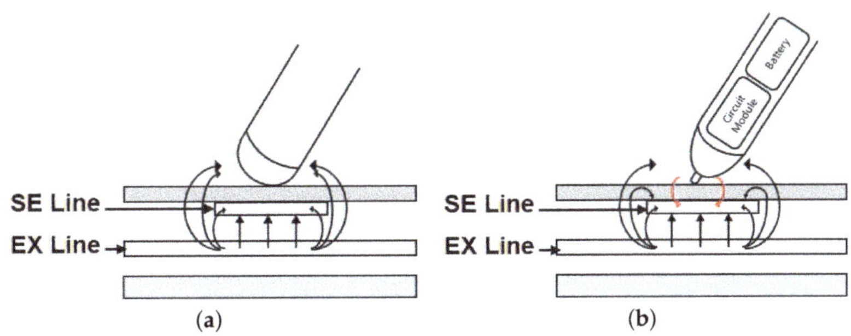

Figure 14. Stylus technologies for capacitive touchscreens. (**a**) Passive stylus. (**b**) Active stylus.

If larger inverted pulses are applied to the tip, the different output voltage from finger-touch as well as no-touch will be obtained at charge amplifiers, which enables the stylus differentiation from the finger [106]. However, this scheme needs much higher voltage amplification at stylus circuits than EX pulses of the touchscreen to give rise to the additional voltage level required for the stylus discrimination. It reduces both dynamic range and SNR for the finger-touch detection when the ADC has the fixed input voltage range. In addition, the boosted voltage levels should be separately adjusted in accordance with touchscreens.

An et al. [108] introduced the other active stylus based on the multiple-frequency driving scheme as presented in Figure 15. It could distinguish the stylus from the finger, because tip pulses of the stylus had different frequencies from EX pulses of the touchscreen. Even though it can achieve the high SNR, it required the high computational power due to the FFT implementation. Its stylus supported pressure and tilt sensing functions by means of additional force gauge and gyro sensor.

Lee et al. [187] proposed an electrically coupled resonance (ECR) stylus. As depicted in Figure 16, when the excitation pulse is asserted, the pulse is transferred to an LC resonance circuit of inductor (L_{ST}) and capacitor (C_{PR}) in the ECR stylus and the signal of the resonance frequency is transmitted from the tip of the stylus to the touch sensing circuits. Since the ECR stylus consists of only passive elements, it does not need any batteries. In addition, the finger touch can be differentiated from the stylus. Because C_{PR} is modified by the pressure, the pressure level can be differentiated by the resonance frequency. However, the large excitation pulse amplitude is necessary to generate the resonance pulse signals that can be detected at the touch sensor circuits. Besides, the frequency detection requires the increased hardware complexity.

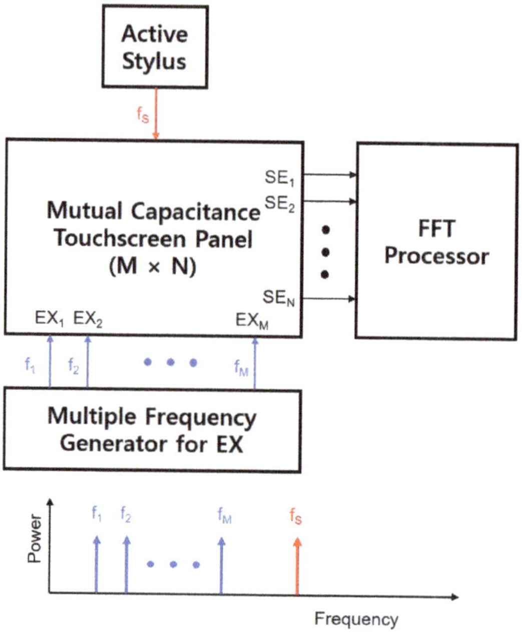

Figure 15. Multiple-frequency driving touchscreen scheme based on an FFT processor. EX lines as well as stylus can be discriminated in the frequency domain at the same time by assigning different frequencies.

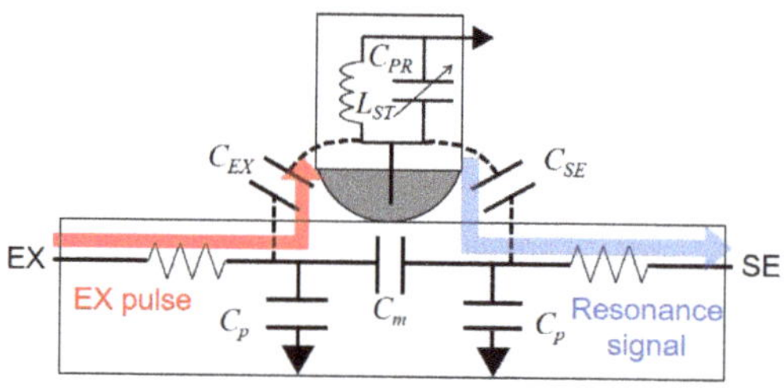

Figure 16. Electrically coupled resonance stylus. EX pulses are transmitted to the inside LC circuit and resonance signals are transferred to the coupled SE line. C_m is the mutual capacitance, C_p is the parasitic capacitor of EX and SE lines, C_{EX} is the coupling capacitance between EX line and stylus, and C_{SE} is the coupling capacitance between SE line and stylus. L_{ST} and C_{PR} are inductance and capacitance of the resonance circuit in the stylus.

Seol et al. [157,163] adopted the active stylus scheme that used the higher frequency pulses to generate different patterns from EX pulses for the finger-touch detection. It discriminated stylus-touches from no-touches and finger-touches without the high complicated FFT functionality by machine-learning-based classifiers as depicted in Figure 17. It also showed that the proposed method allowed for the effective data communication between touchscreens and styli. On the other hand, since the patterns of stylus-touches were placed in between finger-touches and no-touches, the achieved SNR was relatively low compared to the state-of-the-art high SNR touchscreen technologies.

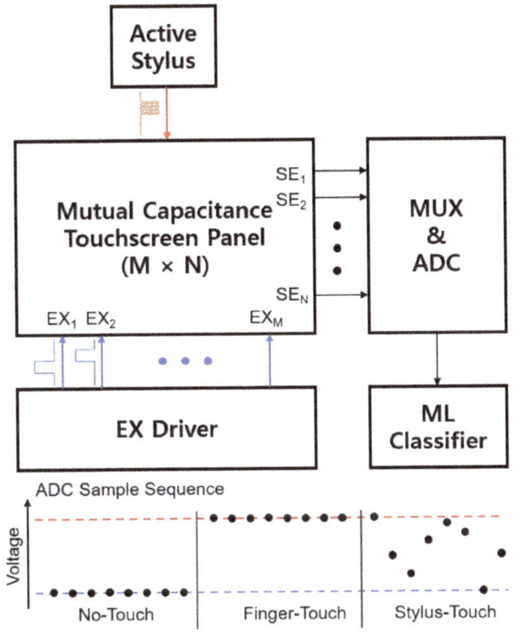

Figure 17. ML-based touchscreen scheme. The higher frequency pulses of a stylus generate the different sequence from finger-touch and no-touch, and three touches are discriminated by ML-based classifiers.

While previous passive and active ways are grounded in the capacitive touchscreen panel, there is a stylus with a different approach that works with an electro-magnetic resonance (EMR) technology [107]. Since EMR responds only to the stylus and the capacitive touchscreen panel senses only the fingers, two separate touch sensing schemes enable the finger to be distinguished from the stylus. However, this technique needs additional layers that increase hardware complexity as well as manufacturing cost.

The stylus technologies that can be implemented on the capacitive touchscreen are summarized in Table 3 in terms of tip size, stylus discrimination support, SNR degradation, computational cost, and hardware complexity.

Table 3. Comparison of stylus technologies for capacitive touchscreens.

Reference	Passive [103,199]	Active [105,106]	FFT [108]	ML [157,163]	EMR [107]	ECR [187]
Tip Size	Large	Small	Small	Small	Small	Small
Stylus Discrimination	No	No ([105]) Yes ([106])	Yes	Yes	Yes	Yes
SNR Degradation	No	No ([105]) Yes ([106])	No	Yes	No	Yes
Pressure Sensing	Yes [199]	-	Yes	No	Yes	Yes
Computational Cost	Low	Low	High	Medium	Low	High
Hardware Complexity	Low	Low	High	Low	High	High

4. Machine Learning Approaches in Mutual Capacitance Capacitive Touchscreen

Several ML algorithms have been employed to the capacitive touchscreen in a variety of applications such as user identification/authentication, gesture detection, accuracy improvement, and input discrimination. While many approaches used traditional ML techniques of decision tree (DT), random forest (RF), naive Bayes (NB), radial basis function network (RBFN), back propagation neural network (BPNN), support vector machine (SVM), and Gaussian process regression (GPR), the latest ML networks, such as convolutional neural network (CNN), anomaly detection (AD), and recurrent neural network (RNN), have been also utilized in the touchscreen field. The following machine learning applications are implemented on the off-the-shelf smartphones and tablet PCs with mutual capacitance capacitive touchscreens.

4.1. User Identification/Authentication

The user identification and authentication is the most active field to apply the machine learning to the touchscreen because the touch behaviors are different according to the users. Kolly et al. [135] used DT, RF, and NB classifiers over touch gestures for the user authentication application. Mean and maximal pressures, the point in the time at the maximal pressure event, minimal and maximal gradients of the pressure, the hold time, mean x-axis and y-axis positions, and the variances in x-axis and y-axis directions were employed as input features that were crowdsourced by designing a quiz game. The resultant identification accuracy for five users was 80% or more. It also proposed the anomaly detection for the user authentication based on the learned distribution of features from 5 button touch events. The equal error rate of false reject ratio (FRR) and false accept ratio (FAR) was estimated as about 30%.

Feng et al. [138] proposed a finger-gesture authentication system using touchscreen (FAST). FAST collected touch gesture information including gesture type, x-axis and y-axis coordinates, directions of the finger motion, finger motion speed, pressure, and the distance between multi-touch points. Totally 53 features for each gesture and six gestures of down to up swipe, up to down swipe, left to right swipe, right to left swipe, zoom-in and zoom-out were put into DT, RF, and NB classifiers. It achieved FAR of 4.66% and FRR of 0.13% for the continuous post-login user authentication.

For the user authentication, Meng et al. [139] constructed 21 features such as average touch movement speeds for eight directions, fractions of touch movements for eight directions, average single-touch time, average multi-touch time, number of touch movements per session, number of single-touch events per session, and number of multi-touch events per session. They evaluated the performance of DT, NB, Kstar, RBFN, and BPNN, leading to the conclusion that RBFN showed the best performance with FAR and FRR of 7.08% and 8.34%. In addition, the particle swarm optimization (PSO) with a RBFN classifier reduced FAR and FRR further to 2.5% and 3.34%, respectively.

Saravanan et al. [140] proposed the authentication scheme based on the user's touch interaction with common user interface elements such as buttons, checkboxes, and sliders. Using SVM and RF, they achieved average accuracies of 97.9% and 96.79% with mobile phone and tablet PC, respectively.

Guo et al. [143] proposed CapAuth that is a user identification and authentication technique based on capacitive touchscreen data combined with machine learning classifiers. It used the capacitive image of the hands-flat pose revealing the more distinguishing features. CapAuth was built based on quadratic-kernel SVM classifiers, a binary classifier for authentication and a multi-class one-to-one classifier for identification. The measure FRR and FAR for authentication were 5.5% and 0.1%, respectively. The accuracy of the identification was 94.0% for 20 users.

Rilvan et al. [148] used four fingers, thumb, as well as ear as the types of biometrics for authentication with the machine learning classifiers such as SVM and RF. It achieved the maximum authentication accuracy of 98.84% over four fingers with SVM and maximum identification accuracy of 97.61% by four fingers with RF.

Meng et al. [152] enhanced touch behavioral authentication by cost-based classifier selection, where the best classifier with the lowest cost value was selected among a set of classifiers including DT, NB, RBFN, and BPNN. It had nine touch features such as the number of touch movements per session, the number of single-touch events per session, the number of multi-touch events per session, the average time duration of touch movements per session, the average time duration of single-touch per session, the average time duration of multi-touch per session, average speed of touch movement, average touch size, and average touch pressure. The average error rate of FAR and FRR was measured to be less than 5% for 15 sessions or more.

4.2. Gesture Detection

The user identification and authentication require various touch data such as location, speed, force, and gestures as features that are of the most importance for the better performance. Therefore, there have been researches to extract the gesture data from the touch data by means of machine learning algorithms. Xiao et al. [146] came up with an approach for estimating 3D finger angles such as pitch and yaw relative to a touchscreen's surface. It used the capacitive image that was the capacitance measured at each point of a touch sensor's capacitive grid. The pitch was estimated by a Gaussian process regression with 42 features and the yaw was computed by the major axis of the ellipsoid of the sensed touch pixels. While mean pitch errors were 9.7 and 14.5, mean yaw errors were 26.8 and 31.7 for phone and watch, respectively. It also proposed the possible applications such as zoom and rotate functions even with a single-touch event which would be very useful in small-size watch displays.

Mayer et al. [151] proposed the neural network approach to estimate the finger orientation of pitch and yaw. They evaluated the performance over separated deep neural networks (DNN) for pitch and yaw, combined DNN with two output neurons, and CNN along with L2 regularization [200,201] and batch normalization (BatchNorm) [202]. The blob detection provided the 15×22 sized data that was fed into neural networks. It achieved the best pitch error of 12.75 with CNN+L2+BatchNorm and the best yaw error of 17.6 with CNN.

Boceck et al. [161] extracted the pressure from capacitive images by using CNN. It used a ReLU function as an activation function, and dropout layers. Final fully connected layers contained LeakyReLU and L1/L2 regularization. They achieved the lower root mean square error of 471.99 g, compared to 583.36 g and 593.51 g of RF and SVM.

Schweigert et al. [162] added knuckle related features by differentiating knuckles from fingers and classified 17 finger and knuckle gestures by CNN and long short-term memory (LSTM). CNN layers extracted the representation from the 15×27 capacitive image and then LSTM layers generated 17 outputs over 50 consecutive images through a softmax activation function. Seventeen gestures contained tap, two tap, swipe left, swipe right, swipe up, swipe down, two swipe up, two swipe down, circle, arrowhead left, arrowhead right, rotate, and five additional gestures. It achieved the accuracies of 97.9% and 86.8% on train and test sets, respectively.

4.3. Accuracy Improvement

Weir et al. [137] adopted GPR to find a mapping between two-dimensional reported touch locations and a corresponding intended two-dimensional touch location on the display. It collected the touch data by randomly displaying the crosshairs that the users had to touch. The resultant average reductions in error rates were 23.79% for 2 mm buttons, 14.79% for 3 mm buttons, and 5.11% for 4 mm buttons.

Fischer et al. [155] presented a system using capacitive sensing to accurately classify hand touches and proximity. Touch data were collected through 50 finger touches with different fingers, angles, locations and speeds, 25 glove touches, and 20 non-valid touches. Then, the collected data were further processed by dimensionality reduction, data augmentation, and normalization. Hidden Markov model (HMM) and RNN of gated recurrent units (GRUs) were evaluated as the classifier over three classes of non-touch, near, and touch. The RNN model showed the better overall accuracy of 97.1% even with gloves while HMM achieved the accuracy performance of 84.21%.

Kumar et al. [158] improved the accuracy of touch locations by CNN. The dataset consisted of capacitive images at the dimension of 15 × 27 as well as the estimated touch positions represented by the centroid of the touch blob. The proposed CNN achieved an average error offset of 41.23 pixels based on a screen resolution of 1920 × 1080 on a 4.94 inch display. It was the improvement of 23.0%, compared to the error offset of 50.7 pixels of the standard touch controller.

Kim et al. [160] introduced a sensor substitution system that generates time-series sensor data based on RNN. It estimated capacitive touch sensor signals by motion and audio signals caused by touch. By other types of multivariate time-series signals, the touch sensor sequences were supplemented even at dynamic and hostile environments that degraded the touch sensor's performance.

4.4. Input Discrimination

There have been studies on the discrimination of inputs such as fingers, palm, and stylus. Schwarz et al. [142] employed the decision tree to distinguish between legitimate stylus and palm based on spatiotemporal touch features and iterative classification. It identified five properties of palms such as the large contact area, the segmentation into a collection of touch points, the clustering, the area change, and the little movements. As a result, min distance to other touches, number of touch events, and min/mean/max/stdev of touch radius were used as the features. The instant classification achieved an accuracy of 98.4% and the continuous iterative classification increased the accuracy to 99.5%.

Le et al. [154] differentiated between touches of fingers and palm to devise an additional input modality. Their PalmTouch showed possible one-handed and two-handed palm interactions by placing flat hand or fist on the screen. It used the capacitive images for the features and CNN for the classification, leading to the accuracy of 99.53%.

Le et al. [153] proposed the finger-aware interaction that identified fingers touching the whole device surface to add the input modalities. In a prototype, front and back side touchscreens were developed by two stacked smartphones and their three edges were attached with 37 capacitive sensors. It used CNN with L2 regularization to obtain 15 outputs that were a three-dimensional coordinate, (x, y, z), for five fingers. The identification accuracy was 95.78% with the position error of 0.74 cm.

Seol et al. [157,163,203] employed the machine learning based classifiers such as SVM and autoencoder-based AD for finger and stylus discrimination. The higher frequency pulses were transmitted from a stylus to a capacitive touchscreen and the outputs of the charge amplifiers were sampled by ADC and classified by the classifier into no-touch, finger-touch, and stylus-touch. While no-touch and finger-touch were the constant level sample sequences, the stylus-touch was the random sequence between two constant levels. Therefore, SVM and AD classifiers achieved lower bit error rates (BERs) than 10^{-6} with the palm rejection. In addition, it was shown that its data communication algorithm could be applied in data transmission and user identification.

5. Conclusions and Future Directions

In this paper, we have provided an extensive review on touchscreen technologies. We mainly dealt with the overview of various touchscreen schemes from resistive to optical methods, and two main research directions of SNR improvement and stylus support as well as machine learning approaches in mutual capacitance capacitive touchscreens that are the most widely adopted scheme at present in smartphones, tablet PCs, notebook PCs, and smartwatches. For the aspect of the SNR improvement, accumulation, differential sensing, TMDS, dual mode of self and mutual capacitance, CDMS, delta-sigma modulation, and multiple-frequency driving have been introduced. High SNRs have been achieved by reduced noises and increased dynamic ranges. For the stylus support, passive, active, multiple frequency driving, ECR, and ML-based schemes have been addressed along with their pressure sensing capabilities. The machine learning applications in capacitive touchscreens have been classified in four categories of user identification/authentication, gesture detection, accuracy improvement, and input discrimination by means of a variety of algorithms such as DT, RF, NB, RBFN, BPNN, SVM, GPR, CNN, AD, and RNN.

Although many advancements have been accomplished in touchscreen technologies, challenges still exist in various fields. We will point out some of these challenges. We hope that this review not only helps the understanding of the touchscreen technologies but also paves the way to future researches on integrating machine learning algorithms into touchscreens for more various applications. As the resolutions of touchscreens are getting larger, fingerprints can be detected on any locations of the screen without additional sensors [204,205]. However, high power consumption and low scan rate should be addressed. One of possible solutions would be the super-resolution (SR) that gives rise to the high resolution capacitive image from the low resolution capacitive image. Many deep-learning-based SR algorithms have been reported [206,207]. As discussed in the previous sections, the SNR is one of the most important performance metrics and the high SNR is required to integrate touch sensing and display driving electronics into one integrated circuit. There exists the research field of de-noising that generates the clean one from the noisy image. Therefore, this de-noising scheme can be applied to enhance the SNR over the acquired capacitive images. Lastly, because latest smartphones, tablet PCs, and notebook PCs contain many sensors such as cameras, IR sensors, microphones, accelerometers, and gyroscopes besides touchscreens, there will keep being many approaches based on multi-sensor fusion technologies in the user interface field.

Author Contributions: Conceptualization, H.N. and K.-H.S.; formal analysis, H.N. and K.-H.S.; resources, H.N., K.-H.S., J.L., H.C., and S.W.J.; writing—original draft preparation, H.N.; writing—review and editing, H.N. and K.-H.S.; visualization, H.N.; supervision, H.N.; project administration, H.N.; funding acquisition, H.N. All authors have read and agreed to the published version of the manuscript.

Funding: This research was supported by IDEC (EDA Tool) and the National Research Foundation of Korea (NRF) funded by the Ministry of Science, ICT & Future Planning (NRF-2016R1A2B4009787) and the Brain Korea 21 Four Program in 2021.

Institutional Review Board Statement: Not applicable.

Informed Consent Statement: Not applicable.

Data Availability Statement: Data sharing not applicable.

Conflicts of Interest: The authors declare no conflict of interest.

References

1. Cundall, C.M. Cathode-ray-tube display systems. *Electron. Power* **1968**, *14*, 115–120. [CrossRef]
2. Anderson, L.K. The Cathode Ray Tube Display: Why Use Anything Else? *J. Vac. Sci. Technol.* **1973**, *10*, 761. [CrossRef]
3. Seats, P. Fundamentals of Cathode-Ray Tubes. In *Digest of Technical Papers of the International Symposium of the SID*; Winner: New York, NY, USA, 1976; pp. 172–173.
4. Lehrer, N.H. *Flat-Panel Displays and CRTs*; Springer: Heidelberg, Germany, 1985.
5. Bitzer, D.L.; Slottow, H.G. The plasma display panel: A digitally addressable display with inherent memory. In Proceedings of the American Federation of Information Processing Societies (Fall), San Francisco, CA, USA, 7–10 November 1966; pp. 541–547.
6. Meunier, J.; Belenguer, P.; Boeuf, J.P. Numerical model of an ac plasma display panel cell in neon-xenon mixtures. *J. Appl. Phys.* **1995**, *78*, 731–745. [CrossRef]
7. Rauf, S.; Kushner, M.J. Dynamics of a coplanar-electrode plasma display panel cell. I. Basic operation. *J. Appl. Phys.* **1999**, *85*, 3460–3469. [CrossRef]
8. Rauf, S.; Kushner, M.J. Dynamics of a coplanar-electrode plasma display panel. II. Cell optimization. *J. Appl. Phys.* **1999**, *85*, 3470–3476. [CrossRef]
9. Shinoda, T.; Wakitani, M.; Nanto, T.; Awaji, N.; Kanagu, S. Development of panel structure for a high-resolution 21-in-diagonal full-color surface-discharge plasma display panel. *IEEE Trans. Electron. Devices* **2000**, *47*, 77–81. [CrossRef]

10. Boeuf, J.P. Plasma display panels: Physics, recent developments and key issues. *J. Phys. D Appl. Phys.* **2003**, *36*, R53–R79. [CrossRef]
11. White, D.L.; Taylor, G.N. New absorptive mode reflective liquid-crystal display device. *J. Appl. Phys.* **1974**, *45*, 4718–4723. [CrossRef]
12. Snell, A.J.; Mackenzie, K.D.; Spear, W.E.; LeComber, P.G.; Hughes, A.J. Application of amorphous silicon field effect transistors in addressable liquid crystal display panels. *Appl. Phys.* **1981**, *24*, 357–362. [CrossRef]
13. Castellano, J.A. Liquid Crystal Display Applications: Past, Present & Future. *Liq. Cryst. Today* **1991**, *1*, 4–6. [CrossRef]
14. Ishii, Y. The World of Liquid-Crystal Display TVs—Past, Present, and Future. *J. Disp. Technol.* **2007**, *3*, 351–360. [CrossRef]
15. Kim, D.H.; Lim, Y.J.; Kim, D.E.; Ren, H.; Ahn, S.H.; Lee, S.H. Past, present, and future of fringe-field switching-liquid crystal display. *J. Inf. Disp.* **2014**, *15*, 99–106. [CrossRef]
16. Tang, C.W.; VanSlyke, S.A. Organic electroluminescent diodes. *Appl. Phys. Lett.* **1987**, *51*, 913–915. [CrossRef]
17. Baldo, M.A.; O'Brien, D.F.; You, Y.; Shoustikov, A.; Sibley, S.; Thompson, M.E.; Forrest, S.R. Highly efficient phosphorescent emission from organic electroluminescent devices. *Nature* **1998**, *395*, 151–154. [CrossRef]
18. Adachi, C.; Baldo, M.A.; Thompson, M.E.; Forrest, S.R. Nearly 100% internal phosphorescence efficiency in an organic light-emitting device. *J. Appl. Phys.* **2001**, *90*, 5048–5051. [CrossRef]
19. Arnold, A.D.; Castro, P.E.; Hatwar, T.K.; Hettel, M.V.; Kane, P.J.; Ludwicki, J.E.; Miller, M.E.; Murdoch, M.J.; Spindler, J.P.; Slyke, S.A.V.; Mameno, K.; Nishikawa, R.; Omura, T.; Matsumoto, S. Full-color AMOLED with RGBW pixel pattern. *J. Soc. Inf. Disp.* **2005**, *13*, 525–535. [CrossRef]
20. Nathan, A.; Chaji, G.; Ashtiani, S. Driving schemes for a-Si and LTPS AMOLED displays. *J. Disp. Technol.* **2005**, *1*, 267–277. [CrossRef]
21. Geffroy, B.; le Roy, P.; Prat, C. Organic light-emitting diode (OLED) technology: Materials, devices and display technologies. *Polym. Int.* **2006**, *55*, 572–582. [CrossRef]
22. Chen, H.W.; Lee, J.H.; Lin, B.Y.; Chen, S.; Wu, S.T. Liquid crystal display and organic light-emitting diode display: Present status and future perspectives. *Light Sci. Appl.* **2018**, *7*, 17168. [CrossRef]
23. Lee, V.W.; Twu, N.; Kymissis, I. Micro-LED Technologies and Applications. *Inf. Disp.* **2016**, *32*, 16–23. [CrossRef]
24. Wu, T.; Sher, C.W.; Lin, Y.; Lee, C.F.; Liang, S.; Lu, Y.; Chen, S.W.H.; Guo, W.; Kuo, H.C.; Chen, Z. Mini-LED and Micro-LED: Promising Candidates for the Next Generation Display Technology. *Appl. Sci.* **2018**, *8*, 1557. [CrossRef]
25. Paranjpe, A.; Montgomery, J.; Lee, S.M.; Morath, C. Micro-LED Displays: Key Manufacturing Challenges and Solutions. *Dig. Tech. Pap. Int. Symp. SID* **2018**, *49*, 597–600. [CrossRef]
26. Huang, Y.; Tan, G.; Gou, F.; Li, M.; Lee, S.; Wu, S. Prospects and challenges of mini-LED and micro-LED displays. *J. Soc. Inf. Disp.* **2019**, *27*, 387–401. [CrossRef]
27. Zhoua, X.; Tiana, P.; Sher, C.W.; Wu, J.; Liu, H.; Liu, R.; Kuo, H.C. Growth, transfer printing and colour conversion techniques towards full-colour micro-LED display. *Prog. Quantum. Electron.* **2020**, *71*, 100263. [CrossRef]
28. Huang, Y.; Hsiang, E.L.; Deng, M.Y.; Wu, S.T. Mini-LED, Micro-LED and OLED displays: Present status and future perspectives. *Light Sci. Appl.* **2020**, *9*, 105. [CrossRef] [PubMed]
29. Schaeffler, J. *Digital Signage: Software, Networks, Advertising, and Displays: A Primer for Understanding the Business*; Taylor & Francis: Oxfordshire, UK, 2008.
30. Nam, H.; Lee, S.W. Low-power liquid crystal display television panel with reduced motion blur. *IEEE Trans. Consum. Electron.* **2010**, *56*, 307–311. [CrossRef]
31. Nam, H. Low power active dimming liquid crystal display with high resolution backlight. *Electron. Lett.* **2011**, *47*, 538–540. [CrossRef]
32. Han, S.M.; Lee, B.W.; Ji, I.H.; Sung, S.D.; Arkhipov, A.; Kim, S.S. Smart Power-Saving Driving Scheme for AMOLEDs Using Dynamic Power Rail Control. *Dig. Tech. Pap. Int. Symp. SID* **2011**, *42*, 183–185. [CrossRef]
33. You, B.H.; Bae, J.S.; Park, D.W.; Hong, S.H.; Saito, S.; Moon, J.T. UD Resolution 240Hz LCD TV Display System with High Speed Driving. *Dig. Tech. Pap. Int. Symp. SID* **2012**, *43*, 395–398. [CrossRef]
34. de Greef, P.; Hulzeu, H.G. Adaptive Dimming and Boosting Backlight for LCD-TV Systems. *Dig. Tech. Pap. Int. Symp. SID* **2012**, *38*, 1332–1335. [CrossRef]
35. Nam, W.; Shim, J.; Shin, H.; Kim, J.; Ha, W.; Park, K.; Kim, H.; Kim, B.; Oh, C.; Ahn, B.; et al. 55-inch OLED TV using InGaZnO TFTs with WRGB Pixel Design. *Dig. Tech. Pap. Int. Symp. SID* **2013**, *44*, 243–246. [CrossRef]
36. Hara, Y.; Kikuchi, T.; Kitagawa, H.; Morinaga, J.; Ohgami, H.; Imai, H.; Daitoh, T.; Matsuo, T. IGZO-TFT technology for large-screen 8K display. *J. Soc. Inf. Disp.* **2018**, *26*, 169–177. [CrossRef]
37. Shin, H.J.; Takasugi, S.; Choi, W.S.; Chang, M.K.; Choi, J.Y.; Jeon, S.K.; Yun, S.H.; Park, H.W.; Kim, J.M.; Kim, H.S.; et al. A Novel OLED Display Panel with High-Reliability Integrated Gate Driver Circuit using IGZO TFTs for Large-Sized UHD TVs. *Dig. Tech. Pap. Int. Symp. SID* **2018**, *49*, 358–361. [CrossRef]
38. Shin, H.J.; Choi, W.S.; Chang, M.K.; Choi, J.Y.; Choi, S.H.; Yun, S.H.; Kim, J.M.; Kim, H.S.; Oha, C.H. A High Image-Quality OLED Display with Integrated Gate Driver using MPRT Enhancement Technology for Large Size Premium TVs. *Dig. Tech. Pap. Int. Symp. SID* **2019**, *50*, 199–202. [CrossRef]
39. Bae, K.S.; Oh, M.; Park, B.; Cho, Y.J.; Cho, S.H.; Kim, D. Novel Pixel Structure for 8K QUHD LCD Panel with the Enhanced Optical Performances. *Dig. Tech. Pap. Int. Symp. SID* **2019**, *50*, 703–706. [CrossRef]

40. Consumer Reports, TV Trends to Watch in 2021. Available online: https://www.consumerreports.org/lcd-led-oled-tvs/tv-trends-to-watch-for/ (accessed on 17 April 2021).
41. PC Magazine, The Best TVs for 2021. Available online: https://www.pcmag.com/picks/the-best-tvs (accessed on 17 April 2021).
42. CNet, CES 2018 Wrap-Up: The TVs of Tomorrow Will Turn Invisible. Available online: https://www.cnet.com/news/ces-2018-wrap-up-the-tvs-of-tomorrow-will-turn-invisible/ (accessed on 17 April 2021).
43. Kim, S.; Kwon, H.; Lee, S.; Shim, H.; Chun, Y.; Choi, W.; Kwack, J.; Han, D.; Song, M.; Kim, S.; et al. Low-Power Flexible Organic Light-Emitting Diode Display Device. *Adv. Mater.* **2011**, *23*, 3511–3516. [CrossRef]
44. Takubo, Y.; Hisatake, Y.; Lizuka, T.; Kawamura, T. Ultra-High Resolution Mobile Displays. *Dig. Tech. Pap. Int. Symp. SID* **2012**, *43*, 869–872. [CrossRef]
45. Sakaigawa, A.; Kabe, M.; Harada, T.; Goto, F.; Takasaki, N.; Mitsui, M.; Nakahara, T.; Ikeda, K.; Seki, K.; Nagatsuma, T.; et al. Low Power Consumption Technology for Ultra-High Resolution Mobile Display by Using RGBW System. *IEICE Trans. Electron.* **2013**, *E96-C*, 1367–1372. [CrossRef]
46. Ohshima, H. Mobile display technologies: Past developments, present technologies, and future opportunities. *Jpn. J. Appl. Phys.* **2014**, *53*. [CrossRef]
47. Yan, J.; Ho, J.; Chen, J. Foldable AMOLED Display Development: Progress and Challenges. *Inf. Disp.* **2015**, *31*, 12–16. [CrossRef]
48. Kim, J.S.; Lee, S.W. Peripheral Dimming: A New Low-Power Technology for OLED Display Based on Gaze Tracking. *IEEE Access* **2020**, *8*, 209064–209073. [CrossRef]
49. Tom's Guide, Best Big Phones of 2021: Top Phablets 6 Inches or Larger. Available online: https://www.tomsguide.com/best-picks/best-big-phones (accessed on 17 April 2021).
50. Bastani, B.; Turner, E.; Vieri, C.; Jiang, H.; Funt, B.; Balram, N. Foveated Pipeline for AR/VR Head-Mounted Displays. *Inf. Disp.* **2017**, *33*, 14–35. [CrossRef]
51. Bhowmik, A.K. Advances in Virtual, Augmented, and Mixed Reality Technologies. *Inf. Disp.* **2018**, *34*, 18–21. [CrossRef]
52. Vieri, C.; Lee, G.; Balram, N.; Jung, S.H.; Yang, J.Y.; Yoon, S.Y.; Kang, I.B. An 18 megapixel 4.3″ 1443 ppi 120 Hz OLED display for wide field of view high acuity head mounted displays. *J. Soc. Inf. Disp.* **2018**, *26*, 314–324. [CrossRef]
53. Park, S.; Kim, Y.I.; Nam, H. Foveation-based reduced resolution driving scheme for immersive virtual reality displays. *Opt. Express* **2019**, *27*, 29594–29605. [CrossRef]
54. Muñoz-Saavedra, L.; Miró-Amarante, L.; Domínguez-Morales, M. Augmented and Virtual Reality Evolution and Future Tendency. *Appl. Sci.* **2020**, *10*, 322. [CrossRef]
55. Zhan, T.; Yin, K.; Xiong, J.; He, Z.; Wu, S.T. Augmented Reality and Virtual Reality Displays: Perspectives and Challenges. *iScience* **2020**, *23*, 101397. [CrossRef]
56. Lee, B.; Yoo, C.; Jeong, J.; Lee, B.; Bang, K. Key issues and technologies for AR/VR head-mounted displays. In Proceedings of the SPIE 11304, Advances in Display Technologies X, San Francisco, CA, USA, 1–6 February 2020; p. 1130402.
57. Jang, H.J.; Lee, J.Y.; Kim, J.; Kwak, J.; Park, J.H. Progress of display performances: AR, VR, QLED, and OLED. *J. Inf. Disp.* **2020**, *21*, 1–9. [CrossRef]
58. Kim, H.; Kwak, B.C.; Lim, H.S.; Kwon, O.K. Pixel Circuit for Organic Light-Emitting Diode-on Silicon Microdisplays Using the Source Follower Structure. *Jpn. J. Appl. Phys.* **2010**, *49*, 03CD05. [CrossRef]
59. Kwak, B.C.; Lim, H.S.; Kwon, O.K. Organic Light-Emitting Diode-on-Silicon Pixel Circuit Using the Source Follower Structure with Active Load for Microdisplays. *Jpn. J. Appl. Phys.* **2011**, *50*, 03CC05. [CrossRef]
60. Hong, S.W.; Kwak, B.C.; Na, J.S.; Hong, S.K.; Kwon, O.K. Simple pixel circuits for high resolution and high image quality organic light emitting diode-on-silicon microdisplays with wide data voltage range. *J. Soc. Inf. Disp.* **2016**, *24*, 110–116. [CrossRef]
61. Liu, B.; Ding, D.; Zhou, T.; Zhang, M. A Novel Pixel Circuit Providing Expanded Input Voltage Range for OLEDoS Microdisplays. *Dig. Tech. Pap. Int. Symp. SID* **2017**, *48*, 1438–1441. [CrossRef]
62. Huo, X.; Liao, C.; Wu, J.; Yi, S.; Wang, Y.; Jiao, H.; Zhang, M.; Zhang, S. An OLEDoS Pixel Circuit with Extended Data Voltage Range for High Resolution Micro-Displays. *Dig. Tech. Pap. Int. Symp. SID* **2018**, *49*, 1373–1376. [CrossRef]
63. Na, J.S.; Hong, S.K.; Kwon, O.K. A 4410-ppi Resolution Pixel Circuit for High Luminance Uniformity of OLEDoS Microdisplays. *IEEE J. Electron Devices Soc.* **2019**, *7*, 1026–1032. [CrossRef]
64. Wadlow, T.A. The Xerox Alto Computer. *Byte Mag.* **1981**, *6*, 58–68.
65. Knight, L.W.; Retter, D. Datahand: Design, Potential Performance, and Improvements in the Computer Keyboard and Mouses. *Proc. Hum. Factors Soc. Annu. Meet.* **1989**, *33*, 450–454. [CrossRef]
66. Wolf, C.G. A comparative study of gestural, keyboard, and mouse interfaces. *Behav. Inf. Technol.* **1992**, *11*, 13–23. [CrossRef]
67. Mahach, K.R. A Comparison of Computer Input Devices: Linus Pen, Mouse, Cursor Keys and Keyboard. *Proc. Hum. Factors Soc. Annu. Meet.* **1989**, *33*, 330–334. [CrossRef]
68. Colla, R.; Ziab, K.; Collc, J.H. A comparison of three computer cursor control devices: Pen on horizontal tablet, mouse and keyboard. *Inf. Manag.* **1994**, *27*, 329–339. [CrossRef]
69. Sharples, M.; Beale, R. A technical review of mobile computational devices. *J. Comput. Assist. Learn.* **2003**, *19*, 392–395. [CrossRef]
70. Walker, G. Tablet Product and Market History. Available online: http://walkermobile.com/Tablet_History.pdf (accessed on 17 April 2021).
71. Lin, C.H.; Schmidt, K.J. User Preference and Performance with Three Different Input Devices: Keyboard, Mouse, or Touchscreen. *Educ. Technol.* **1993**, *33*, 56–59.

72. Bakhtiyari, K.; Taghavi, M.; Husain, H. Hybrid affective computing—Keyboard, mouse and touch screen: From review to experiment. *Neural Comput. Appl.* **2015**, *26*, 1277–1296. [CrossRef]
73. Walker, G. Touch Displays. In *Handbook of Digital Imaging*; Kriss, M., Ed.; John Wiley & Sons, Ltd.: Hoboken, NJ, USA, 2015.
74. Noah, B.; Li, J.; Rothrock, L. An evaluation of touchscreen versus keyboard/mouse interaction for large screen process control displays. *Appl. Ergon.* **2017**, *60*, 1–13. [CrossRef]
75. Vaughan, N.; Steven, J. New Interfaces at the Touch of a Fingertip. *Computer* **2007**, *40*, 12–15. [CrossRef]
76. Bhalla, M.R.; Bhalla, A.V. Comparative Study of Various Touchscreen Technologies. *Int. J. Comput. Appl.* **2010**, *6*, 12–18. [CrossRef]
77. Colegrove, J. The State of the Touch-Screen Market in 2010. *Inf. Disp.* **2010**, *26*, 22–24. [CrossRef]
78. Aguilar, R.N.; Meijer, G.C.M. Fast interface electronics for a resistive touch-screen. In Proceedings of the IEEE Sensors, Orlando, FL, USA, 12–14 June 2002.
79. Downs, R. Texas Instruments Inc., Using Resistive Touch Screens for Human/Machine Interface. Available online: https://www.ti.com/jp/lit/an/slyt209a/slyt209a.pdf (accessed on 23 April 2021).
80. Stetson, J.W. Analog Resistive Touch Panels and Sunlight Readability. *Inf. Disp.* **2006**, *12*, 26–30.
81. Krein, P.T.; Meadows, R.D. The electroquasistatics of the capacitive touch panel. *IEEE Trans. Ind. Appl.* **1990**, *26*, 529–534. [CrossRef]
82. Barrett, G.; Omote, R. Projected-Capacitive Touch Technology. *Inf. Disp.* **2010**, *26*, 16–21. [CrossRef]
83. Yanase, J.; Takatori, K.; Asada, H. Algorithm for Recognizing Pinch Gestures on Surface-Capacitive Touch Screens. *Dig. Tech. Pap. Int. Symp. SID* **2015**, *46*, 899–902. [CrossRef]
84. Walker, G. Part 1: Fundamentals of Projected-Capacitive Touch Technology. Available online: http://www.walkermobile.com/Touch_Technologies_Tutorial_Latest_Version.pdf (accessed on 23 April 2021).
85. Dieulesaint, E.; Royer, D.; Legras, O.; Chaabi, A. Acoustic plate mode touch screen. *Electron. Lett.* **1991**, *27*, 49–51. [CrossRef]
86. Nara, T.; Takasaki, M.; Maeda, T.; Higuchi, T.; Ando, S.; Tachi, S. Surface acoustic wave tactile display. *IEEE Comput. Graph. Appl.* **2001**, *21*, 56–63. [CrossRef]
87. Takasaki, M.; Kotani, H.; Mizuno, T.; Nara, T. Transparent surface acoustic wave tactile display. In Proceedings of the IEEE/RSJ International Conference on Intelligent Robots and Systems, Edmonton, AB, Canada, 2–6 August 2005; pp. 3354–3359.
88. North, K.; D'Souza, H. Acoustic Pulse Recognition Enters Touch-Screen Market. *Inf. Disp.* **2006**, *12*, 22–25.
89. Harris, N. Applications of Bending Wave Technology in Human Interface Devices. *J. Audio Eng. Soc.* **2009**; 7658.
90. Reis, S.; Correia, V.; Martins, M.; Barbosa, G.; Sousa, R.M.; Minas, G.; Lanceros-Mendez, S.; Rocha, J.G. Touchscreen based on acoustic pulse recognition with piezoelectric polymer sensors. In Proceedings of the IEEE International Symposium on Industrial Electronics, Bari, Italy, 4–7 July 2010; pp. 516–520.
91. Hao, X.; Huang, K. A Low-Power Ultra-Light Small and Medium Size Acoustic Wave Touch Screen. *Appl. Mech. Mater.* **2014**, *513–517*, 4072–4075. [CrossRef]
92. Quaegebeur, N.; Masson, P.; Beaudet, N.; Sarret, P. Touchscreen Surface Based on Interaction of Ultrasonic Guided Waves with a Contact Impedance. *IEEE Sens. J.* **2016**, *16*, 3564–3571. [CrossRef]
93. Firouzi, K.; Nikoozadeh, A.; Carver, T.E.; Khuri-Yakub, B.T. Multitouch touchscreen using reverberant lamb waves. In Proceedings of the IEEE International Ultrasonics Symposium, Washington, DC, USA, 6–9 September 2017; pp. 1–4.
94. Yang, Z.; Liu, X.; Wu, B.; Liu, R. Adaptability of Ultrasonic Lamb Wave Touchscreen to the Variations in Touch Force and Touch Area. *Sensors* **2021**, *21*, 1736. [CrossRef]
95. Han, J.Y. Low-cost multi-touch sensing through frustrated total internal reflection. In Proceedings of the 18th annual ACM symposium on User interface software and technology, Seattle, WA, USA, 23–26 October 2005; pp. 115–118.
96. Maxwell, I. An Overview of Optical-Touch Technologies. *Inf. Disp.* **2007**, *12*, 26–30.
97. Zhang, H. Optical touch screen with virtual force. In Proceedings of the IEEE International Conference on Systems, Man and Cybernetics, San Antonio, TX, USA, 11–14 October 2009; pp. 807–811.
98. Sheikh, S.S.A.; Hanana, S.M.; Al-Hosany, Y.; Soudan, B. Design and implementation of an FTIR camera-based multi-touch display. In Proceedings of the 5th IEEE GCC Conference & Exhibition, Kuwait, Kuwait, 17–19 March 2009, pp. 1–6.
99. Ahsanullah.; Mahmood, A.K.B.; Sulaiman, S. Design and implementation of multi-touch system using FTIR technique for optimization of finger touch detection. In Proceedings of the International Symposium on Information Technology, Kuala Lumpur, Malaysia, 15–17 June 2010; Volume 1, pp. 1–7.
100. Pedersen, H.C.; Jakobsen, M.L.; Hanson, S.G.; Mosgaard, M.; Iversen, T.; Korsgaard, J. Optical touch screen based on waveguide sensing. *Appl. Phys. Lett.* **2011**, *99*, 061102. [CrossRef]
101. Han, J.; Choi, S.; Heo, S.; Lee, G. Optical touch sensing based on internal scattering in touch surface. *Electron. Lett.* **2012**, *48*, 1420–1422. [CrossRef]
102. Park, S.M.; Lee, K.; Kyung, K.U. A new stylus for touchscreen devices. In Proceedings of the IEEE International Conference on Consumer Electronics, Las Vegas, NV, USA, 9–12 January 2011, pp. 491–492.
103. Badaye, M.; Schediwy, R.R. Passive Stylus for Capacitive Sensors. U.S. Patent 8,125,469 B2, 28 February 2012.
104. Zachut, R. Digitizer, Stylus and Method of Synchronization Therewith. U.S. Patent 9,524,044 B2, 20 December 2016.
105. Vuppu, S.; Cranfill, D.; Olley, M.; Valentine, M. Active Stylus for Use with Touch-Sensitive Interfaces and Corresponding Method. U.S. Patent 8,766,954 B2, 1 July 2014.
106. Shahparnia, S.; Sundara-Rajan, K.; Ali, Y.; Bentov, I. Active Stylus with High Voltage. U.S. Patent 9,933,866 B2, 3 April 2018.

107. Chen, C.H.; Huang, C.H. Method for Sensing Fast Motion, Controller and Electromagnetic Sensing Apparatus. U.S. Patent 2015/0331504 A1, 19 November 2015.
108. An, J.S.; Han, S.H.; Kim, J.E.; Yoon, D.H.; Kim, Y.H.; Hong, H.H.; Ye, J.H.; Jung, S.J.; Lee, S.H.; Jeong, J.Y.; et al. 9.6 A 3.9 kHz-frame-rate capacitive touch system with pressure/tilt angle expressions of active stylus using multiple-frequency driving method for 65″ 104× 64 touch screen panel. In Proceedings of the IEEE International Solid-State Circuits Conference, San Francisco, CA, USA, 5–9 February 2017, pp. 168–169.
109. Park, S.M.; Lee, K.; Kyung, K.U. Target size study for one-handed thumb use on small touchscreen devices. In Proceedings of the Conference on Human-Computer Interaction with Mobile Devices and Services, Helsinki, Finland, 12–15 September 2006; pp. 203–210.
110. Butler, A.; Izadi, S.; Hodges, S. SideSight: Multi-"touch" interaction around small devices. In Proceedings of the Annual ACM Symposium on User Interface Software and Technology, Monterey, CA, USA, 19–22 October 2008; pp. 201–204.
111. Harrison, C.; Hudson, S.E. Abracadabra: Wireless, high-precision, and unpowered finger input for very small mobile devices. In Proceedings of the Annual ACM Symposium on User Interface Software and Technology, Victoria, BC, Canada, 4–7 October 2009; pp. 121–124.
112. Dybdal, M.L.; Agustin, J.S.; Hansen, J.P. Gaze input for mobile devices by dwell and gestures. In Proceedings of the Symposium on Eye Tracking Research and Applications, Santa Barbara, CA, USA, 28–30 March 2012; pp. 225–228.
113. Stellmach, S.; Dachselt, R. Look & touch: Gaze-supported target acquisition. In Proceedings of the SIGCHI Conference on Human Factors in Computing Systems, Austin, TX, USA, 5–10 May 2008; pp. 2981–2990.
114. Xiao, R.; Laput, G.; Harrison, C. Expanding the input expressivity of smartwatches with mechanical pan, twist, tilt and click. In Proceedings of the SIGCHI Conference on Human Factors in Computing Systems, Toronto, ON, Canada, 26 April–1 May 2014; pp. 193–196.
115. Oakley, I.; Lee, D. Interaction on the edge: Offset sensing for small devices. In Proceedings of the SIGCHI Conference on Human Factors in Computing Systems, Toronto, ON, Canada, 26 April–1 May 2014; pp. 169–178.
116. Lim, S.C.; Shin, J.; Kim, S.C.; Park, J. Expansion of Smartwatch Touch Interface from Touchscreen to Around Device Interface Using Infrared Line Image Sensors. *Sensors* **2015**, *15*, 16642–16653. [CrossRef]
117. Zhou, J.; Zhang, Y.; Laput, G.; Harrison, C. AuraSense: Enabling Expressive Around-Smartwatch Interactions with Electric Field Sensing. In Proceedings of the Annual Symposium on User Interface Software and Technology, Tokyo, Japan, 16–19 October 2016; pp. 81–86.
118. Zhang, C.; Guo, A.; Zhang, D.; Li, Y.; Southern, C.; Arriaga, R.I.; Abowd, G.D. Beyond the Touchscreen: An Exploration of Extending Interactions on Commodity Smartphones. *ACM Trans. Interact. Intell. Syst.* **2016**, *6*, 16. [CrossRef]
119. Weigel, M.; Steimle, J. DeformWear: Deformation Input on Tiny Wearable Devices. *Proc. ACM Interact. Mobile Wearable Ubiquitous Technol.* **2017** , *1*, 1–23. [CrossRef]
120. Lee, J.; Lee, C.; Kim, G.J. Vouch: Multimodal touch-and-voice input for smart watches under difficult operating conditions. *J. Multimodal User Interfaces* **2017**, *11*, 289–299. [CrossRef]
121. Kwon, M.C.; Park, G.; Choi, S. Smartwatch User Interface Implementation Using CNN-Based Gesture Pattern Recognition. *Sensors* **2018**, *18*, 2997. [CrossRef]
122. Song, S.J.; Nam, H. Sound-of-Tapping user interface technology with medium identification. *Displays* **2018**, *53*, 54–64. [CrossRef]
123. Wang, B.; Grossman, T. BlyncSync: Enabling Multimodal Smartwatch Gestures with Synchronous Touch and Blink. In Proceedings of the CHI Conference on Human Factors in Computing Systems, Honolulu, HI, USA, 25–30 April 2020; pp. 1–14.
124. Harrison, C.; Schwarz, J.; Hudson, S.E. TapSense: enhancing finger interaction on touch surfaces. In Proceedings of the Annual ACM Symposium on User Interface Software and Technology, Barbara, CA, USA, 16–19 October 2011; pp. 627–636.
125. Ikematsu, K.; Yamanaka, S. ScraTouch: Extending Touch Interaction Technique Using Fingernail on Capacitive Touch Surfaces. In Proceedings of the Extended Abstracts of the CHI Conference on Human Factors in Computing Systems, Honolulu, HI, USA, 25–30 April 2020, pp. 1–10.
126. Ha, T.; Woo, W. Bare Hand Interface for Interaction in the Video See-Through HMD Based Wearable AR Environment. In Proceedings of the International Conference on Entertainment Computing, Cambridge, UK, 20–22 September 2006.
127. Witt, H.; Nicolai, T.; Kenn, H. Designing a wearable user interface for hands-free interaction in maintenance applications. In Proceedings of the Annual IEEE International Conference on Pervasive Computing and Communications Workshops, Pisa, Italy, 13–17 March 2006.
128. Lee, S.; Ha, G.; Cha, J.; Kim, J.; Lee, H.; Kim, S. CyberTouch—Touch and Cursor Interface for VR HMD. In Proceedings of the Posters' Extended Abstracts of International Conference on Human-Computer Interaction, Los Angeles, CA, USA, 2–7 August 2015.
129. Sidorakis, N.; Koulieris, G.A.; Mania, K. Binocular eye-tracking for the control of a 3D immersive multimedia user interface. In Proceedings of the IEEE Workshop on Everyday Virtual Reality, Arles, France, 23 March 2015; pp. 15–18.
130. Ono, M.; Shizuki, B.; Tanaka, J. Sensing Touch Force using Active Acoustic Sensing. In Proceedings of the International Conference on Tangible, Embedded, and Embodied Interaction, Stanford, CA, USA, 15–19 January 2015; pp. 355–358.
131. Kim, J.; Cha, J.; Lee, H.; Kim, S. Hand-free natural user interface for VR HMD with IR based facial gesture tracking sensor. In Proceedings of the ACM Symposium on Virtual Reality Software and Technology, Gothenburg, Sweden, 8–10 November 2017; pp. 1–2.

132. Kim, M.; Lee, J.; Jeon, C.; Kim, J. A Study on Interaction of Gaze Pointer-Based User Interface in Mobile Virtual Reality Environment. *Symmetry* **2017**, *9*, 189. [CrossRef]
133. Park, K.; Chin, S. Let us guide a smart interface for VR HMD and leap motion. In Proceedings of the ACM Symposium on Virtual Reality Software and Technology, Tokyo, Japan, 28 November–1 December 2018; pp. 1–2.
134. Hansen, J.P.; Alapetite, A.; Thomsen, M.; Wang, Z.; Minakata, K.; Zhang, G. Head and gaze control of a telepresence robot with an HMD. In Proceedings of the ACM Symposium on Eye Tracking Research & Applications, Warsaw, Poland, 14–17 June 2018; pp. 1–3.
135. Kolly, S.M.; Wattenhofer, R.; Welten, S. A personal touch: Recognizing users based on touch screen behavior. In Proceedings of the International Workshop on Sensing Applications on Mobile Phones, Toronto, ON, Canada, 6 November 2012; pp. 1–5.
136. Angulo, J.; Wästlund, E. Exploring Touch-Screen Biometrics for User Identification on Smart Phones. *IFIP Adv. Inf. Commun. Technol.* **2012**, *375*, 130–143. [CrossRef]
137. Weir, D.; Rogers, S.; Murray-Smith, R.; Löchtefeld, M. A user-specific machine learning approach for improving touch accuracy on mobile devices. In Proceedings of the Annual ACM Symposium on User Interface Software and Technology, Cambridge, MA, USA, 7–10 October 2012; pp. 465–476.
138. Feng, T.; Liu, Z.; Kwon, K.A.; Shi, W.; Carbunar, B.; Jiang, Y.; Nguyen, N. Continuous mobile authentication using touchscreen gestures. In Proceedings of the IEEE Conference on Technologies for Homeland Security, Waltham, MA, USA, 13–15 November 2012.
139. Meng, Y.; Wong, D.; Schlegel, R.; Kwok, L. Touch Gestures Based Biometric Authentication Scheme for Touchscreen Mobile Phones. In Proceedings of the Lecture Notes in Computer Science of International Conference on Information Security and Cryptology, Guangzhou, China, 27–30 November 2013; pp. 110–113.
140. Saravanan, P.; Clarke, S.; Chau, D.H.P.; Zha, H. LatentGesture: Active user authentication through background touch analysis. In Proceedings of the International Symposium of Chinese CHI, Toronto, ON, Canada, 26–27 April 2014; pp. 110–113.
141. Meng, Y.; Wong, D.S.; Kwok, L.F. Design of touch dynamics based user authentication with an adaptive mechanism on mobile phones. In Proceedings of the Annual ACM Symposium on Applied Computing, Gyeongju, Korea, 24–28 March 2014; pp. 1680–1687.
142. Schwarz, J.; Xiao, R.; Mankoff, J.; Hudson, S.E.; Harrison, C. Probabilistic palm rejection using spatiotemporal touch features and iterative classification. In Proceedings of the SIGCHI Conference on Human Factors in Computing Systems, Toronto, ON, Canada, 26 April–1 May 2014.
143. Guo, A.; Xiao, R.; Harrison, C. CapAuth: Identifying and Differentiating User Handprints on Commodity Capacitive Touchscreens. In Proceedings of the International Conference on Interactive Tabletops & Surfaces, Madeira, Portugal, 15–18 November 2015; pp. 59–62.
144. Lin, J.W.; Wang, C.; Huang, Y.Y.; Chou, K.T.; Chen, H.Y.; Tseng, W.L.; Chen, M.Y. BackHand: Sensing Hand Gestures via Back of the Hand. In Proceedings of the Annual ACM Symposium on User Interface Software & Technology, Charlotte, NC, USA, 11–15 November 2015; pp. 557–564.
145. Buschek, D.; Alt, F. TouchML: A Machine Learning Toolkit for Modelling Spatial Touch Targeting Behaviour. In Proceedings of the International Conference on Intelligent User Interfaces, Atlanta, GA, USA, 29 March–1 April 2015; pp. 110–114.
146. Xiao, R.; Schwarz, J.; Harrison, C. Estimating 3d finger angle on commodity touchscreens. In Proceedings of the International Conference on Interactive Tabletops & Surfaces, Madeira, Portugal, 15–18 November 2015.
147. Mock, P.; Gerjets, P.; Tibus, M.; Trautwein, U.; Möller, K.; Rosenstiel, W. Using touchscreen interaction data to predict cognitive workload. In Proceedings of the ACM International Conference on Multimodal Interaction, Tokyo, Japan, 12–16 November 2016.
148. Rilvan, M.A.; Lacy, K.I.; Hossain, M.S.; Wang, B. User authentication and identification on smartphones by incorporating capacitive touchscreen. In Proceedings of the IEEE International Performance Computing and Communications Conference, Las Vegas, NV, USA, 9–11 December 2016.
149. Sharma, V.; Enbody, R. User authentication and identification from user interface interactions on touch-enabled devices. In Proceedings of the ACM Conference on Security and Privacy in Wireless and Mobile Networks, Boston, MA, USA, 18–20 July 2017; pp. 1–11.
150. Henze, N.; Mayer, S.; Le, H.V.; Schwind, V. Improving software-reduced touchscreen latency. In Proceedings of the International Conference on Human-Computer Interaction with Mobile Devices and Services, Vienna, Austria, 4–7 September 2017; pp. 1–8.
151. Mayer, S.; Le, H.V.; Henze, N. Estimating the finger orientation on capacitive touchscreens using convolutional neural networks. In Proceedings of the ACM International Conference on Interactive Surfaces and Spaces, Brighton, UK, 17–20 October 2017.
152. Meng, W.; Li, W.; Wong, D.S. Enhancing touch behavioral authentication via cost-based intelligent mechanism on smartphones. *Multimed. Tools Appl.* **2018**, *77*, 30167–30185. [CrossRef]
153. Le, H.V.; Mayer, S.; Henze, N. InfiniTouch: Finger-Aware Interaction on Fully Touch Sensitive. In Proceedings of the Annual ACM Symposium on User Interface Software and Technology, Berlin, Germany, 14 October 2018; pp. 779–792.
154. Le, H.V.; Kosch, T.; Mayer, S.; Henze, N. Demonstrating palm touch: the palm as an additional input modality on commodity smartphones. In Proceedings of the International Conference on Human-Computer Interaction with Mobile Devices and Services Adjunct, Barcelona, Spain, 3–6 September 2018.
155. Fischer, T.; Etchart, M.; Biempica, E. Frame-level proximity and touch recognition using capacitive sensing and semi-supervised sequential modeling. In Proceedings of the IEEE International Workshop on Machine Learning for Signal Processing, Aalborg, Denmark, 17–20 September 2018.

156. Le, H.V.; Kosch, T.; Bader, P.; Mayer, S.; Henze, N. PalmTouch: Using the Palm as an Additional Input Modality on Commodity Smartphones. In Proceedings of the CHI Conference on Human Factors in Computing Systems, Montreal, QC, Canada, 21–26 April 2018.
157. Seol, K.H.; Park, S.; Song, S.J.; Nam, H. Finger and stylus discrimination scheme based on capacitive touch screen panel and support vector machine classifier. *Jpn. J. Appl. Phys.* **2019**, *58*, 074501. [CrossRef]
158. Kumar, A.; Radjesh, A.; Mayer, S.; Le, H.V. Improving the Input Accuracy of Touchscreens using Deep Learning. In Proceedings of the Extended Abstracts of the CHI Conference on Human Factors in Computing Systems, Glasgow Scotland, UK, 4–9 May 2019.
159. Le, H.V.; Mayer, S.; Henze, N. Investigating the feasibility of finger identification on capacitive touchscreens using deep learning. In Proceedings of the International Conference on Intelligent User Interfaces, Marina del Ray, CA, USA, 17–20 March 2019.
160. Kim, S.C.; Han, B.K. Emulating touch signals from multivariate sensor data using gated RNNs. In Proceedings of the ACM/IEEE International Conference on Human-Robot Interaction, Marina del Ray, CA, USA, 17–20 March 2019.
161. Boceck, T.; Sprott, S.; Le, H.V.; Mayer, S. Force Touch Detection on Capacitive Sensors using Deep Neural Networks. In Proceedings of the International Conference on Human-Computer Interaction with Mobile Devices and Services, Taipei, Taiwan, 1–4 October 2019.
162. Schweigert, R.; Leusmann, J.; Hagenmayer, S.; Weiß, M.; Le, H.V.; Mayer, S.; Bulling, A. KnuckleTouch: Enabling Knuckle Gestures on Capacitive Touchscreens using Deep Learning. In Proceedings of the Mensch und Computer, Hamburg, Germany, 8–11 September 2019; pp. 387–397.
163. Seol, K.H.; Park, S.; Lee, J.; Nam, H. Active stylus-touch discrimination scheme based on anomaly detection algorithm. *J. Soc. Inf. Disp.* **2020**, *28*, 831–836. [CrossRef]
164. Bello, A.A.; Chiroma, H.; Gital, A.Y.; Gabralla, L.A.; Abdulhamid, S.M.; Shuib, L. Machine learning algorithms for improving security on touch screen devices: a survey, challenges and new perspectives. *Neural Comput. Appl.* **2020**, *32*, 13651–13678. [CrossRef]
165. Walker, G. A review of technologies for sensing contact location on the surface of a display. *J. Soc. Inf. Disp.* **2012**, *20*, 413–440. [CrossRef]
166. Kwon, O.K.; An, J.S.; Hong, S.K. Capacitive Touch Systems With Styli for Touch Sensors: A Review. *IEEE Sens. J.* **2018**, *18*, 4832–4846. [CrossRef]
167. Hansson, P.R. Multiple input detection for resistive touch panel. WO Patent 2009/156803 A1, 30 December 2009.
168. Bogana, M.P.; Celani, A.; Pastore, N. Method for Determining Multiple Touch Inputs on a Resistive Touch Screen and a Multiple Touch Controller. U.S. Patent 2012/0068969 A1, 22 March 2012.
169. Calpe-Maravilla, J.; Medina, I.; Martínez, M.J.; Carbajo, A. Dual touch and gesture recognition in 4-wire resistive touchscreens. In Proceedings of the IEEE Sensors, Valencia, Spain, 2–5 November 2014.
170. Chi, H.S.; Crockett, T.W.; Makley, A.V. Resistive Scanning Grid Touch Panel. U.S. Patent 7,492,358 B2, 17 February 2009.
171. Wu, C.K.; Wu, M.T. Matrix Resistive Touch Device. U.S. Patent 2010/0164899 A1, 1 July 2010.
172. Fang, W.X. Simultaneous Multiple Location Touch Systems. U.S. Patent 8,310,464 B2, 13 November 2012.
173. An, J.S.; Han, S.H.; Park, K.B.; Kim, J.E.; Ye, J.H.; Lee, S.H.; Jeong, J.Y.; Kim, J.S.; Baek, K.H.; Chung, K.S.; et al. Multi-Way Interactive Capacitive Touch System with Palm Rejection of Active Stylus for 86″ Touch Screen Panels. In Proceedings of the IEEE International Solid-State Circuits Conference, San Francisco, CA, USA, 11–15 February 2018; pp. 182–183.
174. Kim, C.; Lee, D.S.; Kim, J.H.; Kim, H.B.; Shin, S.R.; Jung, J.H.; Song, I.H.; Jang, C.S.; Kwon, K.S.; Kim, S.H.; et al. Advanced In-cell Touch Technology for Large Sized Liquid Crystal Displays. *Dig. Tech. Pap. Int. Symp. SID* **2015**, *46*, 895–898. [CrossRef]
175. Takahashi, S.; Lee, B.J.; Koh, J.H.; Saito, S.; You, B.H.; Kim, N.D.; Kim, S.S. Embedded Liquid Crystal Capacitive Touch Screen Technology for Large Size LCD Applications. *Dig. Tech. Pap. Int. Symp. SID* **2009**, *40*, 563–566. [CrossRef]
176. Park, H.S.; Kim, Y.J.; Han, M.K. Touch-Sensitive Active-Matrix Display with Liquid Crystal Capacitance Detector Arrays. *Jpn. J. Appl. Phys.* **2010**, *49*, 03CC01. [CrossRef]
177. Walker, G.; Fihn, M. LCD In-Cell Touch. *Inf. Disp.* **2010**, *3*, 8–14. [CrossRef]
178. Lee, S.H.; An, J.S.; Hong, S.K.; Kwon, O.K. In-cell Capacitive Touch Panel Structures and Their Readout Circuits. In Proceedings of the International Workshop on Active-Matrix Flatpanel Displays and Devices, Kyoto, Japan, 6–8 July 2016; pp. 258–261.
179. Chen, Y.; Geng, D.; Jang, J. Integrated Active-Matrix Capacitive Sensor Using a-IGZO TFTs for AMOLED. *IEEE J. Electron Devices Soc.* **2018**, *6*, 214–218. [CrossRef]
180. 3M, Dispersive Signal Touch Technology. Available online: https://multimedia.3m.com/mws/media/443866O/3mtm-microtouch-system-sct2270dx-technology-profile.pdf (accessed on 23 April 2021).
181. Hwang, T.H.; Cui, W.H.; Yang, I.S.; Kwon, O.K. A highly area-efficient controller for capacitive touch screen panel systems. *IEEE Trans. Consum. Electron.* **2010**, *56*, 1115–1122. [CrossRef]
182. Luo, C.; Borkar, M.A.; Redfern, A.J.; McClellan, J.H. Compressive sensing for sparse touch detection on capacitive touch screens. *IEEE J. Emerg. Sel. Topics Circuits Syst.* **2012**, *2*, 639–648. [CrossRef]
183. Li, H.; Wei, Y.; Li, H.; Young, S.; Convey, D.; Lewis, J.; Maniar, P. Multitouch Pixilated Force Sensing Touch Screen. *Dig. Tech. Pap. Int. Symp. SID* **2009**, *40*, 455–458. [CrossRef]
184. Hwang, S.; Bianchi, A.; yun Wohn, K. MicPen: Pressure-Sensitive Pen Interaction Using Microphone with Standard Touchscreen. In Proceedings of the Extended Abstracts of the CHI Conference on Human Factors in Computing Systems, Austin, TX, USA, 5–10 May 2012; pp. 1847–1852.

185. Kim, H.K.; Lee, S.; Yun, K.S. Capacitive tactile sensor array for touch screen application. *Sens. Actuator. A Phys.* **2011**, *165*, 2–7. [CrossRef]
186. Reynolds, K.; Shepelev, P.; Graf, A. Touch and Display Integration with Force. *Dig. Tech. Pap. Int. Symp. SID* **2016**, *47*, 617–620. [CrossRef]
187. Lee, K.H.; Nam, S.P.; Lee, J.H.; Choi, M.; Ko, H.J.; Byun, S.H.; Lee, J.C.; Lee, Y.H.; Rhee, Y.C.; Choi, Y.K.; et al. A Noise-Immune Stylus Analog Front-End Using Adjustable Frequency Modulation and Linear Interpolating Data Reconstruction for Both Electrically Coupled Resonance and Active Styluses. In Proceedings of the IEEE International Solid-State Circuits Conference, San Francisco, CA, USA, 11–15 February 2018, pp. 184–185.
188. Ko, S.; Shin, H.; Lee, J.; Jang, H.; So, B.C.; Yun, I.; Lee, K. Low Noise Capacitive Sensor for Multi-touch Mobile Handset's Applications. In Proceedings of the IEEE Asian Solid-State Circuits Conference, Beijing, China, 8–10 November 2010.
189. Yang, I.S.; Kwon, O.K. A touch controller using differential sensing method for on-cell capacitive touch screen panel systems. *IEEE Trans. Consum. Electron.* **2011**, *57*, 1027–1032. [CrossRef]
190. Kim, K.D.; Byun, S.H.; Choi, Y.K.; Baek, J.H.; Cho, H.H.; Park, J.K.; Ahn, H.Y.; Lee, C.J.; Cho, M.S.; Lee, J.H.; et al. A Capacitive Touch Controller Robust to Display Noise for Ultrathin Touch Screen Displays. In Proceedings of the IEEE International Solid-State Circuits Conference, San Francisco, CA, USA, 19–23 February 2012.
191. Yang, J.H.; Park, S.H.; Choi, J.M.; Kim, H.S.; Park, C.B.; Ryu, S.T.; Cho, G.H. A highly noise-immune touch controller using filtered-delta-integration and a charge-interpolation technique for 10.1-inch capacitive touch-screen panels. In Proceedings of the IEEE International Solid-State Circuits Conference, San Francisco, CA, USA, 17–21 February 2013; pp. 390–391.
192. Seo, J.; Nam, H. Low power and low noise shift register for in-cell touch display applications. *IEEE J. Electron Devices Soc.* **2018**, *6*, 726–732. [CrossRef]
193. Miura, N.; Dosho, S.; Tezuka, H.; Miki, T.; Fujimoto, D.; Kiriyama, T.; Nagata, M. A 1 mm Pitch 80 × 80 Channel 322 Hz Frame-Rate Multitouch Distribution Sensor With Two-Step Dual-Mode Capacitance Scan. *IEEE J. Solid-State Circuits* **2015**, *50*, 2741–2749. [CrossRef]
194. Shin, H.; Ko, S.; Jang, H.; Yun, I.; Lee, K. A 55 dB SNR with 240 Hz frame scan rate mutual capacitor 30 × 24 touch-screen panel read-out IC using code-division multiple sensing technique. In Proceedings of the IEEE International Solid-State Circuits Conference, San Francisco, CA, USA, 17–21 February 2013; pp. 388–389.
195. Hamaguchi, M.; Nagao, A.; Miyamoto, M. 12.3 A 240 Hz-reporting-rate 143 × 81 mutual-capacitance touch-sensing analog front-end IC with 37 dB SNR for 1mm-diameter stylus. In Proceedings of the IEEE International Solid-State Circuits Conference, San Francisco, CA, USA, 9–13 February 2014; pp. 214–215.
196. Heo, S.; Ma, H.; Song, J.; Park, K.; Choi, E.H.; Kim, J.J.; Bien, F. 72 dB SNR, 240 Hz frame rate readout IC with differential continuous-mode parallel architecture for larger touch-screen panel applications. *IEEE Trans. Circuits Syst. I Reg. Pap.* **2016**, *63*, 960–971. [CrossRef]
197. Park, J.E.; Lim, D.H.; Jeong, D.K. A Reconfigurable 40-to-67 dB SNR, 50-to-6400 Hz Frame-Rate, Column-Parallel Readout IC for Capacitive Touch-Screen Panels. *IEEE J. Solid-State Circuits* **2014**, *49*, 2305–2318. [CrossRef]
198. An, J.S.; Ra, J.H.; Kang, E.; Pertijs, M.A.P.; Han, S.H. 28.1 A Capacitive Touch Chipset with 33.9 dB Charge- Overflow Reduction Using Amplitude-Modulated Multi-Frequency Excitation and Wireless Power and Data Transfer to an Active Stylus. In Proceedings of the IEEE International Solid-State Circuits Conference, San Francisco, CA, USA, 16–20 February 2020; pp. 430–431.
199. Lin, C.L.; Li, C.S.; Chang, Y.M.; Lin, T.C.; Chen, J.F.; Lin, U.C. Pressure Sensitive Stylus and Algorithm for Touchscreen Panel. *J. Disp. Technol.* **2013**, *9*, 17–23. [CrossRef]
200. Bishop, C. *Pattern Recognition and Machine Learning*; Springer: New York, NY, USA, 2007.
201. Goodfellow, I.; Bengio, Y.; Courville, A. *Deep Learning*; MIT Press: Cambridge, MA, USA, 2016. Available online: http://www.deeplearningbook.org (accessed on 23 April 202).
202. Ioffe, S.; Szegedy, C. Batch Normalization: Accelerating Deep Network Training by Reducing Internal Covariate Shift. In Proceedings of the International Conference on Machine Learning, Lille, France, 7–9 July 2015; pp. 448–456.
203. Seol, K.H.; Park, S.; Nam, H. A Contact-Based Data Communication Technique Using Capacitive Touch Screen Panel and Support Vector Machine Classifier. *IEEE Access* **2019**, *7*, 178596–178604. [CrossRef]
204. Ma, H.; Liu, Z.; Heo, S.; Lee, J.; Na, K.; Jin, H.B.; Jung, S.; Park, K.; Kim, J.J.; Bien, F. On-Display Transparent Half-Diamond Pattern Capacitive Fingerprint Sensor Compatible With AMOLED Display. *IEEE Sens. J.* **2016**, *16*, 8124–8131. [CrossRef]
205. Seo, W.; Pi, J.E.; Cho, S.H.; Kang, S.Y.; Ahn, S.D.; Hwang, C.S.; Jeon, H.S.; Kim, J.U.; Lee, M. Transparent Fingerprint Sensor System for Large Flat Panel Display. *Sensors* **2018**, *18*, 293. [CrossRef]
206. Yang, W.; Zhang, X.; Tian, Y.; Wang, W.; Xue, J.H.; Liao, Q. Deep learning for single image super-resolution: A brief review. *IEEE Trans. Multimed.* **2019**, *21*, 3106–3121. [CrossRef]
207. Wang, Z.; Chen, J.; Hoi, S.C. Deep learning for image super-resolution: A survey. *IEEE Trans. Pattern Anal. Mach. Intell.* **2021**, in press. [CrossRef]

MDPI
St. Alban-Anlage 66
4052 Basel
Switzerland
Tel. +41 61 683 77 34
Fax +41 61 302 89 18
www.mdpi.com

Sensors Editorial Office
E-mail: sensors@mdpi.com
www.mdpi.com/journal/sensors